The
Handbook of
Portfolio
Mathematics

The Handbook of Portfolio Mathematics

Formulas for Optimal Allocation & Leverage

RALPH VINCE

John Wiley & Sons, Inc.

Published by John Wiley & Sons, Inc., Hoboken, New Jersey.
Published simultaneously in Canada.

Chapters 1–10 contain revised material from three of the author's previous books, *Portfolio Management Formulas: Mathematical Trading Methods for the Futures, Options, and Stock Markets* (1990), *The Mathematics of Money Management: Risk Analysis Techniques for Traders* (1992), and *The New Money Management: A Framework for Asset Allocation* (1995), all published by John Wiley & Sons, Inc.

Wiley Bicentennial Logo: Richard J. Pacifico

For general information on our other products and services or for technical support, please contact our Customer Care Department within the United States at (800) 762-2974, outside the United States at (317) 572-3993 or fax (317) 572-4002.

Wiley also publishes its books in a variety of electronic formats. Some content that appears in print may not be available in electronic format. For more information about Wiley products, visit our web site at www.wiley.com

Library of Congress Cataloging-in-Publication Data:

Vince, Ralph, 1958–
    The handbook of portfolio mathematics : formulas for optimal allocation & leverage /
Ralph Vince:
        p.   cm.
    ISBN-13: 978-0-471-75768-9 (cloth)
    ISBN-10: 0-471-75768-3 (cloth)
    1. Portfolio management–Mathematical models.   2. Investments–Mathematical models.
I. Title.
    HG4529.5.V555 2007
    332.601'51 – dc22   2006052577

Printed in the United States of America

10   9   8   7   6   5   4   3   2   1

"You must not be extending your empire while you are at war or run into unnecessary dangers. I am more afraid of our own mistakes than our enemies' designs."

—Pericles, in a speech to the Athenians during the Peloponnesian War, as represented by Thucydides

Contents

Preface

I t's always back there, bubbling away. It seems I cannot shut off my mind from it. Every conversation I ever have, with programmers and traders, engineers and gamblers, Northfield Park Railbirds and Warrensville Workhouse jailbirds—those equations that describe these very things are cast in this book.

Let me say I am averse to gambling. I am averse to the notion of creating risk where none need exist, averse to the idea of attempting to be rewarded in the absence of creating or contributing something (or worse yet, taxing a man's labor!). Additionally, I find amorality in charging or collecting interest, and the absence of this innate sense in others riles me.

This book starts out as a compilation, cleanup, and in some cases, reformulation of the previous books I have written on this subject. I'm standing on big shoulders here. The germ of the idea of those previous books can trace its lineage to my good friend and past employer, Larry Williams. In the dust cloud of his voracious research, was the study of the Kelly Criterion, and how that might be applied to trading. What followed over the coming years then was something of an explosion in that vein, culminating in a better portfolio model than the one which is still currently practiced.

For years now I have been away from the markets—intentionally. In a peculiar irony, it has sharpened my bird's-eye view on the entire industry. People still constantly seek me out, bend my ears, try to pick my hollow, rancid pumpkin about the markets. It has all given me a truly gigantic field of view, a dizzying phantasmagoria, on who is doing what, and how.

I'd like to share some of that with you here.

We are not going to violate anyone's secrets here, realizing that most of these folks work very hard to obtain what they know. What I will speak of is generalizations and commonalities in what people are doing, so that we can analyze, distinguish, compare, and, I hope, arrive at some well-founded conclusions.

But I am not in the markets' trenches anymore. My time has been spent on software for parametric geometry generation of industrial componentry and "smart" robots that understand natural language and can go out and do things like perform research for me, come back, draw inferences, and discuss their findings with me. These are wonderful endeavors for me, allowing me to extend my litany of failures.

Speaking of which, in the final section of this text, we step into the near-silent, blue-lit morgue of failure itself, dissecting it both in a mathematical and abstract sense, as well as the real-world one. In this final chapter, the two are indistinguishable.

When we speak of the *real world*, some may get the mistaken impression that the material is easy. It is not. That has not been a criterion of mine here. What has been a criterion is to address the real-world application of the previous three books that this book incorporates. That means looking at the previous material with regard to failure, with regard to drawdown. Money managers and personal traders alike tend to have utility preference curves that are incongruent with maximizing their returns. Further, I am aware of no one, nor have I ever encountered any trader, fund manager, or institution, who could even tell you what his or her utility preference function was. This is a prime example of the chasm—the disconnect—between theory and real-world application.

Historically, risk has been defined in theoretical terms as the variance (or semivariance) in returns. This, too, is rarely (though in certain situations) a desired proxy for risk. Risk is the chance of getting your head handed to you. It is not, except in rare cases, variance in returns. It is not semivariance in returns; it is not determined by a utility preference function. Risk is the probability of being ruined. Ruin is touching or penetrating a lower barrier on your equity. So we can say to most traders, fund managers, and institutions that risk is the probability of touching a lower barrier on equity, such that it would constitute ruin to someone. Even in the rare cases where variance in returns is a concern, risk is still primarily a drawdown to a lower absorbing barrier.

So what has been needed, and something I have had bubbling away for the past decade or so, is a way to apply the optimal f framework within the real-world constraints of this universally regarded definition of risk. That is, how do we apply optimal f with regard to risk of ruin and its more familiar and real-world-applicable-cousin, risk of drawdown?

Of course, the concepts are seemingly complicated—we're seeking to maximize return for a given level of drawdown, not merely juxtapose returns and variance in returns. Do you want to maximize growth for a given level of drawdown, or do you want to do something easier?

So this book is more than just a repackaging of previous books on this subject. It incorporates new material, including a study of correlations between pairwise components in a portfolio (and *why* that is such a bad idea). Chapter 11 examines what portfolio managers have (not) been doing with regards to the concepts presented in this book, and Chapter 12 takes the new Leverage Space Portfolio Model and juxtaposes it to the probability of a given drawdown to provide a now-superior portfolio model, based on the previous chapters in this book, and applicable to the real world.

I beg the reader to look at everything in this text—as merely my articulation of something, and not an autocratic dictation. Not only am I not infallible, but also my real aim here is to engage you in the study of something I find fascinating, and I want to share that very raw joy with you. Because, you see, as I started out saying, it's always back there, bubbling away—my attraction to those equations on the markets, pertaining to allocation and leverage. It's not a preoccupation with the markets, though—to me it could be the weather or any other dynamic system. It is the allure of nailing masses and motions and relationships with an equation.

Rapture!

That is my motivation, and that is why I can never shut it off. It is that very rapture that I seek to share, which augments that very rapture I find in it. As stated earlier, I stand on big shoulders. My hope is that my shoulders can support those who wish to go further with these concepts.

This book covers my thinking on these subjects for more than two and a half decades. There are a lot of people to thank. I won't mention them, either—they know who they are, and I feel uneasy mentioning the names of others here in one way or another, or others in the industry who wish to remain nameless. I don't know how they might take it.

There is one guilty party, however, whom I *will* mention—Rejeanne. This one, finally, is for you.

RALPH VINCE

Chagrin Falls, Ohio
August 2006

Introduction

This is a book in two distinct parts. Originally, my task was to distill the previous three books on this subject into one book. In effect, Part I comprises that text.

It's been reorganized, rehashed, and reworked to resemble the original texts while creating a contiguous path of reasoning, which takes us from the basic gambling theory and statistics, through the introduction of the Kelly criterion, optimal f, and finally onto the Leverage Space Portfolio Model for multiple-simultaneous positions.

The Leverage Space Portfolio Model addresses allocations and leverage. Often these are two distinct facets, but herein they refer to the same thing. *Allocation* is the *relative* leverage between multiple portfolio components. Thus, when we speak of *leverage*, we are also speaking of *allocation*, and vice versa.

Likewise, *money management* and *portfolio construction*, as practiced, don't necessarily refer to the same exercise, yet in this text, they do. Collectively, whatever the endeavor of risk, be it a bond portfolio, a commodities fund, or a team of blackjack players invading a casino, the collective exercise will be herein referred to as *allocation*.

I have tried to keep the geometric perspective on these concepts, and keep those notions about them intact. The first section is necessarily heavy on math. The first section is purely conceptual. It is about allocation and leverage to maximize returns without respect to anything else.

Everything in Part I was conjured up more than a decade or two ago. I was younger then.

Since that time, I have repeatedly been approached with the question, "How do you apply it?" I used to be baffled by this; the obvious (to me) answer being, "As is."

As used herein, a ln utility preference curve is one that is characteristic of someone who acts so as to maximize the ratio of his or her returns to the risk assumed to do so.

The notion that someone's *utility preference function* could be anything other than ln was evidence of both the person's insanity and weakness.

I saw it as a means for risk takers to enjoy the rush of their compulsive gambling under the ruse of the academic justification of *utility preference*.

I'm older now (seemingly not tempered with age—you see, I still know the guy who wrote those previous books), but I have been able to at least accept the exercise—the rapture—of working to solve the dilemma of optimal allocations and leverage under the constraint of a utility preference curve that is *not* ln.

By the definition of a ln utility preference curve, given a few paragraphs ago, a sane[1] person is therefore one who is levered up to the optimal *f* level in a game favorable to him or minimizes his number of plays in a game unfavorable to him. Anyone who goes to a casino and plunks down all he is willing to lose on that trip in one play is not a compulsive gambler. But who does that? Who has that self-control? Who has a utility preference curve that *is* ln?

That takes us to Part II of the book, the part I call the *real-world application* of the concepts illuminated in Part I, because people's utility preference curves are not ln.

So Part II attempts to tackle the mathematical puzzle posed by attempting to employ the concepts of Part I, given the weakness and insanity of human beings. What could be more fun?

* * *

Many of the people who have approached me with the question of "How do you apply it?" over the years have been professionals in the industry. Since, ultimately, their clients are the very individuals whose utility preference curves are not ln, I have found that these entities have utility preference functions that mirror those of their clients (or they don't have clients for long).

Many of these entities have been successful for many years. Naturally, their procedures pertaining to allocation, leverage, and trading implementation were of great interest to me.

Part II goes into this, into what these entities typically do. The best of them, I find, have not employed the concepts of the last chapter except in very rudimentary and primitive ways. There is a long way to go.

Often, I have been criticized as being "all theory—no practice." Well, Part I is indeed all theory, but it *is* exhaustive in that sense—not on portfolio construction in general and all the multitude of ways of performing that, but rather, on portfolio construction in terms of optimal position sizes (i.e., in the vein of an optimal *f* approach). Further, I did not want Part I to be

---

[1] Academics prefer the nomenclature "rational," versus "sane." The subtle difference between the two is germane to this discussion.

a mere republishing, almost verbatim, of the previous books. Therefore, I have incorporated some new material into Part I. This is material that has become evident to me in the years since the original material was published.

Part II is entirely new. I have been fortunate in that my first exposure to the industry was as a margin clerk. I had an opportunity to observe a sizable universe of ways people go about doing things in this business. Later, thanks to my programming abilities, from which the other books germinated, I had exposure to many professionals in the industry, and was often privy to how they practiced things, or was in a position where I could reverse-engineer it. I have had the good fortune of being on a course that has afforded me a bird's-eye view of the way people practice their allocation, leverage, and trading implementations in this business. Part II is derived from that high-altitude bird's-eye view, and the desire to provide a real-world implementation of the concepts of Part I—that is, to make them applicable to those people whose utility preference functions are not ln.

* * *

Things I have written of in the past have received a good deal of criticism over the years. I welcome it, and a chance to address it. To me, it says people are thinking about these ideas, trying to mold them further, or remold those areas where I may have been wrong (I'm not so much interested in being "right" about any of this as I am about "this"). Though I have not consciously intended that, this book, in many ways, answers some of those criticisms.

The main criticism was that it was too theoretical with no real-world application. The criticism is well founded in the sense that drawdown was all but ignored. For better or worse, people and institutions never seem to have utility functions that are ln. Yet, nearly all utility functions of people and institutions are ln within a drawdown constraint. That is, they seek to maximize the ratio of returns to risk (drawdown) within a certain drawdown. That disconnect between what I have written in the past has now, more than a decade later, been resolved.

A second major criticism is that trading at optimal f is too wild for any mere human. I know of no professional funds that have traded at the optimal f levels. I have known people who have traded at optimal f, usually for short periods of time, in only a single market, before panicking in a drawdown. There it is again: drawdown. You see, it wasn't so much this construct of their utility preference curve (talk about too theoretical!) as it was their drawdown that was incongruent with their trading at the optimal f level.

If you are getting the notion that we will be looking into the nature of drawdown later on in this book, when we discuss what I have been doing in terms of working on this material for the past decade-plus, you're right. We're going to look at drawdown herein beyond what anyone has.

Which takes us to the third major criticism, being that optimal f or the Leverage Space Model allocates without respect to drawdown. This, too, has now been addressed directly in Chapter 12. However, as we will see in that chapter, drawdown is, in a sequence of independent trials, but one permutation of many permutations. Thus, to address drawdown, one must address it in those terms.

The last major criticism has been that regarding the complexity of calculation. People desire a simple solution, a heuristic, something they could perform by hand if need be.

Unfortunately, that was not the case, and that desire of others is now something even more remote. In the final chapter, we can see that one must perform millions of calculations (as a sample to billions of calculations!) in order to derive certain answers.

However, such seemingly complex tasks can be made simple by packaging them up as black-box computer applications. Once someone understands what calculations are performed and why, the machine can do the heavy lifting. Ultimately, that is even simpler than performing a simple calculation by hand.

If one can put in the scenarios, their outcomes, and probability of occurrence—their joint probabilities of occurrence with other scenarios in other scenario spectrums—one can feed the machine and derive that number which satisfies the ideal composition, the optimal allocations and leverage among portfolio components to satisfy that ln utility preference function within a certain drawdown constraint.

To be applicable to the real world, a book like this should, it would seem, be about trading. This is *not* a book on how to trade the markets. (This makes the real-world application section difficult.) It is about how very basic, mathematical laws are working on us—you and me—when we engage in a stream of risk-related outcomes wherein we don't have control over those outcomes. Rather, we have control only over the relative impacts on us. In that sense, the mathematics applies to us in trading.

I don't want to pretend to know a thing about trading, really. Just as I am not an academic, I am also not a trader. I've been around and worked for some amazing traders—but that doesn't mean I am one.

That's *your* domain—and why you are reading this book: To augment the knowledge you have about trading vis-à-vis cross-pollination with these *outside* formulas. And if they are too cumbersome, or too complicated, please don't blame me. I wish they were simply along the lines of $2 + 2$. But they are not.

This is not by my design. When you trade, you are somewhat trying to intuitively carve your way along the paths of these equations, yet you are oblivious to what the equations are. You are, for instance, trying to maximize your returns within a certain probability of a given drawdown over the next period.

But you don't really have the equations to do so. Now you do. Don't blame me if you find them to be too cumbersome. These formulas are what we seek to know—and somehow use—as they apply to us in trading, whether we acknowledge that or not. I have heard ample criticism about the difficulties in applications. In this text, I will attempt to show you what others are doing *compared* to using these formulas. However, these formulas are at work on everyone when they trade. It is in the disparity between the two that your past criticisms of me lie; it is in that very disparity that my criticisms of you lie.

When you step up to the service line and line up to serve to my backhand, say, the fact that gravity operates with an acceleration of 9.8 meters per second squared applies to you. It applies to your serve landing in the box or not (among other things), whether you acknowledge this or not. It is an explanation of how things work more so than how to work things. You are trying to operate within a world defined by certain formulas. It does not mean you can implement them in your work, or that, because you cannot, they are therefore invalid. Perhaps you can implement them in your work. Clearly, if you could, without expense to the other aspects of "your work," wouldn't it be safe to say, then, that you certainly wouldn't be worse off?

And so with the equations in the book. Perhaps you can implement them—and if you can, without expense to the other aspects of your game, then won't you be better off? And if not, does it invalidate their truths any more than a tennis pro who dishes up a first serve, oblivious to the 9.8 m/s^2 at work?

* * *

This is, in its totality, what I know about allocations and leverage in trading. It is the sum of all I have written of it in the past, and what I have savored over the past decade-plus. As with many things, I truly love this stuff. I hope my passion for it rings contagiously herein. However, it sits as dead and cold as any inanimate abstraction. It is only your working with these concepts, your application and your critiques of them, your volley back over the net, that give them life.

PART I

Theory

CHAPTER 1

The Random Process and Gambling Theory

W e will start with the simple coin-toss case. When you toss a coin in the air there is no way to tell for certain whether it will land heads or tails. Yet over many tosses the outcome can be reasonably predicted.

This, then, is where we begin our discussion.

Certain axioms will be developed as we discuss the random process. The first of these is that *the outcome of an individual event in a random process cannot be predicted. However, we can reduce the possible outcomes to a probability statement.*

Pierre Simon Laplace (1749–1827) defined the probability of an event as the ratio of the number of ways in which the event can happen to the total possible number of events. Therefore, when a coin is tossed, the probability of getting tails is 1 (the number of tails on a coin) divided by 2 (the number of possible events), for a probability of .5. In our coin-toss example, we do not know whether the result will be heads or tails, but we do know that the probability that it will be heads is .5 and the probability it will be tails is .5. So, *a probability statement is a number between 0 (there is no chance of the event in question occurring) and 1 (the occurrence of the event is certain).*

Often you will have to convert from a probability statement to odds and vice versa. The two are interchangeable, as the odds imply a probability, and a probability likewise implies the odds. These conversions are given now. The formula to convert to a probability statement, when you know the given odds is:

$$\text{Probability} = \text{odds for}/(\text{odds for} + \text{odds against}) \quad (1.01)$$

3

If the odds on a horse, for example, are 4 to 1 (4:1), then the probability of that horse winning, as implied by the odds, is:

$$\text{Probability} = 1/(1+4)$$
$$= 1/5$$
$$= .2$$

So a horse that is 4:1 can also be said to have a probability of winning of .2. What if the odds were 5 to 2 (5:2)? In such a case the probability is:

$$\text{Probability} = 2/(2+5)$$
$$= 2/7$$
$$= .2857142857$$

The formula to convert from probability to odds is:

$$\text{Odds (against, to one)} = 1/\text{probability} - 1 \qquad (1.02)$$

So, for our coin-toss example, when there is a .5 probability of the coin's coming up heads, the odds on its coming up heads are given as:

$$\text{Odds} = 1/.5 - 1$$
$$= 2 - 1$$
$$= 1$$

This formula always gives you the odds "to one." In this example, we would say the odds on a coin's coming up heads are 1 to 1.

How about our previous example, where we converted from odds of 5:2 to a probability of .2857142857? Let's work the probability statement back to the odds and see if it works out.

$$\text{Odds} = 1/.2857142857 - 1$$
$$= 3.5 - 1$$
$$= 2.5$$

Here we can say that the odds in this case are 2.5 to 1, which is the same as saying that the odds are 5 to 2. So when someone speaks of odds, they are speaking of a probability statement as well.

Most people can't handle the uncertainty of a probability statement; it just doesn't sit well with them. We live in a world of exact sciences, and human beings have an innate tendency to believe they do not understand an event if it can only be reduced to a probability statement. The domain of physics seemed to be a solid one prior to the emergence of quantum physics. We had equations to account for most processes we had observed. These equations were real and provable. They repeated themselves over and over and the outcome could be exactly calculated before the event took place. With the emergence of quantum physics, suddenly a theretofore exact science could only reduce a physical phenomenon to a probability statement. Understandably, this disturbed many people.

I am not espousing the random walk concept of price action nor am I asking you to accept anything about the markets as random. Not yet, anyway. Like quantum physics, the idea that there is or is not randomness in the markets is an emotional one. At this stage, let us simply concentrate on the random process as it pertains to something we are certain is random, such as coin tossing or most casino gambling. In so doing, we can understand the process first, and later look at its applications. Whether the random process is applicable to other areas such as the markets is an issue that can be developed later.

Logically, the question must arise, "When does a random sequence begin and when does it end?" It really doesn't end. The blackjack table continues running even after you leave it. As you move from table to table in a casino, the random process can be said to follow you around. If you take a day off from the tables, the random process may be interrupted, but it continues upon your return. So, when we speak of a random process of X events in length we are arbitrarily choosing some finite length in order to study the process.

INDEPENDENT VERSUS DEPENDENT TRIALS PROCESSES

We can subdivide the random process into two categories. First are those events for which the probability statement is constant from one event to the next. These we will call independent trials processes or sampling with replacement. A coin toss is an example of just such a process. Each toss has a 50/50 probability regardless of the outcome of the prior toss. Even if the last five flips of a coin were heads, the probability of this flip being heads is unaffected, and remains .5.

Naturally, the other type of random process is one where the outcome of prior events *does* affect the probability statement and, naturally, the probability statement is not constant from one event to the next. These types of events are called dependent trials processes or sampling without replacement. Blackjack is an example of just such a process. Once a card is played, the composition of the deck for the next draw of a card is different from what it was for the previous draw. Suppose a new deck is shuffled and a card removed. Say it was the ace of diamonds. Prior to removing this card the probability of drawing an ace was 4/52 or .07692307692. Now that an ace has been drawn from the deck, and not replaced, the probability of drawing an ace on the next draw is 3/51 or .05882352941.

Some people argue that dependent trials processes such as this are really not random events. For the purposes of our discussion, though, we will assume they are—since the outcome still cannot be known beforehand. The best that can be done is to reduce the outcome to a probability statement. Try to think of the difference between independent and dependent trials processes as simply whether the probability statement is *fixed* (independent trials) or *variable* (dependent trials) from one event to the next based on prior outcomes. This is in fact the only difference.

Everything can be reduced to a probability statement. Events where the outcomes can be known prior to the fact differ from random events mathematically only in that their probability statements equal 1. For example, suppose that 51 cards have been removed from a deck of 52 cards and you know what the cards are. Therefore, you know what the one remaining card is with a probability of 1 (certainty). For the time being, we will deal with the independent trials process, particularly the simple coin toss.

MATHEMATICAL EXPECTATION

At this point it is necessary to understand the concept of mathematical expectation, sometimes known as the player's edge (if positive to the player) or the house's advantage (if negative to the player):

$$\text{Mathematical Expectation} = (1 + A) * P - 1 \qquad (1.03)$$

where: P = Probability of winning.
         A = Amount you can win/Amount you can lose.

So, if you are going to flip a coin and you will win $2 if it comes up heads, but you will lose $1 if it comes up tails, the mathematical expectation per flip is:

$$\begin{aligned}
\text{Mathematical Expectation} &= (1 + 2) * .5 - 1 \\
&= 3 * .5 - 1 \\
&= 1.5 - 1 \\
&= .5
\end{aligned}$$

In other words, you would expect to make 50 cents on average each flip.

This formula just described will give us the mathematical expectation for an event that can have two possible outcomes. What about situations where there are more than two possible outcomes? The next formula will give us the mathematical expectation for an unlimited number of outcomes. It will also give us the mathematical expectation for an event with only two possible outcomes such as the 2 for 1 coin toss just described. Hence, it is the preferred formula.

$$\text{Mathematical Expectation} = \sum_{i=1}^{N}(P_i * A_i) \qquad (1.03a)$$

where: P = Probability of winning or losing.
A = Amount won or lost.
N = Number of possible outcomes.

The mathematical expectation is computed by multiplying each possible gain or loss by the probability of that gain or loss, and then summing those products together.

Now look at the mathematical expectation for our 2 for 1 coin toss under the newer, more complete formula:

$$\text{Mathematical Expectation} = .5 * 2 + .5 * (-1)$$
$$= 1 + (-.5)$$
$$= .5$$

In such an instance, of course, your mathematical expectation is to win 50 cents per toss on average.

Suppose you are playing a game in which you must guess one of three different numbers. Each number has the same probability of occurring (.33), but if you guess one of the numbers you will lose \$1, if you guess another number you will lose \$2, and if you guess the right number you will win \$3. Given such a case, the mathematical expectation (ME) is:

$$\text{ME} = .33 * (-1) + .33 * (-2) + .33 * 3$$
$$= -.33 - .66 + .99$$
$$= 0$$

Consider betting on one number in roulette, where your mathematical expectation is:

$$\text{ME} = 1/38 * 35 + 37/38 * (-1)$$
$$= .02631578947 * 35 + .9736842105 * (-1)$$
$$= .9210526315 + (-.9736842105)$$
$$= -.05263157903$$

If you bet $1 on one number in roulette (American double-zero), you would expect to lose, on average, 5.26 cents per roll. If you bet $5, you would expect to lose, on average, 26.3 cents per roll. Notice how *different amounts bet have different mathematical expectations in terms of amounts, but the expectation as a percent of the amount bet is always the same.*

The player's expectation for a series of bets is the total of the expectations for the individual bets. So if you play $1 on a number in roulette, then $10 on a number, then $5 on a number, your total expectation is:

$$ME = (-.0526) * 1 + (-.0526) * 10 + (-.0526) * 5$$
$$= -.0526 - .526 - .263$$
$$= -.8416$$

You would therefore expect to lose on average 84.16 cents.

This principle explains why systems that try to change the size of their bets relative to how many wins or losses have been seen (assuming an independent trials process) are doomed to fail. The sum of negative-expectation bets is always a negative expectation!

EXACT SEQUENCES, POSSIBLE OUTCOMES, AND THE NORMAL DISTRIBUTION

We have seen how flipping one coin gives us a probability statement with two possible outcomes—heads or tails. Our mathematical expectation would be the sum of these possible outcomes. Now let's flip two coins. Here the possible outcomes are:

Coin 1	Coin 2	Probability
H	H	.25
H	T	.25
T	H	.25
T	T	.25

This can also be expressed as there being a 25% chance of getting both heads, a 25% chance of getting both tails, and a 50% chance of getting a head and a tail. In tabular format:

Combination	Probability	
H2	.25	*
T1H1	.50	**
T2	25	*

The asterisks to the right show how many different ways the combination can be made. For example in the above two-coin flip there are two asterisks for T1H1, since there are two different ways to get this combination. Coin A could be heads and coin B tails, or the reverse, coin A tails and coin B heads. The total number of asterisks in the table (four) is the total number of different combinations you can get when flipping that many coins (two).

If we were to flip three coins, we would have:

Combination	Probability	
H3	.125	*
H2T1	375	***
T2H1	.375	***
T3	125	*

for four coins:

Combination	Probability	
H4	.0625	*
H3T1	.25	****
H2T2	.375	******
T3H1	.25	****
T4	.0625	*

and for six coins:

Combination	Probability	
H6	.0156	*
H5T1	.0937	******
H4T2	.2344	***************
H3T3	.3125	********************
T4H2	.2344	***************
T5H1	.0937	******
T6	.0156	*

Notice here that if we were to plot the asterisks vertically we would be developing into the familiar bell-shaped curve, also called the Normal or Gaussian Distribution (see Figure 1.1).[1]

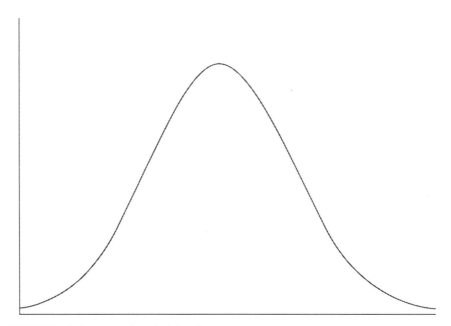

FIGURE 1.1 Normal probability function

---

[1]Actually, the coin toss does not conform to the Normal Probability Function in a pure statistical sense, but rather belongs to a class of distributions called the Binomial Distribution (a.k.a. Bernoulli or Coin-Toss Distributions). However, as N becomes large, the Binomial approaches the Normal Distribution as a limit (provided the probabilities involved are not close to 0 or 1). This is so because the Normal Distribution is continuous from left to right, whereas the Binomial is not, and the Normal is always symmetrical whereas the Binomial needn't be. Since we are treating a finite number of coin tosses and trying to make them representative of the universe of coin tosses, and since the probabilities are always equal to .5, we will treat the distributions of tosses as though they were Normal. As a further note, the Normal Distribution can be used as an approximation of the Binomial if both N times the probability of an event occurring and N times the complement of the probability occurring are both greater than 5. In our coin-toss example, since the probability of the event is .5 (for either heads or tails) and the complement is .5, then so long as we are dealing with N of 11 or more we can use the Normal Distribution as an approximation for the Binomial.

Finally, for 10 coins:

Combination	Probability	
H10	.001	*
H9T1	.01	**********
H8T2	.044	*****(45 different ways)
H7T3	.117	*****(120 different ways)
H6T4	205	*****(210 different ways)
H5T5	.246	*****(252 different ways)
T6H4	.205	*****(210 different ways)
T7H3	.117	*****(120 different ways)
T8H2	.044	*****(45 different ways)
T9H1	.01	**********
T10	.001	*

Notice that *as the number of coins increases, the probability of getting all heads or all tails decreases.* When we were using two coins, the probability of getting all heads or all tails was .25. For three coins it was .125, for four coins .0625; for six coins .0156, and for 10 coins it was .001.

POSSIBLE OUTCOMES AND STANDARD DEVIATIONS

So a coin flipped four times has a total of 16 possible exact sequences:

1.	H	H	H	H
2.	H	H	H	T
3.	H	H	T	H
4.	H	H	T	T
5.	H	T	H	H
6.	H	T	H	T
7.	H	T	T	H
8.	H	T	T	T
9.	T	H	H	H
10.	T	H	H	T
11.	T	H	T	H
12.	T	H	T	T
13.	T	T	H	H
14.	T	T	H	T
15.	T	T	T	H
16.	T	T	T	T

The term "exact sequence" here means the exact outcome of a random process. The set of all possible exact sequences for a given situation is called the *sample space*. Note that the four-coin flip just depicted can be four coins all flipped at once, or it can be one coin flipped four times (i.e., it can be a chronological sequence).

If we examine the exact sequence T H H T and the sequence H H T T, the outcome would be the same for a person flat-betting (i.e., betting 1 unit on each instance). However, to a person not flat-betting, the end result of these two exact sequences can be far different. To a flat bettor there are only five possible outcomes to a four-flip sequence:

> 4 Heads
>
> 3 Heads and 1 Tail
>
> 2 Heads and 2 Tails
>
> 1 Head and 3 Tails
>
> 4 Tails

As we have seen, there are 16 possible exact sequences for a four-coin flip. This fact would concern a person who is not flat-betting. We will refer to people who are not flat-betting as "system" players, since that is most likely what they are doing—betting variable amounts based on some scheme they think they have worked out.

If you flip a coin four times, you will of course see only one of the 16 possible exact sequences. If you flip the coin another four times, you will see another exact sequence (although you could, with a probability of 1/16 = .0625, see the exact same sequence). If you go up to a gaming table and watch a series of four plays, you will see only one of the 16 exact sequences. You will also see one of the five possible end results. *Each exact sequence (permutation) has the same probability of occurring*, that being .0625. *But each end result (combination) does not have equal probability of occurring*:

End Result	Probability
4 Heads	.0625
3 Heads and 1 Tail	.25
2 Heads and 2 Tails	.375
1 Head and 3 Tails	.25
4 Tails	.0625

Most people do not understand the difference between exact sequences (permutation) and end results (combination) and as a result falsely conclude that exact sequences and end results are the same thing. This is a common misconception that can lead to a great deal of trouble. It is the end results (not the exact sequences) that conform to the bell curve—the Normal Distribution, which is a particular type of probability distribution. An interesting characteristic of all probability distributions is a statistic known as the *standard deviation.*

For the Normal Probability Distribution on a simple binomial game, such as the one being used here for the end results of coin flips, the standard deviation (SD) is:

$$D = N * \sqrt{\frac{P * (1 - P)}{N}} \tag{1.04}$$

where: $P =$ Probability of the event (e.g., result of heads).
$N =$ Number of trials.

For 10 coin tosses (i.e., $N = 10$):

$$SD = 10 * \sqrt{.5 * (1 - .5)/10}$$
$$= 10 * \sqrt{.5 * .5/10}$$
$$= 10 * \sqrt{.25/10}$$
$$= 10 * .158113883$$
$$= 1.58113883$$

The center line of a distribution is the peak of the distribution. In the case of the coin toss the peak is at an even number of heads and tails. So for a 10-toss sequence, the center line would be at 5 heads and 5 tails. For the Normal Probability Distribution, approximately 68.26% of the events will be + or − 1 standard deviation from the center line, 95.45% between + and − 2 standard deviations from the center line, and 99.73% between + and − 3 standard deviations from the center line (see Figure 1.2). Continuing with our 10-flip coin toss, 1 standard deviation equals approximately 1.58. We can therefore say of our 10-coin flip that 68% of the time we can expect to have our end result be composed of 3.42 (5 − 1.58) to 6.58 (5 + 1.58) being heads (or tails). So if we have 7 heads (or tails), we would be beyond 1 standard deviation of the expected outcome (the expected outcome being 5 heads and 5 tails).

Here is another interesting phenomenon. Notice in our coin-toss examples that as the number of coins tossed increases, the probability of getting an even number of heads and tails decreases. With two coins the probability of getting H1T1 was .5. At four coins the probability of getting 50% heads and 50% tails dropped to .375. At six coins it was .3125, and at 10 coins .246. Therefore, we can state that *as the number of events increases, the probability of the end result exactly equaling the expected value decreases.*

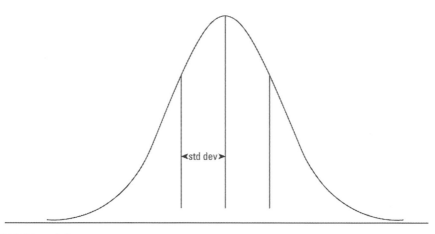

FIGURE 1.2 Normal probability function: Center line and 1 standard deviation in either direction

The mathematical expectation is what we expect to gain or lose, on average, each bet. However, it does not explain the fluctuations from bet to bet. In our coin-toss example we know that there is a 50/50 probability of a toss's coming up heads or tails. We expect that after N trials approximately $1/2 * N$ of the tosses will be heads, and $1/2 * N$ of the tosses will be tails. Assuming that we lose the same amount when we lose as we make when we win, we can say we have a mathematical expectation of 0, regardless of how large N is.

We also know that approximately 68% of the time we will be + or − 1 standard deviation away from our expected value. For 10 trials (N = 10) this means our standard deviation is 1.58. For 100 trials (N = 100) this means we have a standard deviation size of 5. At 1,000 (N = 1,000) trials the standard deviation is approximately 15.81. For 10,000 trials (N = 10,000) the standard deviation is 50.

N	Std Dev	Std Dev/N as%
10	1.58	15.8%
100	5	5.0%
1,000	15.81	1.581%
10,000	50	0.5%

Notice that as N increases, the standard deviation increases as well. This means that contrary to popular belief, *the longer you play, the*

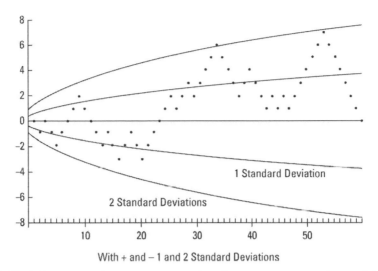

With + and – 1 and 2 Standard Deviations

FIGURE 1.3 The random process: Results of 60 coin tosses, with 1 and 2 standard deviations in either direction

further you will be from your expected value (in terms of units won or lost). However, as N increases, the standard deviation as a percent of N decreases. This means that *the longer you play, the closer to your expected value you will be as a percent of the total action (N).* This is the "Law of Averages" presented in its mathematically correct form. In other words, if you make a long series of bets, N, where T equals your total profit or loss and E equals your expected profit or loss, then T/N tends towards E/N as N increases. Also, the difference between E and T increases as N increases.

In Figure 1.3 we observe the random process in action with a 60-coin-toss game. Also on this chart you will see the lines for + and − 1 and 2 standard deviations. Notice how they bend in, yet continue outward forever. This conforms with what was just said about the Law of Averages.

THE HOUSE ADVANTAGE

Now let us examine what happens when there is a house advantage involved. Again, refer to our coin-toss example. We last saw 60 trials at an even or "fair" game. Let's now see what happens if the house has a 5% advantage. An example of such a game would be a coin toss where if we win, we win $1, but if we lose, we lose $1.10.

Figure 1.4 shows the same 60-coin-toss game as we previously saw, only this time there is the 5% house advantage involved. Notice how, in

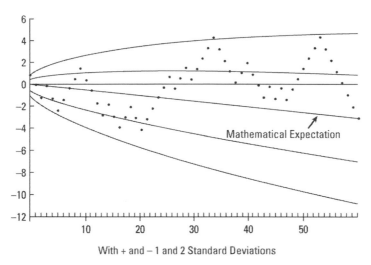

With + and – 1 and 2 Standard Deviations

FIGURE 1.4 Results of 60 coin tosses with a 5% house advantage

this scenario, ruin is inevitable—as the upper standard deviations begin to bend down (to eventually cross below zero).

Let's examine what happens when we continue to play a game with a negative mathematical expectation.

N	Std Dev	Expectation	+ or −1 SD
10	1.58	−.5	+1.08 to −2.08
100	5.00	−5	0 to − 10
1,000	15.81	−50	−34.19 to −65.81
10,000	50.00	−500	−450 to −550
100,000	158.11	−5,000	−4,842 to −5,158
1,000,000	500.00	−50,000	−49,500 to −50,500

The principle of ergodicity is at work here. It doesn't matter if one person goes to a casino and bets $1 one million times in succession or if one million people come and bet $1 each all at once. The numbers are the same. At one million bets, it would take more than 100 standard deviations away from the expectation before the casino started to lose money! Here is the Law of Averages at work. By the same account, if you were to make one million $1 bets at a 5% house advantage, it would be equally unlikely for you to make money. Many casino games have more than a 5% house advantage, as does most sports betting. Trading the markets is a zero-sum game. However, there is a small drain involved in the way of commissions, fees, and slippage. Often these costs can run in excess of 5%.

Next, let's examine the statistics of a 100-coin-toss game with and without a 5% house advantage:

Std. Deviations from Center	Fair 50/50 Game	5% House Advantage Game
+3	+15	+10
+2	+10	+5
+1	+5	0
0	0	−5
−1	−5	−10
−2	−10	−15
−3	−15	−20

As can be seen, at 3 standard deviations, which we can expect to be the outcome 99.73% of the time, we will win or lose between +15 and −15 units in a fair game. At a house advantage of 5%, we can expect our final outcome to be between +10 and −20 units at the end of 100 trials. At 2 standard deviations, which we can expect to occur 95% of the time, we win or lose within + or −10 in a fair game. At a 5% house advantage this is +5 and −15 units. At 1 standard deviation, where we can expect the final outcome to be with 68% probability, we win or lose up to 5 units in a fair game. Yet in the game where the house has the 5% advantage we can expect the final outcome to be between winning nothing and losing 10 units! Note that at a 5% house advantage it is not impossible to win money after 100 trials, but you would have to do better than 1 whole standard deviation to do so. In the Normal Distribution, the probability of doing better than 1 whole standard deviation, you will be surprised to learn, is only .1587!

Notice in the previous example that at 0 standard deviations from the center line (that is, at the center line itself), the amount lost is equal to the house advantage. For the fair 50/50 game, this is equal to 0. You would expect neither to win nor to lose anything. In the game where the house has the 5% edge, you would expect to lose 5%, 5 units for every 100 trials, at 0 standard deviations from the center line. So you can say that *in flat-betting situations involving an independent process, you will lose at the rate of the house advantage.*

MATHEMATICAL EXPECTATION LESS THAN ZERO SPELLS DISASTER

This brings us to another axiom, which can be stated as follows: *In a negative expectancy game, there is no money management scheme that will make you a winner. If you continue to bet, regardless of how you manage your money, it is almost certain that you will be a loser, losing your entire stake regardless of how large it was to start.*

This sounds like something to think about. Negative mathematical expectations (regardless of how negative) have broken apart families and caused suicides and murders and all sorts of other things the bettors weren't bargaining for. I hope you can see what an incredibly losing proposition it is to make bets where there is a negative expectancy, for even a small negative expectancy will eventually take every cent you have. All attempts to outsmart this process are mathematically futile. Don't get this idea confused with whether or not there is a dependent or independent trials process involved; it doesn't matter. If the sum of your bets is a negative expectancy, you are in a losing proposition.

As an example, if you are in a dependent trials process where you have an edge in 1 bet out of 10, then you must bet enough on the bet for which you have an edge so that the sum of all 10 bets is a positive expectancy situation. If you expect to lose 10 cents on average for 9 of the 10 bets, but you expect to make 10 cents on the 1 out of 10 bets where you know you have the edge, then you must bet more than 9 times as much on the bet where you know you have the edge, just to have a net expectation of coming out even. If you bet less than that, you are still in a negative expectancy situation, and complete ruin is all but certain if you continue to play.

Many people have the mistaken impression that if they play a negative expectancy game, they will lose a percentage of their capital relative to the negative expectancy. For example, when most people realize that the mathematical expectation in roulette is 5.26% they seem to think this means that if they go to a casino and play roulette they can expect to lose, on average, 5.26% of their stake. This is a dangerous misconception. The truth is that they can expect to lose 5.26% of their *total action*, not of their entire stake. Suppose they take $500 to play roulette. If they make 500 bets of $20 each, their total action is $10,000, of which they can expect to lose 5.26%, or $526, more than their entire stake.

The only smart thing to do is bet only when you have a positive expectancy. This is not so easily a winning proposition as negative expectancy betting is a losing proposition, as we shall see in a later chapter. You must bet specific quantities, which will be discussed at length. For the time being, though, resolve to bet only on positive expectancy situations.

When it comes to casino gambling, though, the only time you can find a positive expectancy situation is if you keep track of the cards in blackjack, and then only if you are a very good player, and only if you bet your money correctly. There are many good blackjack books available, so we won't delve any further into blackjack here.

BACCARAT

If you want to gamble at a casino but do not want to learn to play blackjack correctly, then baccarat has the smallest negative expectancy of any other casino game. In other words, you'll lose your money at a slower rate. Here are the probabilities in baccarat:

> Banker wins 45.842% of the time.
> Player wins 44.683% of the time.
> A tie occurs 9.547% of the time.

Since a tie is treated as a push in baccarat (no money changes hands, the net effect is the same as if the hand were never played) the probabilities, when ties are eliminated become:

> Banker wins 50.68% of the time.
> Player wins 49.32% of the time.

Now let's look at the mathematical expectations. For the player side:

$$ME = (.4932 * 1) + ((1 - .4932) * (-1))$$
$$= (.4932 * 1) + (.5068 * (-1))$$
$$= .4932 - .5068$$
$$= -.0136$$

In other words, the house advantage over the player is 1.36%.

Now for the banker side, bearing in mind that the banker side is charged a 5% commission on wins only, the mathematical expectation is:

$$ME = (.5068 * .95) + ((1 - .5068) * (-1))$$
$$= (.5068 * .95) + (.4932 * (-1))$$
$$= .48146 - .4932$$
$$= -.01174$$

In other words, the house has an advantage, once commissions on the banker's wins are accounted for, of 1.174%.

As you can see, it makes no sense to bet on the player since the player's negative expectancy is worse than the banker's:

Player's disadvantage	−.0136
Banker's disadvantage	−.01174
Banker's edge over Player	.00186

In other words, after about 538 hands (1/.00186) the banker will be 1 unit ahead of the player. Again, the more hands that are played, the more certain this edge is.

This is not to imply that the banker has a positive mathematical expectation—he doesn't. Both banker and player have negative expectations, but the banker's is not as negative as the player's. Betting 1 unit on the banker on each hand, you can expect to lose 1 unit for approximately every 85 hands (1/.01174); whereas betting 1 unit on the player on each hand, you would expect to lose 1 unit every 74 hands (1/.0136). You will lose your money at a slower *rate*, but not necessarily a slower *pace*. Most baccarat tables have at least a $25 minimum bet. If you are betting banker, 1 unit per hand, after 85 hands you can expect to be down $25.

Let's compare this to betting red/black at roulette, where you have a mathematical expectation of −.0526, but a minimum bet size of at least $2. After 85 spins you would expect to be down about $9 ($2 ∗ 85 ∗ .0526). As you can see, mathematical expectation is also a function of the total amount bet, the action. If, as in baccarat, we were betting $25 per spin in red/black roulette, we would expect to be down $112 after 85 spins, compared with baccarat's expected loss of $25.

NUMBERS

Finally, let's take a look at the probabilities involved in numbers. If baccarat is the game of the rich, numbers is the game of the poor. The probabilities in the numbers game are absolutely pathetic. Here is a game where a player chooses a three-digit number between 0 and 999 and bets $1 that this number will be selected. The number that gets chosen as that day's number is usually some number that (a) cannot be rigged and (b) is well publicized. An example would be to take the first three of the last five digits of the daily stock market volume. If the player loses, then the $1 he bet is lost. If the player should happen to win, then $700 is returned, for a net profit of $699. For numbers, the mathematical expectation is:

$$ME = (699 * (1/1000)) + ((-1) * (1 - (1/1000)))$$
$$= (699 * .001) + ((-1) * (1 - .001))$$
$$= (699 * .001) + ((-1) * .999)$$
$$= .699 + (-.999)$$
$$= -.3$$

In other words your mathematical expectation is to lose 30 cents for every dollar of action. This is far worse than any casino game, including keno. Bad as the probabilities are in a game like roulette, the mathematical expectation in numbers is almost six times worse. The only gambling situations that are worse than this in terms of mathematical expectation are most football pools and many of the state lotteries.

PARI-MUTUEL BETTING

The games that offer seemingly the worst mathematical expectation belong to a family of what are called pari-mutuel games. Pari-mutuel means literally "to bet among ourselves." Pari-mutuel betting was originated in the 1700s by a French perfume manufacturer named Oller. Monsieur Oller, doubling as a bookie, used his perfume bottles as ticket stubs for his patrons while he booked their bets. Oller would take the bets, from this total pool he would take his cut, then he would distribute the remainder to the winners. Today we have different types of games built on this same pari-mutuel scheme, from state lotteries to football pools, from numbers to horse racing. As you have seen, the mathematical expectations on most pari-mutuel games are atrocious. Yet these very games also offer many situations that have a positive mathematical expectancy.

Let's take numbers again, for example. We can approximate how much money is bet in total by taking the average winning purse size and dividing it by 1 minus the take. In numbers, as we have said, the take is 30%, so we have $1 - .3$, or .7. Dividing 1 by .7 yields 1.42857. If the average payout is, say, $1,400, then we can approximate the total purse as 1,400 times 1.42857, or roughly $2,000. So step one in finding positive mathematical expectations in pari-mutuel situations is to know or at least closely approximate the total amount in the pool.

The next step is to take this total amount and divide it by the total number of possible combinations. This gives the average amount bet per combination. In numbers there are 1,000 possible combinations, so in our example we divide the approximate total pool of $2,000 by 1,000, the total number of combinations, to obtain an average bet per combination of $2.

Now we figure the total amount bet on the number we want to play. Here we would need inside information. The purpose here is not to show how to win at numbers or any other gambling situation, but rather to show how to think correctly in approaching a given risk/reward situation. This will be made clearer as we continue with the illustration. For now, let's just assume we can get this information. Now, if we know what the average dollar bet is on any number, and we know the total amount bet on the number we want to play, we simply divide the average bet by the amount bet on our number. This gives us the ratio of what our bet size is relative to the average bet size.

Since the pool can be won by any number, and since the pool is really the average bet times all possible combinations, it stands to reason that naturally we want our bet to be relatively small compared to the average bet. Therefore, if this ratio is 1.5, it means simply that the average bet on a number is 1.5 times the amount bet on our number.

Now this can be converted into an actual mathematical expectation. We take this ratio and multiply it by the quantity (1 − takeout) where the takeout is the pari-mutuel vigorish (also known as the amount that the house skims off the top, and out of the total pool). In the case of numbers, where the takeout is 30%, then 1 minus the takeout equals .7. Multiplying our ratio in our example of 1.5 times .7 gives us 1.05. As a final step, subtracting 1 from the previous step's answer will give us the mathematical expectation, in percent. Since 1.05 − 1 is 5%, we can expect in our example situation to make 5% on our money on average if we make this play over and over.

Which brings us to an interesting proviso here. In numbers, we have probabilities of 1/1000 or .001 of winning. So, in our example, if we bet $1 for each of 1,000 plays, we would expect to be ahead by 5%, or $50, if the given parameters as we just described were always present. Since it is possible to play the number 1,000 times, the mathematical expectation is possible, too.

But let's say you try to do this on a state lottery with over 7 million possible winning combinations. Unless you have a pool together or a lot of money to cover more than one number on each drawing, it is unlikely you will see over 7 million drawings in your lifetime. Since it will take (on average) 7 million drawings until you can mathematically expect your number to have come up, your positive mathematical expectation as we described it in the numbers example is meaningless. You most likely won't be around to collect!

In order for the mathematical expectation to be meaningful (provided it is positive) you must be able to get enough trials off in your lifetime (or the pertinent time period you are considering) to have a fair mathematical chance of winning. The average number of trials needed is the total number of possible combinations divided by the number of combinations you are playing. Call this answer N. Now, if you multiply N by the length of time it takes for 1 trial to occur, you can determine the average length of time needed for you to be able to expect the mathematical expectation to manifest itself. If your chances are 1 in 7 million and the drawing is once a week, you must stick around for 7 million weeks (about 134,615 years) to expect the mathematical expectation to come into play. If you bet 10,000 of those 7 million combinations, you must stick around about 700 weeks (7 million divided by 10,000, or about $13\frac{1}{2}$ years) to expect the mathematical expectation to kick in, since that is about how long, on average, it would take until one of those 10,000 numbers won.

The procedure just explained can be applied to other pari-mutuel gambling situations in a similar manner. There is really no need for inside information on certain games. Consider horse racing, another classic parimutuel situation. We must make one assumption here. We must assume that the money bet on a horse to win divided by the total win pool is an accurate reflection of the true probabilities of that horse winning. For instance, if the total win pool is \$25,000 and there is \$2,500 bet on our horse to win, we must assume that the probability of our horse's winning is .10. We must assume that if the same race were run 100 times with the same horses on the same track conditions with the same jockeys, and so on, our horse would win 10% of the time.

From that assumption we look now for opportunity by finding a situation where the horse's proportion of the show or place pools is much less than its proportion of the win pool. The opportunity is that if a horse has a probability of X of winning the race, then the probability of the horse's coming in second or third should not be less than X (provided, as we already stated, that X is the real probability of that horse winning). If the probability of the horse's coming in second or third is less than the probability of the horse's winning the race, an anomaly is created that we can perhaps capitalize on.

The following formula reduces what we have spoken of here to a mathematical expectation for betting a particular horse to place or show, and incorporates the track takeout. Theoretically, all we need to do is bet only on racing situations that have a positive mathematical expectation. The mathematical expectation of a show (or place) bet is given as:

$$(((W_i / \Sigma W)/(S_i / \Sigma S)) * (1 - \text{takeout}) - 1 \qquad (1.03b)$$

where: W_i = Dollars bet on the ith horse to win.
ΣW = Total dollars in the win pool—i.e., total dollars bet on all horses to win.
S_i = Dollars bet on the ith horse to show (or place).
ΣS = Total dollars in the show (or place) pool—i.e., total dollars on all horses to show (or place).
i = The horse of your choice.

If you've truly learned what is in this book you will use the Kelly formula (more on this in Chapter 4) to maximize the rate of your money's growth. How much to bet, however, becomes an iterative problem, in that the more you bet on a particular horse to show, the more you will change the mathematical expectation and payout—but not the probabilities, since they are dictated by $(W_i/\Sigma W)$. Therefore, when you bet on the horse to place, you alter the mathematical expectation of the bet and you also alter the payout on that horse to place. Since the Kelly formula is affected by the payout, you must be able to iterate to the correct amount to bet.

As in all winning gambling or trading systems, employing the winning formula just shown is far more difficult than you would think. Go to the racetrack and try to apply this method, with the pools changing every 60 seconds or so while you try to figure your formula and stand in line to make your bet and do it within seconds of the start of the race. The real-time employment of any winning system is always more difficult than you would think after seeing it on paper.

WINNING AND LOSING STREAKS IN THE RANDOM PROCESS

We have already seen that in flat-betting situations involving an independent trials process you will lose at the rate of the house advantage. To get around this rule, many gamblers then try various betting schemes that will allow them to win more during hot streaks than during losing streaks, or will allow them to bet more when they think a losing streak is likely to end and bet less when they think a winning streak is about to end. Yet another important axiom comes into play here, which is that *streaks are no more predictable than the outcome of the next event* (this is true whether we are discussing dependent or independent events). In the long run, we can predict approximately how many streaks of a given length can be expected from a given number of chances.

Imagine that we flip a coin and it lands tails. We now have a streak of one. If we flip the coin a second time, there is a 50% chance it will come up tails again, extending the streak to two events. There is also a 50% chance it will come up heads, ending the streak at one. Going into the third flip we face the same possibilities. Continuing with this logic we can construct the following table, assuming we are going to flip a coin 1,024 times:

Length of Streak	No. of Streaks Occurring	How Often Compared to Streak of One	Probability
1	512	1	.50
2	256	1/2	.25
3	128	1/4	.125
4	64	1/8	.0625
5	32	1/16	.03125
6	16	1/32	.015625
7	8	1/64	.0078125
8	4	1/128	.00390625
9	2	1/256	.001953125
10	1	1/512	.0009765625
11+	1	1/1024	.00048828125

The real pattern does not end at this point; rather it continues with smaller and smaller numbers.

Remember that this is the expected pattern. The real-life pattern, should you go out and record 1,024 coin flips, will resemble this, but most likely it won't resemble this exactly. This pattern of 1,024 coin tosses is for a fair 50/50 game. In a game where the house has the edge, you can expect the streaks to be skewed by the amount of the house advantage.

DETERMINING DEPENDENCY

As we have already explained, the coin toss is an independent trials process. This can be deduced by inspection, in that we can calculate the exact probability statement prior to each toss and it is always the same from one toss to the next. There are other events, such as blackjack, that are dependent trials processes. These, too, can be deduced by inspection, in that we can calculate the exact probability statement prior to each draw of a card, and it is not always the same from one draw to the next. For still other events, dependence on prior outcomes cannot be determined upon inspection. Such an event is the profit and loss stream of trades generated by a trading system. For these types of problems we need more tools.

Assume the following stream of coin flips where a plus (+) stands for a win and a minus (−) stands for a loss:

$$+ + - - - - - - - + - + - + - - - + + + - + + + - + ++$$

There are 28 trades, 14 wins and 14 losses. Say there is $1 won on a win and $1 lost on a losing flip. Hence, the net for this series is $0.

Now assume you possess the infant's mind. You do not know if there is dependency or not in the coin-toss situation (although there isn't). Upon seeing such a stream of outcomes you deduce the following rule, which says, "Don't bet after two losers; go to the sidelines and wait for a winner to resume betting." With this new rule, the previous sequence would have been:

$$+ + - - - + - + - - + + - + + + - + ++$$

So, with this new rule the old sequence would have produced 12 winners and 8 losers for a net of $4. You're quite confident of your new rule. You haven't learned to differentiate an exact sequence (which is all that this stream of trades is) from an end result (the end result being that this is a break-even game).

There is a major problem here, though, and that is that you do not know if there is dependency in the sequence of flips. *Unless dependency is proven, no attempt to improve performance based on the stream of profits and losses alone is of any value, and quite possibly you may do more harm than good.*[2] Let us continue with the illustration and we will see why.

---

[2] A distinction must be drawn between a stationary and a nonstationary distribution. A stationary distribution is one where the probability distribution does not change. An example would be a casino game such as roulette, where you are always at a .0526 disadvantage. A nonstationary distribution is one where the expectation changes over time (in fact, the entire probability distribution may change over time). Trading is just such a case. Trading is analogous in this respect to a drunk wandering through a casino, going from game to game. First he plays roulette with $5 chips (for a −.0526 mathematical expectation), then he wanders to a blackjack table, where the deck happens to be running favorable to the player by 2%. His distribution of outcomes curve moves around as he does; the mathematical expectation and distribution of outcomes is dynamic. Contrast this to staying at one table, at one game. In such a case the distribution of outcomes is static. We say it is *stationary*. The outcomes of systems trading appear to be a nonstationary distribution, which would imply that there is perhaps some technique that may be employed to allow the trader to advantageously "trade his equity curve." Such techniques are, however, beyond the mathematical scope of this book and will not be treated here. Therefore, we will not treat nonstationary distributions any differently than stationary ones in the text, but be advised that the two are profoundly different.

Since this was a coin toss, there was in fact no dependency in the trials—that is, the outcome of each successive flip was independent of (unaffected by) the previous flips. Therefore, this exact sequence of 28 flips was totally random. (Remember, each exact sequence has an equal probability of occurring. It is the end results that follow the Normal Distribution, with the peak of the distribution occurring at the mathematical expectation. The end result in this case, the mathematical expectation, is a net profit/loss of zero.) The next exact sequence of 28 flips is going to appear randomly, and there is an equal probability of the following sequence appearing as any other:

$$- - + - - + - - + - - + - - + - - + - - + + + + + + + +$$

Once again, the net of this sequence is nothing won and nothing lost. Applying your rule here, the outcome is:

Fourteen losses and seven wins for a net loss of $7.

As you can see, unless dependency is proven (in a stationary process), no attempt to improve performance based on the stream of profits and losses alone is of any value, and you may do more harm than good.

THE RUNS TEST, Z SCORES, AND CONFIDENCE LIMITS

For certain events, such as the profit and loss stream of a system's trades, where dependency cannot be determined upon inspection, we have the runs test. The runs test is essentially a matter of obtaining the Z scores for the win and loss streaks of a system's trades. Here's how to do it. First, you will need a minimum of 30 closed trades. There is a very valid statistical reason for this. Z scores assume a Normal Probability Distribution (of streaks of wins and losses in this instance). Certain characteristics of the Normal Distribution are no longer valid when the number of trials is less than 30. This is because a minimum of 30 trials are necessary in order to resolve the shape of the Normal Probability Distribution clearly enough to make certain statistical measures valid.

The Z score is simply the number of standard deviations the data is from the mean of the Normal Probability Distribution. For example, a Z score of 1.00 would mean that the data you are testing is within 1 standard deviation from the mean. (Incidentally, this is perfectly normal.) The Z score is then converted into a confidence limit, sometimes also called a degree of certainty. We have seen that the area under the curve of the Normal Probability Function at 1 standard deviation on either side of the mean equals 68% of the total area under the curve. So we take our Z score and convert it to a confidence limit, the relationship being that the Z score is how many standard deviations and the confidence limit is the percentage of area under the curve occupied at so many standard deviations.

Confidence Limit	Z Score
99.73%	3.00
99%	2.58
98%	2.33
97%	2.17
96%	2.05
95.45%	2.00
95%	1.96
90%	1.64
85%	1.44
80%	1.28
75%	1.15
70%	1.04
68.27%	1.00
65%	.94
60%	.84
50%	.67

With a minimum of 30 closed trades we can now compute our Z scores. We are trying to determine how many streaks of wins/losses we can expect from a given system. Are the win/loss streaks of the system we are testing in line with what we could expect? If not, is there a high enough confidence limit that we can assume dependency exists between trades, that is, the outcome of a trade dependent on the outcome of previous trades?

Here, then, is how to perform the runs test, how to find a system's Z score:

1. You will need to compile the following data from your run of trades:

   A. The total number of trades, hereafter called N.

   B. The total number of winning trades and the total number of losing trades. Now compute what we will call X. $X = 2 *$ Total Number of Wins $*$ Total Number of Losses.

   C. The total number of runs in a sequence. We'll call this R.

Let's construct an example to follow along with. Assume the following trades:

$$-3 \quad +2 \quad +7 \quad -4 \quad +1 \quad -1 \quad +1 \quad +6 \quad -1 \quad 0 \quad -2 \quad +1$$

The net profit is +7. The total number of trades is 12; therefore, N = 12 (we are violating the rule that there must be at least 30 trades only to keep the example simple). Now we are not concerned here with how big the wins and losses are, but rather how many wins and losses there are and how many streaks. Therefore, we can reduce our run of trades to a simple sequence of pluses and minuses. Note that a trade with a profit and loss (P&L) of 0 is regarded as a loss. We now have:

$$- + + - + - + + - - - +$$

As can be seen, there are six profits and six losses. Therefore, X = 2 * 6 * 6 = 72. As can also be seen, there are eight runs in this sequence, so R = 8. We will define a *run* as any time we encounter a sign change when reading the sequence as shown above from left to right (i.e., chronologically). Assume also that we start at 1. Therefore, we would count this sequence as follows:

$$- + + - + - + + - - - +$$
$$1 \ 2 \quad 3\ 4\ 5\ 6 \quad 7 \qquad 8$$

2. Solve for the equation:

$$N * (R - .5) - X$$

For our example this would be:

$$12 * (8 - .5) - 72$$
$$12 * 7.5 - 72$$
$$90 - 72$$
$$18$$

3. Solve for the equation:

$$X * (X - N)/(N - 1)$$

So for our example this would be:

$$72 * (72 - 12)/(12 - 1)$$
$$72 * 60/11$$
$$4{,}320/11$$
$$392.727272$$

4. Take the square root of the answer in number 3. For our example this would be:

$$\sqrt{392.727272} = 19.81734777$$

5. Divide the answer in number 2 by the answer in number 4. This is the Z score. For our example this would be: $18/19.81734777 = .9082951063$

6.   Confidence Limit $= 1 - (2 * (X * .31938153 - Y * .356563782$
$$+ (X * Y * 1.781477937 - Y^2 * 1.821255978$$
$$+ 1.821255978 + Y^2 * X * 1.330274429) * 1$$
$$/\sqrt{\text{EXP}(Z^2) * 6.283185307}))$$

where: $X = 1.0/(((\text{ABS}(Z)) * .2316419) + 1.0)$.
$Y = X \wedge 2$.
$Z = $ The Z score you are converting from.
$\text{EXP}(\ ) = $ The exponential function.
$\text{ABS}(\ ) = $ The absolute value function.

This will give you the confidence limit for the so-called "two-tailed" test. To convert this to a confidence limit for a "one-tailed" test:

$$\text{Confidence Limit} = 1 - (1 - A)/2$$

where: $A = $ The "two-tailed" confidence limit.

If the Z score is negative, simply convert it to positive (take the absolute value) when finding your confidence limit. A negative Z score implies positive dependency, meaning fewer streaks than the Normal Probability Function would imply, and hence that wins beget wins and losses beget losses. A positive Z score implies negative dependency, meaning more streaks than the Normal Probability Function would imply, and hence that wins beget losses and losses beget wins.

As long as the dependency is at an acceptable confidence limit, you can alter your behavior accordingly to make better trading decisions, even though you do not understand the underlying cause of the dependency. Now, if you could know the cause, you could then better estimate when the dependency was in effect and when it was not, as well as when a change in the degree of dependency could be expected.

The runs test will tell you if your sequence of wins and losses contains more or fewer streaks (of wins or losses) than would ordinarily be expected in a truly random sequence, which has no dependence between trials. Since we are at such a relatively low confidence limit, we can assume that there is no dependence between trials in this particular sequence.

What would be an acceptable confidence limit then? Dependency can never be proved nor disproved beyond a shadow of a doubt in this test; therefore, what constitutes an acceptable confidence limit is a personal choice. Statisticians generally recommend selecting a confidence limit at least in the high nineties. Some statisticians recommend a confidence limit in excess of 99% in order to assume dependency; some recommend a less stringent minimum of 95.45% (2 standard deviations).

Rarely, if ever, will you find a system that shows confidence limits in excess of 95.45%. Most frequently, the confidence limits encountered are less than 90%. Even if you find one between 90 and 95.45%, this is not exactly a nugget of gold, either. You really need to exceed 95.45% as a bare minimum to assume that there is dependency involved that can be capitalized upon to make a substantial difference.

For example, some time ago a broker friend of mine asked me to program a money management idea of his that incorporated changes in the equity curve. Before I even attempted to satisfy his request, I looked for dependency between trades, since we all know now that unless dependency is proven (in a stationary process) to a very high confidence limit, all attempts to change your trading behavior based on changes in the equity curve are futile and may even be harmful.

Well, the Z score for this system (of 423 trades) clocked in at −1.9739! This means that there is a confidence limit in excess of 95%, a very high reading compared to most trading systems, but hardly an acceptable reading for dependency in a statistical sense. The negative number meant that wins beget wins and losses beget losses in this system. Now this was a great system to start with. I immediately went to work having the system pass all trades after a loss, and continue to pass trades until it passed what would have been a winning trade, then to resume trading. Here are the results:

	Before Rule	After Rule
Total Profits	$71,800	$71,890
Total Trades	423	360
Winning Trades	358	310
Winning Percentage$	84.63%	86.11%
Average Trade	$169.74	$199.69
Maximum Drawdown	$4,194	$2,880
Max. Losers in Succession	4	2
4 losers in a row	2	0
3 losers in a row	1	0
2 losers in a row	7	4

All of the above is calculated with $50 commissions and slippage taken off of each trade. As you can see, this was a terrific system before this rule. So good, in fact, that it was difficult to improve upon it in any way. Yet, once the dependency was found and exploited, the system was materially improved. It was with a confidence limit of slightly over 95%. It is rare to find a confidence limit this high in futures trading systems. However, from a statistical point of view, it is hardly high enough to assume that dependency exists. Ideally, yet rarely you will find systems that have confidence limits in the high nineties.

So far we have only looked at dependency from the point of view of whether the last trade was a winner or a loser. We are trying to determine if the sequence of wins and losses exhibit dependency or not. The runs test for dependency automatically takes the percentage of wins and losses into account. However, in performing the runs test on runs of wins and losses, we have accounted for the sequence of wins and losses but not their size. For the system to be truly independent, not only must the sequence of wins and losses be independent; the sizes of the wins and losses within the sequence must also be independent. It is possible for the wins and losses to be independent, while their sizes are dependent (or vice versa).

One possible solution is to run the runs test on only the winning trades, segregating the runs in some way (e.g., those that are greater than the median win versus those that are less). Then look for dependency among the size of the winning trades; then do the same for the losing trades.

THE LINEAR CORRELATION COEFFICIENT

There is, however, a different, possibly better way to quantify this possible dependency between the size of the wins and losses. The technique to be discussed next looks at the sizes of wins and losses from an entirely different mathematical perspective than does the runs test, and when used in conjunction with the latter, measures the relationship of trades with more depth than the runs test alone could provide. This technique utilizes the linear correlation coefficient, r, sometimes called Pearson's r, to quantify the dependency/independency relationship.

Look at Figure 1.5. It depicts two sequences that are perfectly correlated with each other. We call this effect "positive" correlation.

Now look at Figure 1.6. It shows two sequences that are perfectly uncorrelated with each other. When one line is zigging, the other is zagging. We call this effect "negative" correlation.

The formula for finding the linear correlation coefficient (r) between two sequences, X and Y, follows. (A bar over the variable means the mean

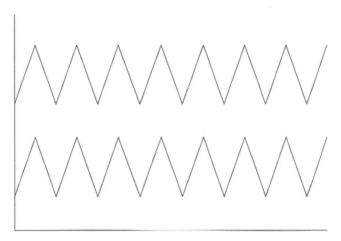

FIGURE 1.5 Perfect positive correlation (r = +1.00)

of the variables; for example, $\overline{X} = ((X_1 + X_2 + \ldots X_n)/n.)$

$$r = \frac{\sum_a(X_a - \overline{X}) * \sum_a(Y_a - \overline{Y})}{\sqrt{\sum_a(X_a - \overline{X})^2} * \sqrt{\sum_a(Y_a - \overline{Y})^2}} \qquad (1.05)$$

Here is how to perform the calculation as shown in the table on page 34:

1. Average the Xs and the Ys.
2. For each period, find the difference between each X and the average X and each Y and the average Y.

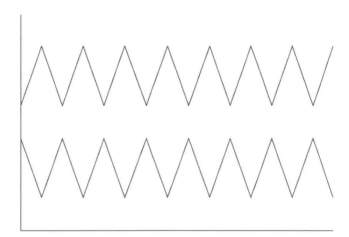

FIGURE 1.6 Perfect negative correlation (r = −1.00)

3. Now calculate the numerator. To do this, for each period, multiply the answers from step 2. In other words, for each period, multiply the difference between that period's X and the average X times the difference between that period's Y and the average Y.

4. Total up all of the answers to step 3 for all of the periods. This is the numerator.

5. Now find the denominator. To do this, take the answers to step 2 for each period, for both the X differences and the Y differences, and square them (they will now all be positive numbers).

6. Sum up the squared X differences for all periods into one final total. Do the same with the squared Y differences.

7. Take the square root of the sum of the squared X differences you just found in step 7. Now do the same with the Ys by taking the square root of the sum of the squared Y differences.

8. Multiply together the two answers you just found in step 7. That is, multiply the square root of the sum of the squared X differences by the square root of the sum of the squared Y differences. This product is your denominator.

9. Divide the numerator you found in step 4 by the denominator you found in step 8. This is your linear correlation coefficient, r.

The value for r will always be between +1.00 and −1.00. A value of 0 indicates no correlation whatsoever.

Look at Figure 1.7. It represents the following sequence of 21 trades:

$$1, 2, 1, \quad -1, 3, 2, \quad -1, -2, -3, 1, \quad -2, 3, 1, 1, 2, 3, 3, \quad -1, 2, \quad -1, 3$$

Now, here is how we use the linear correlation coefficient to see if there is any correlation between the previous trade and the current trade. The idea is to treat the trade P&Ls as the X values in the formula for r. Superimposed over that, we duplicate the same trade P&Ls, only this time we skew them by one trade, and use these as the Y values in the formula for r. In other words the Y value is the previous X value (see Figure 1.8).

The averages are different because you average only those Xs and Ys that have a corresponding X or Y value—that is, you average only those values that overlap; therefore, the last Y value (3) is not figured in the Y average, nor is the first X value (1) figured in the X average.

The numerator is the total of all entries in column E (.8). To find the denominator we take the square root of the total in column F, which is 8.555699, and we take the square root of the total in column G, which is 8.258329, and multiply them together to obtain a denominator of 70.65578. Now we divide our numerator of .8 by our denominator of 70.65578 to obtain 0.011322. This is our linear correlation coefficient, r. If you're really on top of this, you would also compute your Z score on these trades,

FIGURE 1.7 Individual outcomes of 21 bets/trades

which (if you want to check your work) is .5916 to four decimal places, or less than a 50% confidence limit that like begets unlike (since the Z score was positive).

The linear correlation coefficient of .011322 in this case is hardly indicative of anything, but it is pretty much in the range you can expect for most trading systems. A high correlation coefficient in futures trading systems would be one that was greater than .25 to .30 on the positive side, or less than −.25 to −.30 on the negative side. High positive correlation generally suggests that big wins are seldom followed by big losses and

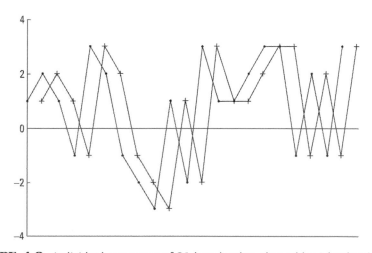

FIGURE 1.8 Individual outcomes of 21 bets/trades, skewed by 1 bet/trade

A	B	C	D	E col C times col D	F col C squared	G col D squared
X	Y	X–X avg	Y–Y avg			
1						
2	1	1.2	0.3	0.36	1.44	0.09
1	2	0.2	1.3	0.26	0.04	1.69
−1	1	−1.8	0.3	−0.54	3.24	0.09
3	−1	2.2	−1.7	−3.74	4.54	2.89
2	3	1.2	2.3	2.76	1.44	5.29
−1	2	−1.8	1.3	−2.34	3.24	1.69
−2	−1	−2.8	−1.7	4.76	7.84	2.89
−3	−2	−3.8	−2.7	10.26	14.44	7.29
1	−3	0.2	−3.7	−0.74	0.04	13.69
−2	1	−2.8	0.3	−0.84	7.84	0.09
3	−2	2.2	−2.7	−5.94	4.84	7.29
1	3	0.2	2.3	0.46	0.04	5.29
1	1	0.2	0.3	0.06	0.04	0.09
2	1	1.2	0.3	0.36	1.44	0.09
3	2	2.2	1.3	2.86	4.84	1.69
3	3	2.2	2.3	5.06	4.84	5.29
−1	3	−1.8	2.3	−4.14	3.24	5.29
2	−1	1.2	−1.7	−2.04	1.44	2.89
−1	2	−1.8	1.3	−2.34	3.24	1.69
3	−1	2.2	−1.7	−3.74	4.84	2.89
	3					
avg = 0.8	avg = 0.7	Totals =		0.8	73.2	68.2

vice versa. Negative correlation readings below −.25 to −.30 imply that big losses tend to be followed by big wins and vice versa.

There are a couple of reasons why it is important to use both the runs test and the linear correlation coefficient together in looking for dependency/correlation between trades. The first is that futures trading system trades (i.e., the profits and losses) do not necessarily conform to a Normal Probability Distribution. Rather, they conform pretty much to whatever the distribution is that futures prices conform to, which is as yet undetermined. Since the runs test assumes a Normal Probability Distribution, the runs test is only as accurate as the degree to which the system trade P&Ls conform to the Normal Probability Distribution.

The second reason for using the linear correlation coefficient in conjunction with the runs test is that the linear correlation coefficient is affected by the size of the trades. It not only interprets to what degree like begets like or like begets unlike, it also attempts to answer questions such as, "Are big winning trades generally followed by big losing trades?" "Are big losing trades generally followed by little losing trades?" And so on.

Negative correlation is just as helpful as positive correlation. For example, if there appears to be negative correlation, and the system has just suffered a large loss, we can expect a large win, and would therefore have more contracts on than ordinarily. Because of the negative correlation, if the trade proves to be a loss, the loss will most likely not be large.

Finally, in determining dependency you should also consider out-of-sample tests. That is, break your data segment into two or more parts. If you see dependency in the first part, then see if that dependency also exists in the second part, and so on. This will help eliminate cases where there appears to be dependency when in fact no dependency exists.

Using these two tools (the runs test and the linear correlation coefficient) can help answer many of these questions. However, they can answer them only if you have a high enough confidence limit and/or a high enough correlation coefficient (incidentally, the system we used earlier in this chapter, which had a confidence limit greater than 95%, had a correlation coefficient of only .0482). Most of the time, these tools are of little help, since all too often the universe of futures system trades is dominated by independence.

Recall the system mentioned in the discussion of Z scores that showed dependency to the 95% confidence limit. Based upon this statistic, we were able to improve this system by developing rules for passing trades. Now here is an interesting but disturbing fact. That system had one optimizeable parameter. When the system was run with a different value for that parameter, the dependency vanished! Was this saying that the appearance of dependency in our cited example was an illusion? Was it saying that only if you keep the value of this parameter within certain bounds can you have any dependency? If so, then isn't it possible that the appearance of dependency can be deceiving? To an extent this seems to be true.

Unfortunately, as traders, we most often must assume that dependency does not exist in the marketplace for the majority of market systems. That is, when trading a given market system, we will usually be operating in an environment where the outcome of the next trade is not predicated upon the outcome(s) of the preceding trade(s). This is not to say that there is never dependency between trades for some market systems (because for some market systems dependency does exist), only that we should act as though dependency does not exist unless there is very strong evidence to the contrary. Such would be the case if the Z score and the linear correlation coefficient indicated dependency, and the dependency held up across markets and across optimizeable parameter values. If we act as though there is dependency when the evidence is not overwhelming, we may well just be fooling ourselves and cause more self-inflicted harm than good.

Even if a system showed dependency to a 95% confidence limit for all values of a parameter, that confidence limit is hardly high enough for us to assume that dependency does in fact exist between the trades of a given market/system.

Yet the confidence limits and linear correlation coefficients are tools that should be used, because on rare occasions they may turn up a diamond in the rough, which can then possibly be exploited. Furthermore, and perhaps more importantly, they increase our understanding of the environment in which we are trying to operate.

On occasion, particularly in longer-term trading systems, you will encounter cases where the Z score and the linear correlation coefficient indicate dependency, and the dependency holds up across markets and across optimizeable parameter values. In such rare cases, you can take advantage of this dependency by either passing certain trades or altering your commitment on certain trades.

By studying these examples, you will better understand the subject matter.

$-10, 10, -1, 1$
Linear Correlation $= -.9172$
Z score $= 1.8371$ or 90 to 95% confidence limit that like begets unlike.

$10, -1, 1, -10$
Linear Correlation $= .1796$
Z score $= 1.8371$ or 90 to 95% confidence limit that like begets unlike.

$10, -10, 10, -10$
Linear Correlation $= -1.0000$
Z score $= 1.8371$ or 90 to 95% confidence limit that like begets unlike.

$-1, 1, -1, 1$
Linear Correlation $= -1.0000$
Z score $= 1.8371$ or 90 to 95% confidence limit that like begets unlike.

$1, 1, -1, -1$
Linear Correlation $= .5000$
Z score $= -.6124$ or less than 50% confidence limit that like begets like.

$100, -1, 50, -100, 1, -50$
Linear Correlation $= -.2542$
Z score $= 2.2822$ or more than 97% confidence limit that like begets unlike.

The *turning points test* is an altogether different test for dependency. Going through the stream of trades, a turning point is counted if a trade is for a greater P&L value than both the trade before it and the trade after it. A trade can also be counted as a turning point if it is for a lesser P&L value than both the trade before it and the trade after it. Notice that we are using the individual trades, not the equity curve (the cumulative values of the trades). The number of turning points is totaled up for the entire stream of trades. Note that we must start with the second trade and end with the next to last trade, as we need a trade on either side of the trade we are considering as a turning point.

Consider now three values (1, 2, 3) in a random series, whereby each of the six possible orderings are equally likely:

$$1, 2, 3 \qquad 2, 3, 1 \qquad 1, 3, 2 \qquad 3, 1, 2 \qquad 2, 1, 3 \qquad 3, 2, 1$$

Of these six, four will result in a turning point. Thus, for a random stream of trades, the expected number of turning points is given as:

$$\text{Expected number of turning points} = 2/3 * (N - 2) \qquad (1.06)$$

where: N = The total number of trades

We can derive the variance in the number of turning points of a random series as:

$$\text{Variance} = (16 * N - 29)/90 \qquad (1.07)$$

The standard deviation is the square root of the variance. Taking the difference between the actual number of turning points counted in the stream of trades and the expected number and then dividing the difference by the standard deviation will give us a Z score, which is then expressed as a confidence limit. The confidence limit is discerned from Equation (2.22) for two-tailed Normal probabilities. Thus, if our stream of trades is very far away (very many standard deviations from the expected number), it is unlikely that our stream of trades is random; rather, dependency is present. If dependency appears to a high confidence limit (at least 95%) with the turning points test, you can determine from inspection whether like begets like (if there are fewer actual turning points than expected) or whether like begets unlike (if there are more actual turning points than expected).

Another test for dependence is the *phase length test*. This is a statistical test similar to the turning points test. Rather than counting up the number of turning points between (but not including) trade 1 and the last trade, the phase length test looks at how many trades have elapsed between turning points. A "phase" is the number of trades that elapse between a turning point high and a turning point low, or a turning point low and a turning point high. It doesn't matter which occurs first, the high turning point or the low turning point. Thus, if trade number 4 is a turning point (high or low) and trade number 5 is a turning point (high or low, so long as it's the opposite of what the last turning point was), then the phase length is 1, since the difference between 5 and 4 is 1.

With the phase length test you add up the number of phases of length 1, 2, and 3 or more. Therefore, you will have three categories: 1, 2, and 3+. Thus, phase lengths of 4 or 5, and so on, are all totaled under the group of 3+. It doesn't matter if a phase goes from a high turning point to a low turning point or from a low turning point to a high turning point; the only thing that matters is how many trades the phase is comprised of. To figure the phase length, simply take the trade number of the latter phase (what number it is in sequence from 1 to N, where N is the total number of trades) and subtract the trade number of the prior phase. For each of the three categories you will have the total number of complete phases that occurred between (but not including) the first and the last trades.

Each of these three categories also has an expected number of trades for that category. The expected number of trades of phase length D is:

$$E(D) = 2 * (N - D - 2) * (D \wedge 2 * 3 * D + 1)/(D + 3)! \qquad (1.08)$$

where: D = The length of the phase.
E(D) = The expected number of counts.
N = The total number of trades.

Once you have calculated the expected number of counts for the three categories of phase length (1, 2, and 3+), you can perform the chi-square test. According to Kendall and colleagues,[3] you should use 2.5 degrees of freedom here in determining the significance levels, as the lengths of the phases are not independent. Remember that the phase length test doesn't tell you about the dependence (like begetting like, etc.), but rather whether or not there is dependence or randomness.

Lastly, this discussion of dependence addresses converting a correlation coefficient to a confidence limit. The technique employs what is known as *Fisher's Z transformation*, which converts a correlation coefficient, r, to a Normally distributed variable:

$$F = .5 * \ln(1 + r)/(1 - r)) \qquad (1.09)$$

where: F = The transformed variable, now Normally distributed.
r = The correlation coefficient of the sample.
ln( ) = The natural logarithm function.

[3]Kendall, M. G., A. Stuart, and J. K. Ord. *The Advanced Theory of Statistics*, Vol. III. New York: Hafner Publishing, 1983.

The distribution of these transformed variables will have a variance of:

$$V = 1/(N - 3) \tag{1.10}$$

where: V = The variance of the transformed variables.
         N = The number of elements in the sample.

The mean of the distribution of these transformed variables is discerned by Equation (1.09), only instead of being the correlation coefficient of the sample, r is the correlation coefficient of the population. Thus, since our population has a correlation coefficient of 0 (which we assume, since we are testing deviation from randomness), then Equation (1.09) gives us a value of 0 for the mean of the population.

Now we can determine how many standard deviations the adjusted variable is from the mean by dividing the adjusted variable by the square root of the variance, Equation (1.10). The result is the Z score associated with a given correlation coefficient and sample size. For example, suppose we had a correlation coefficient of .25, and this was discerned over 100 trades. Thus, we can find our Z score as Equation (1.9) divided by the square root of Equation (1.10), or:

$$Z = .5 * \ln((1 + r)/(1 - r))/\sqrt{1/(N - 3)} \tag{1.11}$$

Which, for our example, is:

$$
\begin{aligned}
Z &= (.5 * \ln((1 + .25)/(1 - .25)))/(1/(100 - 3)) \wedge .5 \\
&= (.5 * \ln(1.25/.75))/(1/97) \wedge .5 \\
&= (.5 * \ln(1.6667))/.010309 \wedge .5 \\
&= (.5 * .51085)/.1015346165 \\
&= .25541275/.1015346165 \\
&= 2.515523856
\end{aligned}
$$

Now we can translate this into a confidence limit by using Equation (2.22) for a Normal Distribution two-tailed confidence limit. For our example this works out to a confidence limit in excess of 98.8%. If we had had 30 trades or less, we would have had to discern our confidence limit by using the Student's Distribution with $N - 1$ degrees of freedom.

CHAPTER 2

Probability Distributions

THE BASICS OF PROBABILITY DISTRIBUTIONS

Imagine if you will that you are at a racetrack and you want to keep a log of the position in which the horses in a race finish. Specifically, you want to record whether the horse in the pole position came in first, second, and so on for each race of the day. You will record only 10 places. If the horse came in worse than in tenth place, you will record it as a tenth-place finish. If you do this for a number of days, you will have gathered enough data to see the *distribution* of finishing positions for a horse starting out in the pole position. Now you take your data and plot it on a graph. The horizontal axis represents where the horse finished, with the far left being the worst finishing position (tenth) and the far right being a win. The vertical axis will record how many times the pole-position horse finished in the position noted on the horizontal axis. You would begin to see a bell-shaped curve develop.

Under this scenario, there are 10 possible finishing positions for each race. We say that there are 10 *bins* in this distribution. What if, rather than using 10 bins, we used five? The first bin would be for a first- or second-place finish, the second bin for a third- or fourth-place finish, and so on. What would have been the result?

Using fewer bins on the same set of data would have resulted in a probability distribution with the same profile as one determined on the same

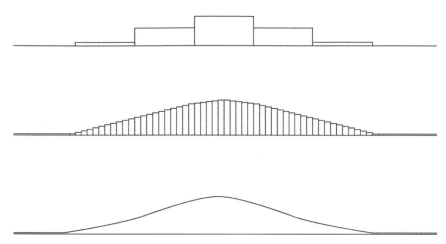

FIGURE 2.1 A continuous distribution is a series of infinitely thin bins

data with more bins. That is, they would look pretty much the same graphically. However, using fewer bins does reduce the information content of a distribution. Likewise, using more bins increases the information content of a distribution. If, rather than recording the finishing position of the pole-position horse in each race, we record the time the horse ran in, rounded to the nearest second, we will get more than 10 bins, and thus the information content of the distribution obtained will be greater.

If we recorded the exact finish time, rather than rounding finish times to use the nearest second, we would be creating what is called a *continuous* distribution. In a continuous distribution, there are no bins. Think of a continuous distribution as a series of infinitely thin bins (see Figure 2.1). A continuous distribution differs from a *discrete* distribution, the type we discussed first, in that a discrete distribution is a binned distribution. Although binning does reduce the information content of a distribution, in real life it is often necessary to bin data. Therefore, in real life it is often necessary to lose some of the information content of a distribution, while keeping the profile of the distribution the same, so that you can process the distribution. Finally, you should know that it is possible to take a continuous distribution and make it discrete by binning it, but it is not possible to take a discrete distribution and make it continuous.

When we are discussing the profits and losses of trades, we are essentially discussing a continuous distribution. A trade can take a multitude of values (although we could say that the data is binned to the nearest cent). In order to work with such a distribution, you may find it necessary to bin the data into, for example, $100-wide bins. Such a distribution would have a bin for trades that made nothing to $99.99, the next bin would be for trades that made $100 to $199.99, and so on. There is a loss of information content in binning this way, yet the profile of the distribution of the trade profits and losses remains relatively unchanged.

DESCRIPTIVE MEASURES OF DISTRIBUTIONS

Most people are familiar with the average, or more specifically the *arithmetic mean.* This is simply the sum of the data points in a distribution divided by the number of data points:

$$A = \left(\sum_{i=1}^{N} X_i\right) \Big/ N \qquad (2.01)$$

where: A = The arithmetic mean.

X$_i$ = The ith data point.

N = The total number of data points in the distribution.

The arithmetic mean is the most common of the types of measures of *location,* or *central tendency* of a body of data, a distribution. However, you should be aware that the arithmetic mean is not the only available measure of central tendency and often it is not the best. The arithmetic mean tends to be a poor measure when a distribution has very broad tails. Suppose you randomly select data points from a distribution and calculate their mean. If you continue to do this, you will find that the arithmetic means thus obtained converge poorly, if at all, when you are dealing with a distribution with very broad tails.

Another important measure of location of a distribution is the *median.* The median is described as the middle value when data are arranged in an array according to size. The median divides a probability distribution into two halves such that the area under the curve of one half is equal to the area under the curve of the other half. The median is frequently a better measure of central tendency than the arithmetic mean. Unlike the arithmetic mean, the median is *not* distorted by extreme outlier values. Further, the median can be calculated even for *open-ended* distributions. An open-ended distribution is a distribution in which all of the values in excess of a certain bin are thrown into one bin. An example of an open-ended distribution is the one we were compiling when we recorded the finishing position in horse racing for the horse starting out in the pole position. Any finishes worse than tenth place were recorded as a tenth-place finish. Thus, we had an open distribution.

The third measure of central tendency is the *mode*—the most frequent occurrence. The mode is the peak of the distribution curve. In some distributions there is no mode and sometimes there is more than one mode. Like the median, the mode can often be regarded as a superior measure of central tendency. The mode is completely independent of extreme outlier values, and it is more readily obtained than the arithmetic mean or the median.

We have seen how the median divides the distribution into two equal areas. In the same way a distribution can be divided by three *quartiles* (to give four areas of equal size or probability), or nine *deciles* (to give 10 areas of equal size or probability) or 99 *percentiles* (to give 100 areas of equal size or probability). The 50th percentile is the median, and along with the 25th and 75th percentiles give us the quartiles. Finally, another term you should become familiar with is that of a *quantile*. A quantile is any of the $N - 1$ variate-values that divide the total frequency into N equal parts.

We now return to the mean. We have discussed the arithmetic mean as a measure of central tendency of a distribution. You should be aware that there are other types of means as well. These other means are less common, but they do have significance in certain applications.

First is the *geometric mean*, which we saw how to calculate in the first chapter. The geometric mean is simply the Nth root of all the data points multiplied together.

$$G = \left(\prod_{i=1}^{N} X_i \right)^{1/N} \tag{2.02}$$

where: $G =$ The geometric mean.

$X_i =$ The ith data point.

$N =$ The total number of data points in the distribution.

The geometric mean cannot be used if any of the variate-values is zero or negative.

Another type of mean is the *harmonic mean*. This is the reciprocal of the mean of the reciprocals of the data points.

$$1/H = 1/N \sum_{i=1}^{N} 1/X_i \tag{2.03}$$

where: $H =$ The harmonic mean.

$X_i =$ The ith data point.

$N =$ The total number of data points in the distribution.

The final measure of central tendency is the *quadratic mean* or *root mean square.*

$$R^2 = 1/N \sum_{i=1}^{N} X_i^2 \qquad (2.04)$$

where: R = The root mean square.

X$_i$ = The ith data point.

N = The total number of data points in the distribution.

You should realize that the arithmetic mean (A) is always greater than or equal to the geometric mean (G), and the geometric mean is always greater than or equal to the harmonic mean (H):

$$H <= G <= \Lambda \qquad (2.05)$$

where: H = The harmonic mean.

G = The geometric mean.

G = The arithmetic mean.

MOMENTS OF A DISTRIBUTION

The central value or location of a distribution is often the first thing you want to know about a group of data, and often the next thing you want to know is the data's variability or "width" around that central value. We call the measures of a distribution's central tendency the *first moment* of a distribution. The variability of the data points around this central tendency is called the *second moment* of a distribution. Hence, the second moment measures a distribution's dispersion about the first moment.

As with the measure of central tendency, many measures of dispersion are available. We cover seven of them here, starting with the least common measures and ending with the most common.

The *range* of a distribution is simply the difference between the largest and smallest values in a distribution. Likewise, the *10–90 percentile range* is the difference between the 90th and 10th percentile points. These first two measures of dispersion measure the spread from one extreme to the other. The remaining five measures of dispersion measure the departure from the central tendency (and hence measure the half-spread).

The *semi-interquartile range or quartile deviation* equals one half of the distance between the first and third quartiles (the 25th and 75th percentiles). This is similar to the 10–90 percentile range, except that with this measure the range is commonly divided by 2.

The *half-width* is an even more frequently used measure of dispersion. Here, we take the height of a distribution at its peak, the mode. If we find the point halfway up this vertical measure and run a horizontal line through it perpendicular to the vertical line, the horizontal line will touch the distribution at one point to the left and one point to the right. The distance between these two points is called the half-width.

Next, the *mean absolute deviation* or *mean deviation* is the arithmetic average of the absolute value of the difference between the data points and the arithmetic average of the data points. In other words, as its name implies, it is the average distance that a data point is from the mean. Expressed mathematically:

$$M = 1/N \sum_{i=1}^{N} ABS(X_i - A) \qquad (2.06)$$

where: M = The mean absolute deviation.

N = The total number of data points.

X_i = The ith data point.

A = The arithmetic average of the data points.

$ABS(\ )$ = The absolute value function.

Equation (2.06) gives us what is known as the *population* mean absolute deviation. You should know that the mean absolute deviation can also be calculated as what is known as the *sample* mean absolute deviation. To calculate the sample mean absolute deviation, replace the term $1/N$ in Equation (2.06) with $1/(N - 1)$. You use the sample version when you are making judgments about the population based on a sample of that population.

The next two measures of dispersion, variance and standard deviation, are the two most commonly used. Both are used extensively, so we cannot say that one is more common than the other; suffice to say they are both the most common. Like the mean absolute deviation, they can be calculated two different ways, for a population as well as a sample. The population version is shown, and again it can readily be altered to the sample version by replacing the term $1/N$ with $1/(N - 1)$.

The *variance* is the same thing as the mean absolute deviation except that we square each difference between a data point and the average of the data points. As a result, we do not need to take the absolute value of each difference, since multiplying each difference by itself makes the result positive whether the difference was positive or negative. Further, since each distance is squared, extreme outliers will have a stronger effect on the variance than they would on the mean absolute deviation. Mathematically expressed:

$$V = 1/N \sum_{i=1}^{N} (X_i - A)^2 \qquad (2.07)$$

where: $V =$ The variance.

$N =$ The total number of data points.

$X_i =$ The ith data point.

$A =$ The arithmetic average of the data points.

Finally, the *standard deviation* is related to the variance (and hence the mean absolute deviation) in that the *standard deviation is simply the square root of the variance.*

The *third moment* of a distribution is called *skewness*, and it describes the extent of asymmetry about a distribution's mean (Figure 2.2). Whereas the first two moments of a distribution have values that can be considered *dimensional* (i.e., having the same units as the measured quantities), skewness is defined in such a way as to make it *nondimensional.* It is a pure number that represents nothing more than the shape of the distribution.

A positive value for skewness means that the tails are thicker on the positive side of the distribution and vice versa. A perfectly symmetrical distribution has a skewness of 0.

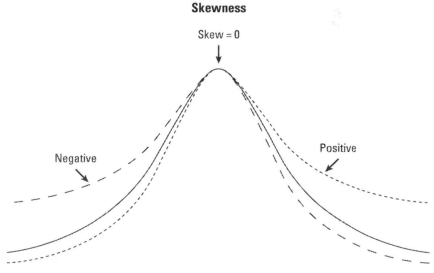

FIGURE 2.2   Skewness

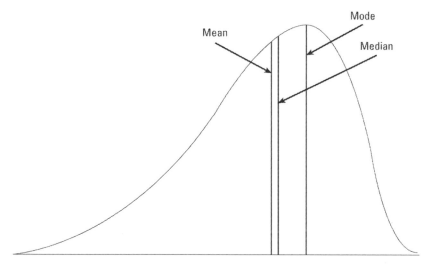

FIGURE 2.3 Skewness alters location

In a symmetrical distribution the mean, median, and mode are all at the same value. However, when a distribution has a nonzero value for skewness, this changes as depicted in Figure 2.3. The relationship for a skewed distribution (any distribution with a nonzero skewness) is:

$$\text{Mean} - \text{Mode} = 3 * (\text{Mean} - \text{Median}) \tag{2.08}$$

As with the first two moments of a distribution, there are numerous measures for skewness, which most frequently will give different answers. These measures now follow:

$$S = (\text{Mean} - \text{Mode})/\text{Standard Deviation} \tag{2.09}$$

$$S = (3 * (\text{Mean} - \text{Median}))/\text{Standard Deviation} \tag{2.10}$$

These last two equations, (2.09) and (2.10), are often referred to as Pearson's first and second coefficients of skewness, respectively. Skewness is also commonly determined as:

$$S = 1/N \sum_{i=1}^{N}((X_i - A)/D)^3 \tag{2.11}$$

where: $S = $ The skewness.

          $N = $ The total number of data points.

          $X_i = $ The ith data point.

          $A = $ The arithmetic average of the data points.

          $D = $ The population standard deviation of the data points.

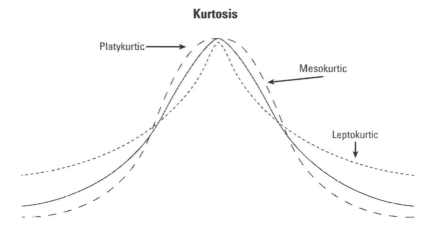

FIGURE 2.4 Kurtosis

Finally, the *fourth moment* of a distribution, *kurtosis* (see Figure 2.4), measures the peakedness or flatness of a distribution (relative to the Normal Distribution). Like skewness, it is a nondimensional quantity. A curve less peaked than the Normal is said to be *platykurtic* (kurtosis will be negative), and a curve more peaked than the Normal is called *leptokurtic* (kurtosis will be positive). When the peak of the curve resembles the Normal Distribution curve, kurtosis equals zero, and we call this type of peak on a distribution *mesokurtic*.

Like the preceding moments, kurtosis has more than one measure. The two most common are:

$$K - Q/P \tag{2.12}$$

where: $K =$ The kurtosis.
$Q =$ The semi-interquartile range.
$P =$ The 10–90 percentile range.

$$K = \left(1/N \sum_{i=1}^{N} ((X_i - A)/D)^4 \right) - 3 \tag{2.13}$$

where: $K =$ The kurtosis.
$N =$ The total number of data points.
$X_i =$ The ith data point.
$A =$ The arithmetic average of the data points.
$D =$ The population standard deviation of the data points.

Finally, it should be pointed out there is a lot more "theory" behind the moments of a distribution than is covered here. The depth of discussion about the moments of a distribution presented here will be more than adequate for our purposes throughout this text.

Thus far, we have covered data distributions in a general sense. Now we will cover the specific distribution called the Normal Distribution.

THE NORMAL DISTRIBUTION

Frequently, the Normal Distribution is referred to as the Gaussian distribution, or de Moivre's distribution, after those who are believed to have discovered it—Karl Friedrich Gauss (1777–1855) and, about a century earlier and far more obscurely, Abraham de Moivre (1667–1754).

The Normal Distribution is considered to be the most useful distribution in modeling. This is due to the fact that the Normal Distribution accurately models many phenomena. Generally speaking, we can measure heights, weights, intelligence levels, and so on from a population, and these will very closely resemble the Normal Distribution.

Let's consider what is known as Galton's board (Figure 2.5). This is a vertically mounted board in the shape of an isosceles triangle. The board is studded with pegs, one on the top row, two on the second, and so on. Each row down has one more peg than the previous row. The pegs are arranged in a triangular fashion such that when a ball is dropped in, it has a 50/50 probability of going right or left with each peg it encounters. At the base of the board is a series of troughs to record the exit gate of each ball.

The balls falling through Galton's board and arriving in the troughs will begin to form a Normal Distribution. The "deeper" the board is (i.e., the more rows it has) and the more balls are dropped through, the more closely the final result will resemble the Normal Distribution.

The Normal is useful in its own right, but also because it tends to be the limiting form of many other types of distributions. For example, if X is distributed binomially, then as N tends toward infinity, X tends to be Normally distributed. Further, the Normal Distribution is also the limiting form of a number of other useful probability distributions such as the Poisson, the Student's, or the T distribution. In other words, as the data (N) used in these other distributions increases, these distributions increasingly resemble the Normal Distribution.

THE CENTRAL LIMIT THEOREM

One of the most important applications for statistical purposes involving the Normal Distribution has to do with the distribution of averages. The

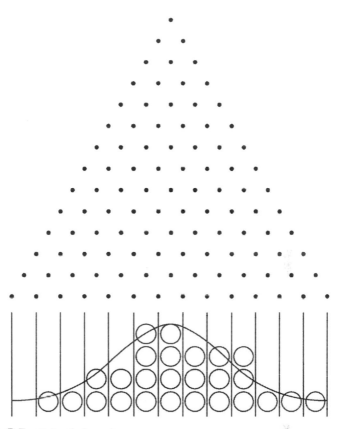

FIGURE 2.5 Galton's board

averages of samples of a given size, taken such that each sampled item is selected independently of the others, will yield a distribution that is close to Normal. This is an extremely powerful fact, for it means that you can generalize about an actual random process from averages computed using sample data.

Thus, we can state that *if N random samples are drawn from a population, then the sums (or averages) of the samples will be approximately Normally distributed, regardless of the distribution of the population from which the samples are drawn. The closeness to the Normal Distribution improves as N (the number of samples) increases.*

As an example, consider the distribution of numbers from 1 to 100. This is what is known as a *uniform distribution*: All elements (numbers in this case) occur only once. The number 82 occurs once and only once, as does 19, and so on. Suppose now that we take a sample of five elements and we take the average of these five sampled elements (we can just as well take their sums). Now, we replace those five elements back into the population, and we take another sample and calculate the sample mean. If we keep on repeating this process, we will see that the sample means are Normally distributed, even though the population from which they are drawn is uniformly distributed.

Furthermore, this is true *regardless* of how the population is distributed! The Central Limit Theorem allows us to treat the distribution of sample means as being Normal without having to know the distribution of the population. This is an enormously convenient fact for many areas of study.

If the population itself happens to be Normally distributed, then the distribution of sample means will be exactly (not approximately) Normal. This is true because how quickly the distribution of the sample means approaches the Normal, as N increases, is a function of how close the population is to Normal. As a general rule of thumb, if a population has a *unimodal distribution*—any type of distribution where there is a concentration of frequency around a single mode, and diminishing frequencies on either side of the mode (i.e., it is convex)—or is uniformly distributed, using a value of 20 for N is considered sufficient, and a value of 10 for N is considered *probably* sufficient. However, if the population is distributed according to the Exponential Distribution (Figure 2.6), then it may be necessary to use an N of 100 or so.

The Central Limit Theorem, this amazingly simple and beautiful fact, validates the importance of the Normal Distribution.

WORKING WITH THE NORMAL DISTRIBUTION

In using the Normal Distribution, we most frequently want to find the percentage of area under the curve at a given point along the curve. In the parlance of calculus this would be called the integral of the function for the curve itself. Likewise, we could call the function for the curve itself the derivative of the function for the area under the curve. Derivatives are often noted with a prime after the variable for the function. Therefore, if we have a function, N(X), that represents the percentage of area under the curve at a given point, X, we can say that the derivative of this function, N′(X) (called N prime of X), is the function for the curve itself at point X.

We will begin with the formula for the curve itself, N′(X). This function is represented as:

$$N'(X) = 1/(S * \sqrt{2 * 3.1415926536}$$
$$* EXP(-(X - U)^2/2 * S^2) \tag{2.14}$$

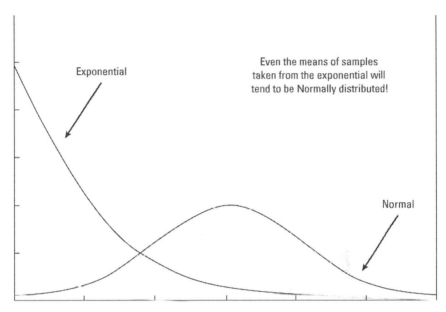

FIGURE 2.6 The Exponential Distribution and the Normal

where: U = The mean of the data.

S = The standard deviation of the data.

X = The observed data point.

EXP( ) = The exponential function.

This formula will give us the Y axis value, or the height of the curve if you will, at any given X axis value.

Often, it is easier to refer to a point along the curve with reference to its X coordinate in terms of how many standard deviations it is away from the mean. Thus, a data point that was 1 standard deviation away from the mean would be said to be one *standard unit* from the mean.

Further, it is often easier to subtract the mean from all of the data points, which has the effect of shifting the distribution so that it is centered over zero rather than over the mean. Therefore, a data point that was 1 standard deviation to the right of the mean would now have a value of 1 on the X axis.

When we make these conversions, subtracting the mean from the data points, then dividing the difference by the standard deviation of the data points, we are converting the distribution to what is called the *standardized normal*, which is the Normal Distribution with mean = 0 and variance $= 1$. Now, $N'(Z)$ will give us the Y axis value (the height of the curve) for any value of Z:

$$N'(Z) = 1/\sqrt{2 * 3.1415926536 * EXP(-(Z^2/2))}$$
$$= .398942 * EXP(-(Z^2/2)) \qquad (2.15a)$$

where: $Z = (X - U)/S$
and $U =$ The mean of the data.
             $S =$ The standard deviation of the data.
             $X =$ The observed data point.
$EXP\ ()=$ The exponential function.

Equation (2.16) gives us the number of *standard units* that the data point corresponds to—in other words, how many standard deviations away from the mean the data point is. When Equation (2.16) equals 1, it is called the *standard normal deviate*. A standard deviation or a standard unit is sometimes referred to as a *sigma*. Thus, when someone speaks of an event's being a "five sigma event," they are referring to an event whose probability of occurrence is the probability of being beyond 5 standard deviations.

Consider Figure 2.7, which shows this equation for the Normal curve. Notice that the height of the standard Normal curve is .39894. From Equation (2.15a), the height is:

$$N'(Z) = .398942 * EXP(-(Z^2/2))$$
$$N'(0) = .398942 * EXP(-(0^2/2))$$
$$N'(0) = .398942$$

Notice that the curve is *continuous*—that is, there are no "breaks" in the curve as it runs from minus infinity on the left to positive infinity on the right. Notice also that the curve is symmetrical, the side to the right of the peak being the mirror image of the side to the left of the peak.

Suppose we had a group of data where the mean of the data was 11 and the standard deviation of the group of data was 20. To see where a data point in that set would be located on the curve, we could first calculate it as a standard unit. Suppose the data point in question had a value of -9. To calculate how many standard units this is, we first must subtract the mean from this data point:

$$-9 - 11 = -20$$

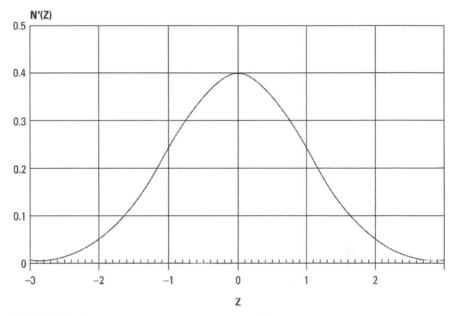

FIGURE 2.7 The Normal Probability density function

Next, we need to divide the result by the standard deviation:

$$-20/20 = -1$$

We can therefore say that the number of standard units is 1, when the data point equals -9, and the mean is 11, and the standard deviation is 20. In other words, we are 1 standard deviation away from the peak of the curve, the mean, and since this value is negative we know that it means we are 1 standard deviation to the left of the peak. To see where this places us on the curve itself (i.e., how high the curve is at 1 standard deviation left of center, or what the Y axis value of the curve is for a corresponding X axis value of -1), we need to now plug this into Equation (2.15a):

$$
\begin{aligned}
N'(Z) &= .398942 * \mathrm{EXP}(-(Z^2/2)) \\
&= .398942 * 2.7182818285(-(-1^2/2)) \\
&= .398942 * 2.7182818285^{-1/2} \\
&= .398942 * .6065307 \\
&= .2419705705
\end{aligned}
$$

Thus, we can say that the height of the curve at $X = -1$ is .2419705705. The function $N'(Z)$ is also often expressed as:

$$\begin{aligned}
N'(Z) &= EXP(-(Z^2/2))/\sqrt{8 * ATN(1)} \\
&= EXP(-(Z^2/2))/\sqrt{8 * .7853983} \qquad (2.15b) \\
&= EXP(-(Z^2/2))/2.506629
\end{aligned}$$

where: $\qquad Z = (X - U)/S$ (2.16)

and $\qquad ATN(\ ) = $ The arctangent function.

$\qquad\qquad U = $ The mean of the data.

$\qquad\qquad S = $ The standard deviation of the data.

$\qquad\qquad X = $ The observed data point.

$\qquad EXP(\ ) = $ The exponential function.

Nonstatisticians often find the concept of the standard deviation (or its square, *variance*) hard to envision. A remedy for this is to use what is known as the mean absolute deviation and convert it to and from the standard deviation in these equations. The mean absolute deviation is exactly what its name implies. The mean of the data is subtracted from each data point. The absolute values of each of these differences are then summed, and this sum is divided by the number of data points. What you end up with is the average distance each data point is away from the mean. The conversion for mean absolute deviation and standard deviation are given now:

$$\begin{aligned}
\text{Mean Absolute Deviation} &= S * \sqrt{2/3.1415926536} \\
&= S * .7978845609 \qquad (2.17)
\end{aligned}$$

where: $\quad M = $ The mean absolute deviation.

$\qquad\quad S = $ The standard deviation.

Thus, we can say that in the Normal Distribution, the mean absolute deviation equals the standard deviation times .7979. Likewise:

$$\begin{aligned}
S &= M * 1/.7978845609 \\
&= M * 1.253314137 \qquad (2.18)
\end{aligned}$$

where: $\quad S = $ The standard deviation.

$\qquad\quad M = $ The mean absolute deviation.

So we can also say that in the Normal Distribution the standard deviation equals the mean absolute deviation times 1.2533. Since the variance is always the standard deviation squared (and standard deviation is always the square root of variance), we can make the conversion between variance and mean absolute deviation.

$$M = \sqrt{V} * \sqrt{2/3.1415926536}$$
$$= \sqrt{V} * .7978845609 \tag{2.19}$$

where: M = The mean absolute deviation.
     V = The variance.

$$V = (M * 1.253314137)^2 \tag{2.20}$$

where: V = The variance.
     M = The mean absolute deviation.

Since the standard deviation in the standard normal curve equals 1, we can state that the mean absolute deviation in the standard normal curve equals .7979.

Further, in a bell-shaped curve like the Normal, the semi-interquartile range equals approximately two-thirds of the standard deviation, and therefore the standard deviation equals about 1.5 times the semi-interquartile range. This is true of most bell-shaped distributions, not just the Normal, as are the conversions given for the mean absolute deviation and standard deviation.

NORMAL PROBABILITIES

We now know how to convert our raw data to standard units and how to form the curve $N'(Z)$ itself (i.e., how to find the height of the curve, or Y coordinate for a given standard unit) as well as $N'(X)$ (Equation (2.14), the curve itself without first converting to standard units). To really use the Normal Probability Distribution, though, we want to know what the probabilities of a certain outcome's happening are. This is *not* given by the height of the curve. Rather, the probabilities correspond to the area under the curve. These areas are given by the integral of this $N'(Z)$ function that we have thus far studied. We will now concern ourselves with $N(Z)$, the integral to $N'(Z)$, to find the areas under the curve (the probabilities).[1]

---

[1]The actual integral to the Normal probability density does not exist in closed form, but it can very closely be approximated by Equation (2.21).

$$N(Z) = 1 - N'(Z) * ((1.330274429 * Y^\wedge 5) - (1.821255978 * Y^\wedge 4)$$
$$+ (1.781477937 * Y^\wedge 3) - (.356563782 * Y^\wedge 2)$$
$$+ (.31938153 * Y)) \tag{2.21}$$

If $Z < 0$, then $N(Z) = 1 - N(Z)$. Now recall Equation (2.15a):

$$N'(Z) = .398942 * EXP(-(Z^2/2))$$

where: $Y = 1/(1 + 2316419 * ABS(Z))$

and $ABS() = $ The absolute value function.

             $EXP() = $ The exponential function.

We will always convert our data to standard units when finding probabilities under the curve. That is, we will not describe an $N(X)$ function, but rather we will use the $N(Z)$ function where:

$$Z = (X - U)/S$$

and $U = $ The mean of the data.

       $S = $ The standard deviation of the data.

       $X = $ The observed data point.

Refer now to Equation (2.21). Suppose we want to know what the probability is of an event's not exceeding $+2$ standard units ($Z = +2$).

$$Y = 1/(1 + 2316419 * ABS(+2))$$
$$= 1/1.4632838$$
$$= .68339443311$$

$$N'(Z) = .398942 * EXP(-(Z^2/2))$$
$$= .398942 * EXP(-2)$$
$$= .398942 * .1353353$$
$$= .05399093525$$

Notice that this tells us the height of the curve at -2 standard units. Plugging these values for Y and $N'(Z)$ into Equation (2.21) we can obtain the probability of an event's not exceeding $+2$ standard units:

$$N(Z) = 1 - N'(Z) * ((1.330274429 * Y^\wedge 5) - (1.821255978 * Y^\wedge 4)$$
$$+ (1.781477937 * Y^\wedge 3) - (.356563782 * Y^\wedge 2)$$
$$+ (.31938153 * Y))$$

$$= 1 - .05399093525 * \begin{pmatrix} 1.330274429 * .68339443311^5 \\ -1.821255978 * .68339443311^4 \\ +1.781477937 * .68339443311^3 \\ -.356563782 * .68339443311^2 \\ +.31928153 * .68339443311 \end{pmatrix}$$

$$= 1 - .05399093525 * \begin{pmatrix} 1.330274429 * .1490587 - 1.821255978* \\ .2181151 + 1.781477937 * .3191643 \\ -.356563782 * .467028 + .31928153* \\ .68339443311 \end{pmatrix}$$

$$= 1 - .05399093525 * (.198299977 - .3972434298$$
$$+ .5685841587 - .16652527 + .2182635596)$$
$$= 1 - .05399093525 * .4213679955$$
$$= 1 - .02275005216$$
$$= .9772499478$$

Thus, we can say that we can expect 97.72% of the outcomes in a Normally distributed random process to fall shy of $+2$ standard units. This is depicted in Figure 2.8.

If we wanted to know what the probabilities were for an event's equaling or exceeding a prescribed number of standard units (in this case $+2$), we would simply amend Equation (2.21), taking out the $1-$ in the beginning of the equation and doing away with the $-Z$ provision (i.e., doing away with

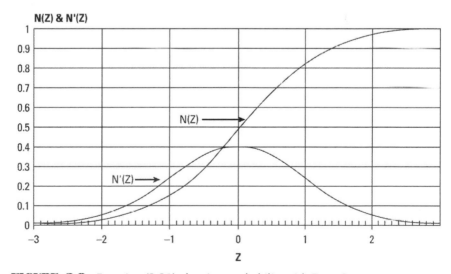

FIGURE 2.8 Equation (2.21) showing probability with $Z = +2$

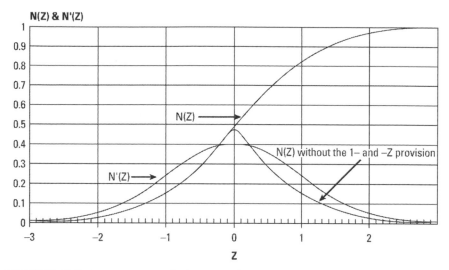

FIGURE 2.9 Doing away with the 1– and –Z provision in Equation (2.21)

"If $Z < 0$ then $N(Z) = 1 - N(Z)$"). Therefore, the second to last line in the last computation would be changed from

$$= 1 - .02275005216$$

to simply

$$.02275005216$$

We would therefore say that there is about a 2.275% chance that an event in a Normally distributed random process would equal or exceed $+2$ standard units. This is shown in Figure 2.9.

Thus far we have looked at areas under the curve (probabilities) where we are dealing only with what are known as "one-tailed" probabilities. That is to say we have thus far looked to solve such questions as, "What are the probabilities of an event's being less (more) than such-and-such standard units from the mean?" Suppose now we were to pose the question as, "What are the probabilities of an event's being *within* so many standard units of the mean?" In other words, we wish to find out what the "2-tailed" probabilities are.

Consider Figure 2.10. This represents the probabilities of being within 2 standard units of the mean. Unlike Figure 2.8, this probability computation does not include the extreme left tail area, the area of less than -2 standard units. To calculate the probability of being within Z standard units of the mean, you must first calculate the one-tailed probability of the absolute value of Z with Equation (2.21). This will be your input to the next

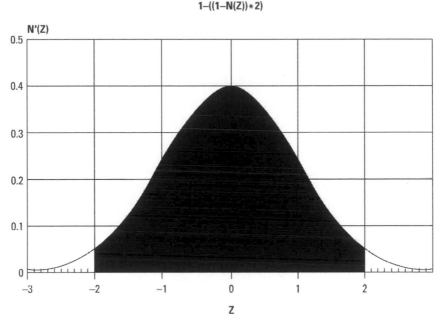

FIGURE 2.10 A two-tailed probability of an event's being + or −2 sigma

Equation, (2.22), which gives us the two-tailed probabilities (i.e., the probabilities of being within ABS(Z) standard units of the mean):

$$\text{Two-tailed probability} = 1 - ((1 - N(ABS(Z))) * 2) \qquad (2.22)$$

If we are considering what our probabilities of occurrence within 2 standard deviations are (Z = 2), then from Equation (2.21) we know that N(2) = .9772499478, and using this as input to Equation (2.22):

$$\begin{aligned}
\text{Two-tailed probability} &= 1 - ((1 - .9772499478) * 2) \\
&= 1 - (.02275005216 * 2) \\
&= 1 - .04550010432 \\
&= .9544998957
\end{aligned}$$

Thus, we can state from this equation that the probability of an event in a Normally distributed random process falling within 2 standard units of the mean is about 95.45%.

Just as with Equation (2.21), we can eliminate the leading 1− in Equation (2.22) to obtain (1 – N(ABS(Z))) * 2, which represents the probabilities of an event's falling outside of ABS(Z) standard units of the mean. This is depicted in Figure 2.11. For the example where Z = 2, we can state that the

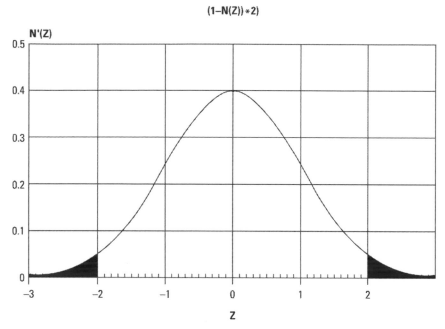

FIGURE 2.11 Two-tailed probability of an event's being beyond 2 sigma

probabilities of an event in a Normally distributed random process falling *outside* of 2 standard units is:

$$\text{Two-tailed probability (outside)} = (1 - .9772499478) * 2$$
$$= .02275005216 * 2$$
$$= .04550010432$$

Finally, we come to the case where we want to find what the probabilities (areas under the $N'(Z)$ curve) are for two different values of Z.

Suppose we want to find the area under the $N'(Z)$ curve between -1 standard unit and $+2$ standard units. There are a couple of ways to accomplish this. To begin with, we can compute the probability of not exceeding $+2$ standard units with Equation (2.21), and from this we can subtract the probability of not exceeding -1 standard units (see Figure 2.12). This would give us:

$$.9772499478 - .1586552595 = .8185946883$$

Another way we could have performed this is to take the number 1, representing the entire area under the curve, and then subtract the sum of

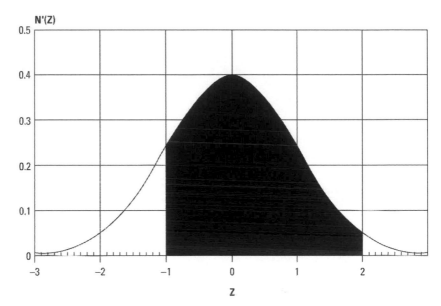

FIGURE 2.12 The area between −1 and +2 standard units

the probability of not exceeding −1 standard unit and the probability of exceeding 2 standard units:

$$= 1 - (.022750052 + .1586552595)$$
$$= 1 - .1814053117$$
$$- .8185946883$$

With the basic mathematical tools regarding the Normal Distribution thus far covered in this chapter, you can now use your powers of reasoning to figure any probabilities of occurrence for Normally distributed random variables.

FURTHER DERIVATIVES OF THE NORMAL

Sometimes you may want to know the second derivative of the N(Z) function. Since the N(Z) function gives us the area under the curve at Z, and the N'(Z) function gives us the height of the curve itself at Z, then the N''(Z) function gives us the *instantaneous slope* of the curve at a given Z:

$$N''(Z) = -Z/2.506628274 * EXP(-(Z^2)/2) \qquad (2.23)$$

where: EXP( ) = The exponential function.

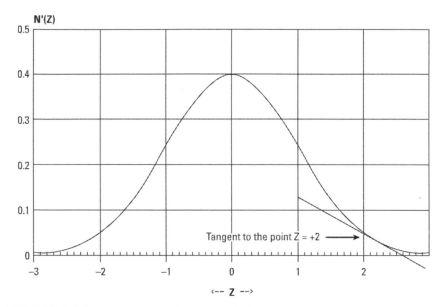

FIGURE 2.13 N''(Z) giving the slope of the line tangent tangent to N'(Z) at $Z = +2$

To determine what the slope of the N'(Z) curve is at $+2$ standard units:

$$N''(Z) = -2/2.506628274 * \text{EXP}(-(2^2)/2)$$
$$= -2/2.506628274 * \text{EXP}(-2)$$
$$= -2/2.506628274 * .1353353$$
$$= -.1079968336$$

Therefore, we can state that the instantaneous rate of change in the N'(Z) function when $Z = +2$ is $-.1079968336$. This represents rise/run, so we can say that when $Z = +2$, the N'(Z) curve is rising $-.1079968336$ for every 1 unit run in Z. This is depicted in Figure 2.13.

For the reader's own reference, further derivatives are now given. These will not be needed throughout the remainder of this text, but are provided for the sake of completeness:

$$N'''(Z) = (Z^2 - 1)/2.506628274 * \text{EXP}(-(Z^2)/2)) \tag{2.24}$$

$$N''''(Z) = ((3 * Z) - Z^3)/2.506628274 * \text{EXP}(-(Z^2)/2)) \tag{2.25}$$

$$N'''''(Z) = (Z^4 - (6 * Z^2) + 3)/2.506628274 * \text{EXP}(-(Z^2)/2)) \tag{2.26}$$

As a final note regarding the Normal Distribution, you should be aware that the distribution is nowhere near as "peaked" as the graphic examples

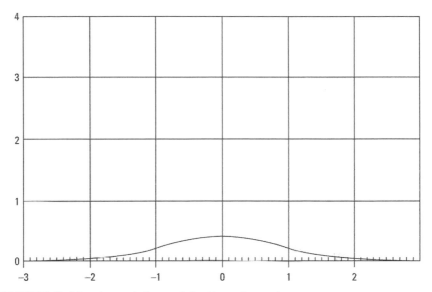

FIGURE 2.14 The real shape of the Normal Distribution

presented in this chapter imply. The real shape of the Normal Distribution is depicted in Figure 2.14.

Notice that here the scales of the two axes are the same, whereas in the other graphic examples they differ so as to exaggerate the shape of the distribution.

THE LOGNORMAL DISTRIBUTION

Many of the real-world applications in trading require a small but crucial modification to the Normal Distribution. This modification takes the Normal, and changes it to what is known as the Lognormal Distribution.

Consider that the price of any freely traded item has zero as a lower limit.[2] Therefore, as the price of an item drops and approaches zero, it

---

[2]This idea that the lowest an item can trade for is zero is not always entirely true. For instance, during the stock market crash of 1929 and the ensuing bear market, the shareholders of many failed banks were held liable to the depositors in those banks. Persons who owned stock in such banks not only lost their full investment, they also realized liability *beyond* the amount of their investment. The point here isn't to say that such an event can or cannot happen again. Rather, we cannot always say that zero is the absolute low end of what a freely traded item can be priced at, although it usually is.

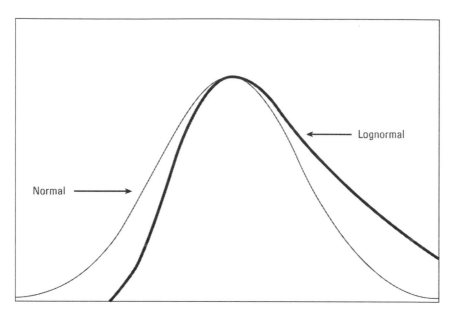

FIGURE 2.15 The Normal and Lognormal Distributions

should in theory become progressively more difficult for the item to get lower. For example, consider the price of a hypothetical stock at $10 per share. If the stock were to drop $5, to $5 per share, a 50% loss, then according to the Normal Distribution it could just as easily drop from $5 to $0. However, under the Lognormal, a similar drop of 50% from a price of $5 per share to $2.50 per share would be about as probable as a drop from $10 to $5 per share.

The Lognormal Distribution, Figure 2.15, works exactly like the Normal Distribution except that with the Lognormal we are dealing with percentage changes rather than absolute changes.

Consider now the upside. According to the Lognormal, a move from $10 per share to $20 per share is about as likely as a move from $5 to $10 per share, as both moves represent a 100% gain.

That isn't to say that we won't be using the Normal Distribution. The purpose here is to introduce you to the Lognormal, show you its relationship to the Normal (the Lognormal uses percentage price changes rather than absolute price changes), and point out that it usually is used when talking about price moves, or anytime that the Normal would apply but be bounded on the low end at zero.

To use the Lognormal Distribution, you simply convert the data you are working with to natural logarithms.[3] Now the converted data will be Normally distributed if the raw data was Lognormally distributed.

For instance, if we are discussing the distribution of price changes as being Lognormal, we can use the Normal distribution on it. First, we must divide each closing price by the previous closing price. Suppose in this instance we are looking at the distribution of monthly closing prices (we could use any time period—hourly, daily, yearly, or whatever). Suppose we now see $10, $5, $10, $10, then $20 per share as our first five months closing prices. This would then equate to a loss of 50% going into the second month, a gain of 100% going into the third month, a gain of 0% going into the fourth month, and another gain of 100% into the fifth month. Respectively, then, we have quotients of .5, 2, 1, and 2 for the monthly price changes of months 2 through 5. We must now convert to natural logarithms in order to study their distribution under the math for the Normal Distribution. Thus, the natural log of .5 is $-.6931473$, of 2 it is .6931471, and of 1 it is 0. We are now able to apply the mathematics pertaining to the Normal Distribution to this converted data.[4]

THE UNIFORM DISTRIBUTION

The *Uniform Distribution*, sometimes referred to as the *Rectangular Distribution* from its shape, occurs when all items in a population have equal frequency. A good example is the 10 digits 0 through 9. If we were to randomly select one of these digits, each possible selection has an equal chance of occurrence. Thus, the Uniform Distribution is used to model truly random events. A particular type of Uniform Distribution where $A = 0$ and $B = 1$ is called the *Standard Uniform Distribution*, and it is used extensively in generating random numbers.

---

[3]The distinction between common and natural logarithms is reiterated here. A common log is a log base 10, while a natural log is a log base e, where $e = 2.7182818285$. The common log of X is referred to mathematically as log(X) while the natural log is referred to as ln(X). The distinction gets blurred when we observe BASIC programming code, which often utilizes a function LOG(X) to return the *natural* log.

[4]This is diametrically opposed to mathematical convention. BASIC does not have a provision for common logs, but the natural log can be converted to the common log by multiplying the natural log by .4342917. Likewise, we can convert common logs to natural logs by multiplying the common log by 2.3026.

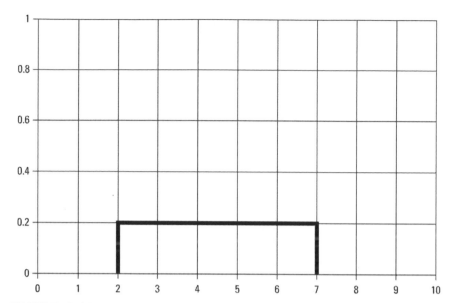

FIGURE 2.16 Probability density functions for the Uniform Distribution (A = 2, B = 7)

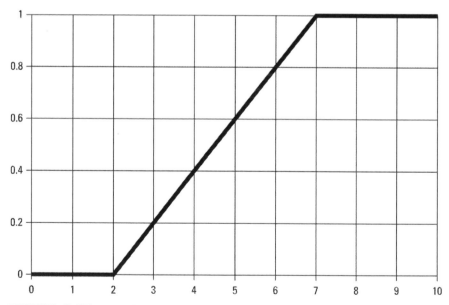

FIGURE 2.17 Cumulative probability functions for the Uniform Distribution (A = 2, B = 7)

The Uniform Distribution is a *continuous* distribution. The probability density function, $N'(X)$, is described as:

$$N'(X) = 1/(B - A) \quad \text{for } A <= X <= B \tag{2.27}$$

else

$$N'(X) = 0$$

where: $B =$ The rightmost limit of the interval AB.
   $A =$ The leftmost limit of the interval AB.

The cumulative density of the Uniform is given by:

$$N(X) = 0 \qquad \text{for } X < A \tag{2.28}$$

else

$$N(X) = (X - A)/(B - A) \qquad \text{for } A <= X <= B$$

else

$$N(X) - 1 \quad \text{for } X > B$$

where: $B =$ The rightmost limit of the interval AB.
   $A =$ The leftmost limit of the interval AB.

Figures 2.16 and 2.17 illustrate the probability density and cumulative probability (i.e., cdf) respectively of the Uniform Distribution.
Other qualities of the Uniform Distribution are:

$$\text{Mean} = (A + B)/2 \tag{2.29}$$
$$\text{Variance} = (B - A)^2/12 \tag{2.30}$$

where: $B =$ The rightmost limit of the interval AB.
   $A =$ The leftmost limit of the interval AB.

THE BERNOULLI DISTRIBUTION

Another simple, common distribution is the *Bernoulli Distribution*. This is the distribution when the random variable can have only two possible values. Examples of this are heads and tails, defective and nondefective

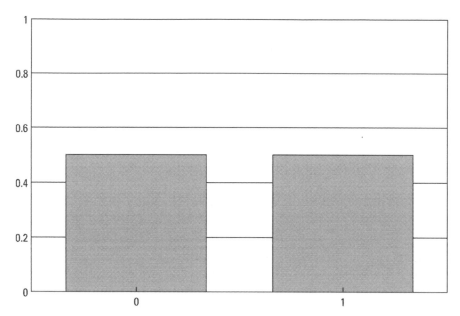

FIGURE 2.18 Probability density functions for the Binomial Distribution (P = .5)

articles, success or failure, hit or miss, and so on. Hence, we say that the Bernoulli Distribution is a *discrete distribution* (as opposed to being a continuous distribution). The distribution is completely described by one parameter, P, which is the probability of the first event occurring. The variance in the Bernoulli is:

$$\text{Variance} = P * Q \tag{2.31}$$

where:

$$Q = P - 1 \tag{2.32}$$

Figure 2.18 and 2.19 illustrate the probability density and cumulative probability (i.e., cdf) respectively of the Bernoulli Distribution.

THE BINOMIAL DISTRIBUTION

The *Binomial Distribution* arises naturally when sampling from a Bernoulli Distribution. The probability density function, N'(X), of the Binomial (the probability of X successes in N trials or X defects in N items or

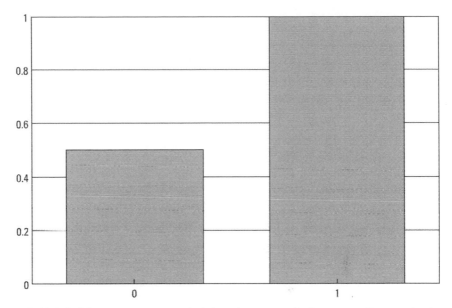

FIGURE 2.19 Cumulative probability function of the Bernoulli Distribution (P = .5)

X heads in N coin tosses, etc.) is:

$$N'(X) = (N!/(X! * (N - X)!)) * P^X * Q^{(N-X)} \qquad (2.33)$$

where: N = The number of trials.

X = The number of successes.

P = The probability of a success on a single trial.

Q = 1 − P.

$$X! = X * (X - 1) * (X - 2) * \ldots * 1 \qquad (2.34)$$

which can be also written as:

$$X! = \prod_{J=0}^{X-1} X - J \qquad (2.34a)$$

Further, by convention:

$$0! = 1 \qquad (2.34b)$$

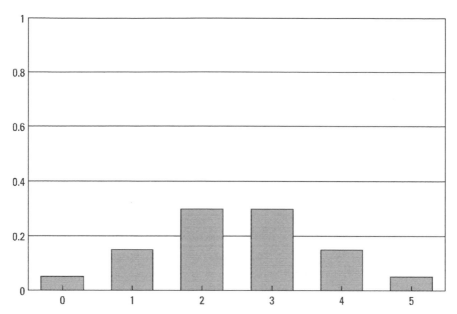

FIGURE 2.20 Probability density functions for the Binomial Distribution (N = 5, P = .5)

The cumulative density function for the Binomial is:

$$N(X) = \sum_{J=0}^{X} (N!/(J! * (N - J)!)) * P^J * Q^{(N-J)} \qquad (2.35)$$

where: N = The number of trials.

X = The number of successes.

P = The probability of a success on a single trial.

Q = 1 − P.

Figures 2.20 and 2.21 illustrate the probability density and cumulative probability (i.e., cdf), respectively, of the Bionomial Distribution.

The Binomial is also a discrete distribution. Other properties of the Binomial Distribution are:

$$\text{Mean} = N * P \qquad (2.36)$$

$$\text{Variance} = N * P * Q \qquad (2.37)$$

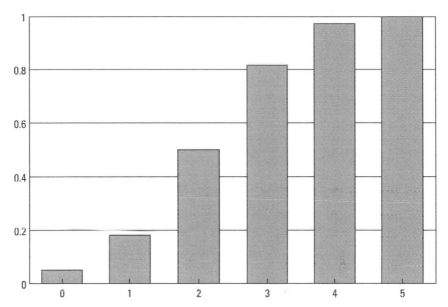

FIGURE 2.21 Cumulative probability functions for the Binomial Distribution (N = 5, P = .5)

where: N = The number of trials.

P = The probability of a success on a single trial.

Q = 1 − P.

As N becomes large, the Binomial tends to the Normal Distribution, with the Normal being the limiting form of the Binomial. Generally, if N * P and N * Q are both greater than 5, you could use the Normal in lieu of the Binomial as an approximation.

The Binomial Distribution is often used to statistically validate a gambling system. An example will illustrate. Suppose we have a gambling system that has won 51% of the time. We want to determine what the winning percentage would be if it performs in the future at a level of 3 standard deviations worse. Thus, the variable of interest here, X, is equal to .51, the probability of a winning trade. The variable of interest need not always be for the probability of a win. It can be the probability of an event being in one of two mutually exclusive groups. We can now perform the first necessary equation in the test:

$$L = P - Z * \sqrt{(P * (P - 1))/(N - 1)} \qquad (2.38)$$

where: L = The lower boundary for P to be at Z standard deviations.

P = The variable of interest representing the probability of being in one of two mutually exclusive groups.

Z = The selected number of standard deviations.

N = The total number of events in the sample.

Suppose our sample consisted of 100 plays. Thus:

$$L = .51 - 3 * \sqrt{(.51 * (1 - .51))/(100 - 1)}$$
$$= .51 - 3 * \sqrt{(.51 * .49)/99}$$
$$= .51 - 3 * \sqrt{.2499/99}$$
$$= .51 - 3 * \sqrt{.0025242424}$$
$$= .51 - 3 * .05024183938$$
$$= .51 - .1507255181$$
$$= .3592744819$$

Based on our history of 100 plays which generated a 51% win rate, we can state that it would take a three-sigma event for the population of plays (the future if we play an infinite number of times into the future) to have less than 35.92744819% winners.

What kind of a confidence level does this represent? That is a function of N, the total number of plays in the sample. We can determine the confidence level of achieving 35 or 36 wins in 100 tosses by Equation (2.35). However, (2.35) is clumsy to work with as N gets large because of all of the factorial functions in (2.35). Fortunately, the Normal distribution, Equation (2.21) for one-tailed probabilities, can be used as a very close approximation for the Binomial probabilities. In the case of our example, using Equation (2.21), 3 standard deviations translates into a 99.865% confidence. Thus, if we were to play this gambling system over an infinite number of times, we could be 99.865% sure that the percentage of wins would be greater than or equal to 35.92744819%.

This technique can also be used for statistical validation of trading systems. However, this method is only valid when the following assumptions are true. First, the N events (trades) are all independent and randomly selected. This can easily be verified for any trading system. Second, the N events (trades) can all be classified into two mutually exclusive groups (wins and losses, trades greater than or less than the median trade, etc.). This assumption, too, can easily be satisfied. The third assumption is that the probability of an event being classified into one of the two mutually exclusive groups is constant from one event to the next. This is not necessarily true in trading, and the technique becomes inaccurate to the degree that this assumption is false. Be that as it may, the technique still can have value for traders.

Not only can it be used to determine the confidence level for a certain method's being profitable; the technique can also be used to determine the confidence level for a given market indicator. For instance, if you have an indicator that will forecast the direction of the next day's close, you then have two mutually exclusive groups: correct forecasts and incorrect forecasts. You can now express the reliability of your indicator to a certain confidence level.

This technique can also be used to discern how many trials are necessary for a system to be profitable to a given confidence level. For example, suppose we have a gambling system that wins 51% of the time on a game that pays 1 to 1. We want to know how many trials we must observe to be certain to a given confidence level that the system will be profitable in an asymptotic sense. Thus, we can restate the problem as, "If the system wins 51% of the time, how many trials must I witness, and have it show a 51% win rate, to know that it will be profitable to a given confidence level?"

Since the payoff is 1:1, the system must win in excess of 50% of the time to be considered profitable. Let's say we want the given confidence level to again be 99.865, or 3 standard deviations (although we are using 3 standard deviations in this discussion, we aren't restricted to that amount; we can use any number of standard deviations that we want). How many trials must we now witness to be 99.865% confident that at least 51% of the trials will be winners?

If $.51 - X = .5$, then $X = .01$. Therefore, the right factors of Equation (2.38), $Z * \sqrt{P * (P - 1)/(N - 1)}$, must equal .01. Since $Z = 3$ in this case, and $.01/3 = .0033$, then:

$$\sqrt{P * (1 - P)/(N - 1)}$$

We know that P equals .51, thus:

$$.51 * \sqrt{(1 - .51)/(N - 1)}$$

Squaring both sides gives us:

$$((.51 * (1 - .51))/(N - 1)) = .00001111$$

To continue:

$$(.51 * .49)/(N - 1) = .00001111$$

$$.2499/(N - 1) = .00001111$$

$$.2499/.00001111 = N - 1$$

$$.2499/.00001111 + 1 = N$$

$$22,491 + 1 = N$$

$$N = 22,492$$

Thus, we need to witness a 51% win rate over 22,492 trials to be 99.865% certain that we will see at least 51% wins.

THE GEOMETRIC DISTRIBUTION

Like the Binomial, the *Geometric Distribution*, also a discrete distribution, occurs as a result of N independent Bernoulli trials. The Geometric Distribution measures the number of trials before the first success (or failure). The probability density function, $N'(X)$, is:

$$N'(X) = Q^{(X-1)} * P \tag{2.39}$$

where: $P =$ The probability of success for a given trial.
$Q =$ The probability of failure for a given trial.

In other words, $N'(X)$ here measures the number of trials until the first success. The cumulative density function for the Geometric is therefore:

$$N(X) = \sum_{J=1}^{X} Q^{(J-1)} * P \tag{2.40}$$

where: $P =$ The probability of success for a given trial.
$Q =$ The probability of failure for a given trial.

Figures 2.22 and 2.23 illustrate the probability density and cumulative probability ability (i.e., cdf), respectively, of the Geometric Distribution. Other properties of the Geometric are:

$$\text{Mean} = 1/P \tag{2.41}$$
$$\text{Variance} = Q/P^2 \tag{2.42}$$

where: $P =$ The probability of success for a given trial.
$Q =$ The probability of failure for a given trial.

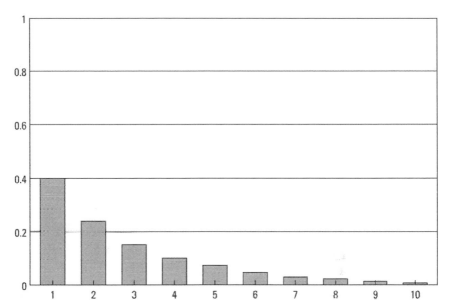

FIGURE 2.22 Probability density functions for the Geometric Distribution (P = .6)

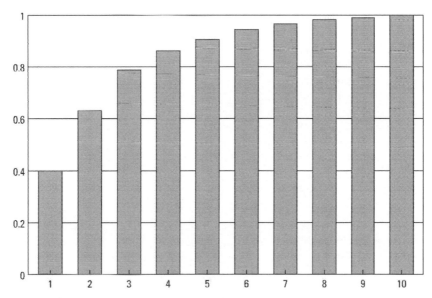

FIGURE 2.23 Cumulative probability functions for the Geometric Distribution (P = .6)

Suppose we are discussing tossing a single die. If we are talking about having the outcome of 5, how many times will we have to toss the die, on average, to achieve this outcome? The mean of the Geometric Distribution tells us this. If we know the probability of throwing a 5 is 1/6 (.1667), then the mean is 1/.1667 = 6. Thus, we would expect, on average, to toss a die six times in order to get a 5. If we kept repeating this process and recorded how many tosses it took until a 5 appeared, plotting these results would yield the Geometric Distribution function formulated in (2.39).

THE HYPERGEOMETRIC DISTRIBUTION

Another type of discrete distribution related to the preceding distributions is termed the *Hypergeometric Distribution*. Recall that in the Binomial Distribution it is assumed that each draw in succession from the population has the same probabilities. That is, suppose we have a deck of 52 cards; 26 of these cards are black and 26 are red. If we draw a card and record whether it is black or red, we then put the card back into the deck for the next draw. This "sampling with replacement" is what the Binomial Distribution assumes. Now, for the next draw, there is still a .5 (26/52) probability of the next card's being black (or red).

The Hypergeometric Distribution assumes almost the same thing, except there is no replacement after sampling. Suppose we draw the first card and it is red, and we *do not* replace it back into the deck. Now, the probability of the next draw's being red is reduced to 25/51 or .4901960784. In the Hypergeometric Distribution there is *dependency*, in that the probabilities of the next event are dependent on the outcome(s) of the prior event(s). Contrast this to the Binomial Distribution, where an event is *independent* of the outcome(s) of the prior event(s).

The basic functions $N'(X)$ and $N'(X)$ of the Hypergeometric are the same as those for the Binomial, (2.33) and (2.35), respectively, except that with the Hypergeometric the variable P, the probability of success on a single trial, changes from one trial to the next.

It is interesting to note the relationship between the Hypergeometric and Binomial Distributions. As N becomes larger, the differences between the computed probabilities of the Hypergeometric and the Binomial draw closer to each other. Thus, we can state that as N approaches infinity, the Hypergeometric approaches the Binomial as a limit.

If you want to use the Binomial probabilities as an approximation of the Hypergeometric, as the Binomial is far easier to compute, how big must the population be? It is not easy to state with any certainty, since the desired accuracy of the result will determine whether the approximation is successful or not. Generally, though, a population to sample size of 100 to 1 is usually sufficient to permit approximating the Hypergeometric with the Binomial.

THE POISSON DISTRIBUTION

The *Poisson Distribution* is another important discrete distribution. This distribution is used to model arrival distributions and other seemingly random events that occur repeatedly yet haphazardly. These events can occur at points in time or at points along a wire or line (one dimension), along a plane (two dimensions), or in any N-dimensional construct. Figure 2.24 shows the arrival of events (the Xs) along a line, or in time.

The Poisson Distribution was originally developed to model incoming telephone calls to a switchboard. Other typical situations that can be modeled by the Poisson are the breakdown of a piece of equipment, the completion of a repair job by a steadily working repairman, a typing error, the growth of a colony of bacteria on a Petri plate, a defect in a long ribbon or chain, and so on.

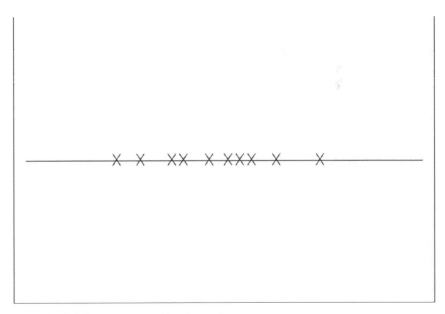

FIGURE 2.24 Sequence of haphazard events in time

The main difference between the Poisson and the Binomial distributions is that the Binomial is not appropriate for events that can occur more than once within a given time frame. Such an example might be the probability of an automobile accident over the next 6 months. In the Binomial we would be working with two distinct cases: Either an accident occurs, with probability P, or it does not, with probability Q (i.e., 1 – P). However, in the Poisson Distribution we can also account for the fact that more than one accident can occur in this time period.

The probability density function of the Poisson, $N'(X)$, is given by:

$$N'(X) = (L^X * EXP(-L))/X! \qquad (2.43)$$

where: $L = $ The parameter of the distribution.

$EXP() = $ The exponential function.

Note that X must take discrete values.

Suppose that calls to a switchboard average four calls per minute ($L = 4$). The probability of three calls ($X = 3$) arriving in the next minute is:

$$N'(3) = (4^3 * EXP(-4))/3!$$
$$= (64 * EXP(-4))/(3 * 2)$$
$$= (64 * .01831564)/6$$
$$= 1.17220096/6$$
$$= .1953668267$$

So we can say there is about a 19.5% chance of getting three calls in the next minute. Note that this is not cumulative—that is, this is not the probability of getting three calls or fewer, it is the probability of getting exactly three calls. If we wanted to know the probability of getting three calls or fewer, we would have had to use the N(3) formula [which is given in (2.46)].

Other properties of the Poisson Distribution are:

$$Mean = L \qquad (2.44)$$
$$Variance = L \qquad (2.45)$$

where: $L = $ The parameter of the distribution.

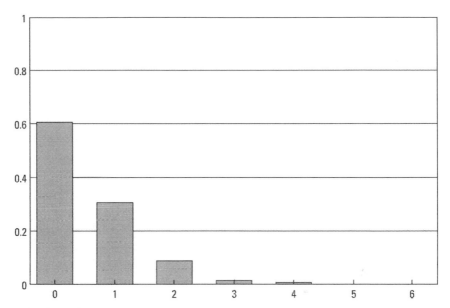

FIGURE 2.25 Probability density functions for the Poisson Distribution (L = .5)

In the Poisson Distribution, both the mean and the variance equal the parameter L. Therefore, in our example case we can say that the mean is four calls and the variance is four calls (or, the standard deviation is 2 calls—the square root of the variance, 4).

When this parameter, L, is small, the distribution is shaped like a reversed J, and when L is large, the distribution is not dissimilar to the Binomial. Actually, the Poisson is the limiting form of the Binomial as N approaches infinity and P approaches O. Figures 2.25 through 2.28 show the Poisson Distribution with parameter values of .5 and 4.5.

The cumulative density function of the Poisson, N(X), is given by:

$$N(X) = \sum_{J=0}^{X}(L^{J} * EXP(-L))/J! \qquad (2.46)$$

where:
L = The parameter of the distribution.

$EXP(\ )$ = The exponential function.

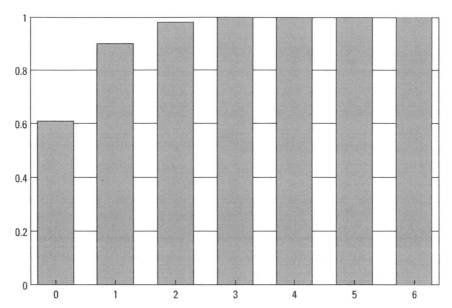

FIGURE 2.26 Cumulative probability functions for the Poisson Distribution (L = .5)

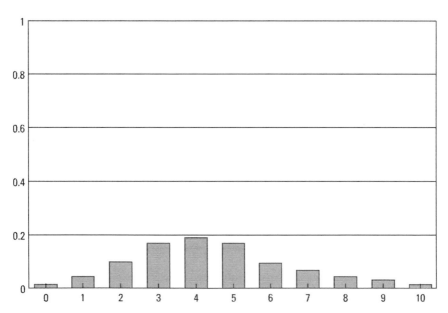

FIGURE 2.27 Probability density functions for the Poisson Distribution (L = 4.5)

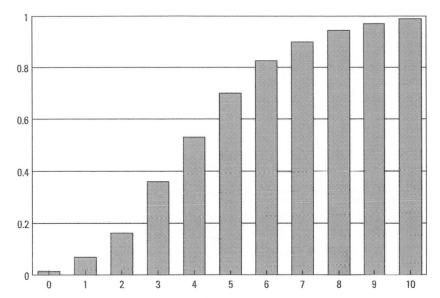

FIGURE 2.28 Cumulative probability functions for the Poisson Distribution (L = 4.5)

THE EXPONENTIAL DISTRIBUTION

Related to the Poisson Distribution is a continuous distribution with a wide utility called the *Exponential Distribution*, sometimes also referred to as the *Negative Exponential Distribution*. This distribution is used to model interarrival times in queuing systems; service times on equipment; and sudden, unexpected failures such as equipment failures due to manufacturing defects, light bulbs burning out, the time that it takes for a radioactive particle to decay, and so on. (There is a very interesting relationship between the Exponential and the Poisson Distributions. The arrival of calls to a queuing system follows a Poisson Distribution, with arrival rate L. The interarrival distribution (the time between the arrivals) is Exponential with parameter 1/L.)

The probability density function $N'(X)$ for the Exponential Distribution is given as:

$$N'(X) = A * EXP(-A * X) \qquad (2.47)$$

where: $A =$ The single parametric input, equal to 1/L in the Poisson Distribution. A must be greater than 0.

$EXP(\ ) =$ The exponential function.

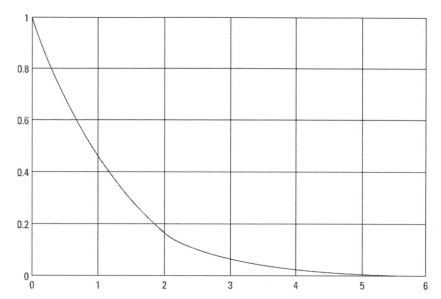

FIGURE 2.29 Probability density functions for the Exponential Distribution (A = 1)

The integral of (2.47), N(X), the cumulative density function for the Exponential Distribution is given as:

$$N(X) = 1 - EXP(-A * X) \qquad (2.48)$$

where: A = The single parametric input, equal to 1/L in the Poisson Distribution. A must be greater than 0.

     EXP( ) = The exponential function.

Figures 2.29 and 2.30 show the functions of the Exponential Distribution. Note that once you know A, the distribution is completely determined.

where: A = The single parametric input, equal to 1/L in the Poisson Distribution. A must be greater than 0.

     EXP( ) = The exponential function.

Figures 2.29 and 2.30 show the functions of the Exponential Distribution. Note that once you know A, the distribution is completely determined. The mean and variance of the Exponential Distribution are:

$$\text{Mean} = 1/A \qquad (2.49)$$
$$\text{Variance} = 1/A^2 \qquad (2.50)$$

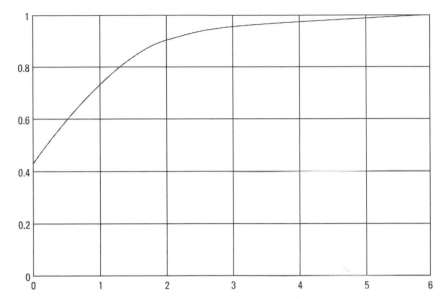

FIGURE 2.30 Cumulative probability functions for the Exponential Distribution (A = 1)

Again Λ is the single parametric input, equal to 1/L in the Poisson Distribution, and must be greater than 0.

Another interesting quality about the Exponential Distribution is that it has what is known as the "forgetfulness property." In terms of a telephone switchboard, this property states that the probability of a call in a given time interval is not affected by the fact that no calls may have taken place in the preceding interval(s).

THE CHI-SQUARE DISTRIBUTION

A distribution that is used extensively in goodness-of-fit testing is the *Chi-Square Distribution* (pronounced ki square, from the Greek letter X (chi) and hence often represented as the X^2 distribution).

Assume that K is a standard normal random variable (i.e., it has mean 0 and variance 1). If we say that K equals the square root of J ($J = K^2$) , then we know that K will be a continuous random variable. However, we know that K will not be less than zero, so its density function will differ from the Normal. The Chi-Square Distribution gives us the density function of K:

$$N'(K) = K^{V/2-1} * EXP(-V/2)/2^{V/2} * GAM(V/2) \qquad (2.51)$$

where: K = The chi-square variable X^2.

V = The number of degrees of freedom, which is the single input parameter.

EXP( ) = The exponential function.

GAM( ) = The standard gamma function.

A few notes on the gamma function are in order. This function has the following properties:

1. $GAM(O) \neq 1$
2. $GAM(1/2) \neq$ The square root of pi, or 1.772453851
3. $GAM(N) \neq (N - 1) * GAM(N - 1)$; therefore, if N is an integer, $GAM(N) = (N-1)!$

Notice in Equation (2.51) that the only input parameter is V, the number of degrees of freedom. Suppose that rather than just taking one independent random variable squared (K^2), we take M independent random variables squared, and take their sum:

$$JM = K_1^2 + K_2^2 \ldots K_M^2$$

Now J_M is said to have the Chi-Square Distribution with M degrees of freedom. It is the number of degrees of freedom that determines the shape of a particular Chi-Square Distribution. When there is one degree of freedom, the distribution is severely asymmetric and resembles the Exponential Distribution (with $A = 1$). At two degrees of freedom the distribution begins to look like a straight line going down and to the right, with just a slight concavity to it. At three degrees of freedom, a convexity starts taking shape and we begin to have a unimodal-shaped distribution. As the number of degrees of freedom increases, the density function gradually becomes more and more symmetric. As the number of degrees of freedom becomes very large, the Chi-Square Distribution begins to resemble the Normal Distribution per the Central Limit Theorem.

THE CHI-SQUARE "TEST"

Do not confuse the Chi-Square "Test" with the Chi-Square Distribution. The former is a hypothesis testing procedure (one of many such procedures). Mention is made of it here, but it should not be confused with the distributional form of the same name.

There exist a number of statistical tests designed to determine if two samples come from the same population. Essentially, we want to know if two distributions are different. Perhaps the most well known of these tests is the chi-square test, devised by Karl Pearson around 1900. It is perhaps the most popular of all statistical tests used to determine whether two distributions are different.

The chi-square statistic, X^2, is computed as:

$$X^2 = \sum_{i=1}^{N}(O_i - E_i)^2/E_i \tag{2.52}$$

where: $N = $ The total number of bins.

$O_i = $ The number of events observed in the ith bin.

$E_i = $ The number of events expected in the ith bin.

A large value for the chi-square statistic indicates that it is unlikely that the two distributions are the same (i.e., the two samples are not drawn from the same population). Likewise, the smaller the value for the chi-square statistic, the more likely it is that the two distributions are the same (i.e., the two samples were drawn from the same population).

Note that the observed values, the O_i's, will always be integers. However, the expected values, the E_i's, can be nonintegers. Equation (2.52) gives the chi-square statistic when both the expected and observed values are integers. When the expected values, the E_i's, are permitted to be nonintegers, we must use a different equation, known as *Yates' correction*, to find the chi-square statistic:

$$X^2 = \sum_{i=1}^{N}(\text{ABS}\,(O_i - E_i) - .5)^2/E_i \tag{2.53}$$

where: $N = $ The total number of bins.

$O_i = $ The number of events observed in the ith bin.

$E_i = $ The number of events expected in the ith bin.

$\text{ABS}() = $ The absolute value function.

We can convert a chi-square statistic such as 37.5336 to a *sigfinicance level*. In the sense we are using here, a significance level is a number between 0, representing that the two distributions are different, and 1, meaning that the two distributions are the same. We can never be 100% certain that two distributions are the same (or different), but we can determine how alike or different two distributions are to a certain significance level. There are two ways in which we can find the significance level. This first and by far the simplest way is by using tables. The second way to convert a chi-square statistic to a significance level is to perform the math yourself (which is how the tables were drawn up in the first place). However, the math requires the use of incomplete gamma functions, not be treated in this text. However, most readers who would want to know how to calculate a significance level from a given chi-square statistic would want to know this because tables are rather awkward to use from a programming standpoint. Therefore, what follows is a snippet of Java code to convert from a given chi-square statistic to a significance level.

```java
Public void ChiSquareTest(int nmbrOfBins, double chiSquareStatistic){
    double confidenceLevel = 1.0;
    double a = 0.0, b = 0.0, c=1.0, d = 0.0, e = 0.0,
    f = 1.0;
    int nbins = nmbrOfBins -3;
    System.out.println("Chi-Square Statistic at " + nbins + "degrees of
freedom is "+chiSquareStatistic);
    if(chiSquareStatistic < 31.0 || nbins > 2){
      e = nbins/2 -1;
      a = 1;
      for(int i = 1;i <=nbons/2 - .5; i++){
        a *=e;
        e -=1.0;
      }
      if(nbins% 2 !=0){
        a *= 1.77245374942627;
      }
      b = Math.pow((chiSquareStatistic/2.0), (double) (nbins/2)) * 2.0/
(Math.exp(chiSquareStatistic/2.0) * a * nbins);
      d = nbins + 2;
      do{
      c *=chiSquareStatistic/d;
      f+=c;
      d+=2.0;
      }while(c > 0.0);
      confidenceLevel = 1.0 - b *f;
    }
    System.out.println("For a Significance level of "+confidenceLevel);
}
```

Whether you determine your significance levels via a table or calculate them yourself, you will need two parameters to determine a significance level. The first of these parameters is, of course, the chi-square statistic itself. The second is the number of *degrees of freedom*. Generally, the number of degrees of freedom is equal to the number of bins minus 1 minus the number of population parameters that have to be estimated for the sample statistics. What follows is a small table for converting between chi-square values and degrees of freedom to significance levels:

Values of X^2

Degrees of Freedom	Significance Level			
	.20	.10	.05	.01
1	1.6	2.7	3.8	6.6
2	3.2	4.6	6.0	9.2
3	4.6	6.3	7.8	11.3
4	6.0	7.8	9.5	13.3
5	7.3	9.2	11.1	15.1
10	13.4	16.0	18.3	23.2
20	25.0	28.4	31.4	37.6

You should be aware that the chi-square test can do a lot more than is presented here. For instance, you can use the chi-square test on a 2×2 contingency table (actually on any $N \times M$ contingency table).

Finally, there is the problem of the arbitrary way we have chosen our bins as regards both their number and their range. Recall that binning data involves a certain loss of information about that data, but generally the profile of the distribution remains relatively the same. If we choose to work with only three bins, or if we choose to work with 30, we will likely get somewhat different results. It is often a helpful exercise to bin your data in several different ways when conducting statistical tests that rely on binned data. In so doing, you can be rather certain that the results obtained were not due solely to the arbitrary nature of how you chose your bins.

In a purely statistical sense, in order for our number of degrees of freedom to be valid, it is necessary that the number of elements in each of the expected bins, the E_i's, be at least five. When there is a bin with less than five expected elements in it, theoretically the number of bins should be reduced until all of the bins have at least five expected elements in them. Often, when only the lowest and/or highest bin has less than five expected elements in it, the adjustment can be made by making these groups "all less than" and "all greater than" respectively.

THE STUDENT'S DISTRIBUTION

The *Student's Distribution,* sometimes called the *t Distribution* or *Student's t*, is another important distribution used in hypothesis testing that is related to the Normal Distribution. When you are working with less than 30 samples of a near-Normally distributed population, the Normal Distribution can no longer be accurately used. Instead, you must use the Student's Distribution. This is a symmetrical distribution with one parametric input, again the degrees of freedom. The degrees of freedom usually equals the number of elements in a sample minus one (N − 1).

The shape of this distribution closely resembles the Normal except that the tails are thicker and the peak of the distribution is lower. As the number of degrees of freedom approaches infinity, this distribution approaches the Normal in that the tails lower and the peak increases to resemble the Normal Distribution. When there is one degree of freedom, the tails are at their thickest and the peak at its smallest. At this point, the distribution is called *Cauchy.*

It is interesting that if there is only one degree of freedom, then the mean of this distribution is said not to exist. If there is more than one degree of freedom, then the mean does exist and is equal to zero, since the distribution is symmetrical about zero. The variance of the Student's Distribution is infinite if there are fewer than three degrees of freedom.

The concept of *infinite variance* is really quite simple. Suppose we measure the variance in daily closing prices for a particular stock for the last month. We record that value. Now we measure the variance in daily closing prices for that stock for the next year and record that value. Generally, it will be greater than our first value, of simply last month's variance. Now let's go back over the last five years and measure the variance in daily closing prices. Again, the variance has gotten larger. The farther back we go—that is, the more data we incorporate into our measurement of variance—the greater the variance becomes. Thus, the variance increases without bound as the size of the sample increases. This is infinite variance. The distribution of the log of daily price changes appears to have infinite variance, and thus the Student's Distribution is sometimes used to model the log of price changes. (That is, if C_0 is today's close and C_1 yesterday's close, then $\ln(C_0/C_1)$ will give us a value symmetrical about 0. The distribution of these values is sometimes modeled by the Student's distribution).

If there are three or more degrees of freedom, then the variance is finite and is equal to:

$$\text{Variance} = V/(V-2) \qquad \text{for } V > 2 \qquad (2.54)$$

$$\text{Mean} = 0 \qquad \text{for } V > 1 \qquad (2.55)$$

where:　　$V =$ The degrees of freedom.

Suppose we have two independent random variables. The first of these, Z, is standard normal (mean of 0 and variance of 1). The second of these, which we call J, is Chi-Square distributed with V degrees of freedom. We can now say that the variable T, equal to Z/(J/V), is distributed according to the Student's Distribution. We can also say that the variable T will follow the Student's Distribution with N − 1 degrees of freedom if:

$$T = \sqrt{N} * (X - U)/S$$

where: X = A sample mean.

   S = A sample standard deviation.

   N = The size of a sample.

   U = The population mean.

The probability density function for the Student's Distribution, $N'(X)$, is given as:

$$N'(X) = \frac{GAM((V + 1)/2)}{\sqrt{V * P} * GAM(V/2)} * (1 + X^2/V)^{-(V+1)/2} \qquad (2.56)$$

where: P = pi, or 3.1415926536.

   V = The degrees of freedom.

  GAM( ) = The standard gamma function.

The mathematics of the Student's Distribution are related to the incomplete beta function. Since we aren't going to plunge into functions of mathematical physics such as the incomplete beta function, we will leave the Student's Distribution at this point. Before we do, however, you still need to know how to calculate probabilities associated with the Student's Distribution for a given number of standard units (Z score) and degrees of freedom. As the following snippet of java code to discern the probabilities. You'll note that as the degrees of freedom variable, DEGFDM, approaches infinity, the values returned, the probabilities, converge to the Normal as given by Equation (2.22):

```
public void StudentsT2TailProbs(double zScore, int degreesOfFreedom){
    double confidenceLevel = 1.0;
    double st = Math.abs(zScore);
    double r8 = Math.atan(st/Math.sqrt((double)degreesOfFreedom));
    double rc8 = Math.cos(r8);
    double x8 = 1.0;
```

```
double r28 = rc8 * rc8;
double rs8 = Math.sin(r8);
double y8 = r8;
double y8 = r8;
if(degreesOfFreedom %2 !=0 ){
   if(degreesOfFreedom !=1){
      y8 = rc8;
      for(int i =3;i<=degreesOfFreedom - 2; i+=2){
         x8 = x8 * r28 * (double)((i - 1)/i);
         y8 = y8 + x8 * rc8;
      }
      y8 = r8 + rs8 * y8;
   }
   confidenceLevel = y8 * 0.6366197723657157;
}else{
   y8=1.0;
   for(int i =2;i <=degreesOfFreedom -2; i+=2){
      x8 = x8 * r28 * (double)((i - 1)/i);
      y8 += x8;
   }
   confidenceLevel = y8 *rs8;
}
System.out.println("The two-tailed probabilities associated with
the T distribution for a Z score of "+zScore+" and "+degreesOfFreedom+
"degrees freedom is "+confidenceLevel);
}
```

Next, we come to another distribution, related to the Chi-Square Distribution, that also has important uses in statistics. The *F Distribution*, sometimes referred to as *Snedecor's Distribution* or *Snedecor's F*, is useful in hypothesis testing. Let A and B be independent chi-square random variables with degrees of freedom of M and N, respectively. Now the random variable:

$$F = (A/M)/(B/N)$$

can be said to have the F Distribution with M and N degrees of freedom. The density function, $N'(X)$, of the F Distribution is given as:

$$N'(X) = \frac{GAM((M+N)/2) * (M/N)^{M/2}}{(GAM(M/2) * GAM(N/2) * (1+M/N))^{(M+N)/2}} \qquad (2.57)$$

where: M = The number of degrees of freedom of the first parameter.

N = The number of degrees of freedom of the second parameter.

GAM( ) = The standard gamma function.

THE MULTINOMIAL DISTRIBUTION

The *Multinomial Distribution* is related to the Binomial, and likewise is a discrete distribution. Unlike the Binomial, which assumes two possible outcomes for an event, the Multinomial assumes that there are M different outcomes for each trial. The probability density function, $N'(X)$, is given as:

$$N'(X) = \left(N! \Big/ \prod_{i=1}^{M} N_i! \right) * \prod_{i=1}^{M} P_i^{N_i} \qquad (2.58)$$

where: N = The total number of trials.

N_i = The number of times the ith trial occurs.

P_i = The probability that outcome number i will be the result of any one trial. The summation of all P_i's equals 1.

M = The number of possible outcomes on each trial.

For example, consider a single die where there are six possible outcomes on any given roll (M = 6). What is the probability of rolling a 1 once, a 2 twice, and a 3 three times out of 10 rolls of a fair die? The probabilities of rolling a 1, a 2, or a 3 are each 1/6. We must consider a fourth alternative to keep the sum of the probabilities equal to 1, and that is the probability of not rolling a 1, 2, or 3, which is 3/6. Therefore, $P_1 = P_2 = P_3 = 1/6$, and $P_4 = 3/6$. Also, $N_1 = 1$, $N_2 = 2$, $N_3 = 3$, and $N_4 = 10 - 3 - 2 - 1 = 4$. Therefore, Equation (2.58) can be worked through as:

$$N'(X) = (10!/(1! * 2! * 3! * 4!)) * (1/6)^1 * (1/6)^2 * (1/6)^3 * (3/6)^4$$
$$= (3628800/(1 * 2 * 6 * 24)) * .1667 * .0278 * .00463 * .0625$$
$$= (3628800/288) * .000001341$$
$$= 12600 * .000001341$$
$$= .0168966 \qquad (2.58)$$

Note that this is the probability of rolling exactly a 1 once, a 2 twice, and a 3 three times, not the cumulative density. This is a type of distribution that uses more than one random variable; hence, its cumulative density cannot be drawn out nicely and neatly in two dimensions as you could with the other distributions discussed thus far. We will not be working with other distributions that have more than one random variable, but you should be aware that such distributions and their functions do exist.

THE STABLE PARETIAN DISTRIBUTION

The *Stable Paretian Distribution* is actually an entire class of distributions, sometimes referred to as "Pareto-Levy" distributions. The probability density function $N'(U)$ is given as:

$$\ln(N''(U)) = i * D * U - V * ABS(U)^A * Z \tag{2.59}$$

where: $U = $ The variable of the stable distribution.

               $A = $ The kurtosis parameter of the distribution.

               $B = $ The skewness parameter of the distribution.

               $D = $ The location parameter of the distribution.

               $V = $ This is also called the scale parameter.

               $i = $ The imaginary unit, $\sqrt{-1}$

               $Z = 1 - i * B * (U/ABS(U)) * \tan(A * 3.1415926536/2)$ when $A > < 1$ and $1 + i * B * (U/ABS(U)) * 2/3.1415926536 * \log(ABS(U))$ when $A = 1$.

     $ABS() = $ The absolute value function.

     $\tan() = $ The tangent function.

      $\ln() = $ The natural logarithm function.

The limits on the parameters of Equation (2.59) are:

$$0 < A <= 2 \tag{2.60}$$

$$-1 <= B <= 1 \tag{2.61}$$

$$0 <= V \tag{2.62}$$

The four parameters of the distribution—A, B, D, and V—allow the distribution to assume a great many different shapes.

The variable A measures the height of the tails of the distribution. Thus, we can say that A represents the kurtosis variable of the distribution. A is also called the *characteristic exponent* of the distribution. When A equals 2, the distribution is Normal, and when A equals 1 the distribution is Cauchy. For values of A that are less than 2, the tails of the distribution are higher than with the Normal Distribution. The total probability in the tails increases as A decreases. When A is less than 2, the variance is infinite. The mean of the distribution exists only if A is greater than 1.

The variable B is the *index of skewness*. When B equals zero, the distribution is perfectly symmetrical. The degree of skewness is larger the larger the absolute value of B. Notice that when A equals 2, W(U,A) equals 0; hence, B has no effect on the distribution. In this case, when A equals 2, no matter what B is, we still have the perfectly symmetrical Normal Distribution. The *scale parameter*, V, is sometimes written as a function of A, in that $V = C^A$, therefore $C = V^{1/A}$. When A equals 2, V is one-half the variance. When A equals 1, the Cauchy Distribution, V is equal to the semi-interquartile range. D is the *location parameter*. When A is equal to 2, the arithmetic mean is an unbiased estimator of D; when A is equal to 1, the median is.

The cumulative density functions for the stable Paretian are not known to exist in closed form. For this reason, evaluation of the parameters of this distribution is complex, and work with this distribution is made more difficult. It is interesting to note that the stable Paretian parameters A, B, C, and D correspond to the fourth, third, second, and first moments of the distribution, respectively. This gives the stable Paretian the power to model many types of real-life distributions—in particular, those where the tails of the distribution are thicker than they would be in the Normal, or those with infinite variance (i.e., when A is less than 2). For these reasons, the stable Paretian is an extremely powerful distribution with applications in economics and the social sciences, where data distributions often have those characteristics (fatter tails and infinite variance) that the stable Paretian addresses.

This infinite variance characteristic makes the Central Limit Theorem inapplicable to data that is distributed per the stable Paretian distribution when A is less than 2. This is a very important fact if you plan on using the Central Limit Theorem.

One of the major characteristics of the stable Paretian is that it is invariant under addition. This means that the sum of independent stable variables with characteristic exponent A will be stable, with approximately the same characteristic exponent. Thus, we have the Generalized Central Limit Theorem, which is essentially the Central Limit Theorem, except that the limiting form of the distribution is the stable Paretian rather than the Normal, and the theorem applies even when the data has infinite variance (i.e., A < 2), which is when the Central Limit Theorem does not apply. For example, the heights of people have finite variance. Thus, we could model the heights of people with the Normal Distribution. The distribution of people's incomes, however, does not have finite variance and is therefore modeled by the stable Paretian distribution rather than the Normal Distribution.

It is because of this Generalized Central Limit Theorem that the stable Paretian Distribution is believed by many to be representative of the distribution of price changes.

There are many more probability distributions that we could still cover (Negative Binomial Distribution, Gamma Distribution, Beta Distribution, etc.); however, they become increasingly more obscure as we continue from here. The distributions we have covered thus far are, by and large, the main common probability distributions.

Efforts have been made to catalogue the many known probability distributions. Undeniably, one of the better efforts in this regard has been done by Karl Pearson, but perhaps the most comprehensive work done on cataloguing the many known probability distributions has been presented by Frank Haight.[5] Haight's "Index" covers almost all of the known distributions on which information was published prior to January, 1958. Haight lists most of the mathematical functions associated with most of the distributions. More important, references to books and articles are given so that a user of the index can find what publications to consult for more in-depth matter on the particular distribution of interest. Haight's index categorizes distributions into ten basic types: (1) Normal; (2) Type III; (3) Binomial; (4) Discrete; (5) Distributions on (A, B); (6) Distributions on (0, infinity); (7) Distributions on (–infinity, infinity); (8) Miscellaneous Univariate; (9) Miscellaneous Bivariate; (10) Miscellaneous Multivariate.

Of the distributions we have covered in this Chapter, the Chi-Square and Exponential (Negative Exponential) are categorized by Haight as Type III. The Binomial, Geometric, and Bernoulli are categorized as Binomial. The Poisson and Hypergeometric are categorized as Discrete. The Rectangular is under Distributions on (A, B), the F Distribution as well as the Pareto are under Distributions on (0, infinity), the Student's Distribution is regarded as a Distribution on (–infinity, infinity), and the Multinomial as a Miscellaneous Multivariate. It should also be noted that not all distributions fit cleanly into one of these ten categories, as some distributions can actually be considered subclasses of others. For instance, the Student's distribution is catalogued as a Distribution on (–infinity, infinity), yet the Normal can be considered a subclass of the Student's, and the Normal is given its own category entirely. As you can see, there really isn't any "clean" way to categorize distributions. However, Haight's index is quite thorough. Readers interested in learning more about the different types of distributions should consult Haight as a starting point.

---

[5]Haight, F. A., "Index to the Distributions of Mathematical Statistics," *Journal of Research of the National Bureau of Standards-B. Mathematics and Mathematical Physics* 65B No. 1, pp. 23–60, January–March 1961.

CHAPTER 3

Reinvestment of Returns and Geometric Growth Concepts

TO REINVEST TRADING PROFITS OR NOT

Let's call the following system "System A." In it we have two trades—the first making 50%, the second losing 40%. Therefore, if we do not reinvest our returns, we make 10%. If we do reinvest, the same sequence of trades loses 10%.

System A

	No Reinvestment		With Reinvestment	
Trade No.	P&L	Accum.	P&L	Accum.
		100		100
1	50	150	50	150
2	−40	110	−60	90

Now let's look at System B, a gain of 15% and a loss of 5%, which also nets out 10% over two trades on a nonreinvestment basis, just like System A. But look at the results of System B with reinvestment. Unlike System A, it makes money.

System B

	No Reinvestment		With Reinvestment	
Trade No.	P&L	Accum.	P&L	Accum.
		100		100
1	15	115	15	115
2	−5	110	−5.75	109.25

An important characteristic of trading with reinvestment that must be realized is that *reinvesting trading profits can turn a winning system into a losing system but not vice versa!* A winning system is turned into a losing system in trading with reinvestment if the returns are not consistent enough. Further, *changing the order or sequence of trades does not affect the final outcome.* This is not only true on a nonreinvestment basis, but also true on a reinvestment basis (contrary to most people's misconception).

System A

	No Reinvestment		With Reinvestment	
Trade No.	P&L	Accum.	P&L	Accum.
		100		100
1	−40	60	−40	60
2	50	110	30	90

System B

	No Reinvestment		With Reinvestment	
Trade No.	P&L	Accum.	P&L	Accum.
		100		100
1	−5	95	−5	95
2	15	110	14.25	109.25

This is not just an aberration caused by a two-trade example. Let's take system A and add two more trades and then examine the results under all four possible sequences of trades.

First Sequence
(System A)

Trade No.	No Reinvestment		With Reinvestment	
	P&L	Accum.	P&L	Accum.
		100		100
1	−40	60	−40	60
2	50	110	30	90
3	−40	70	−36	54
4	50	120	27	81

Second Sequence
(System A)

Trade No.	No Reinvestment		With Reinvestment	
	P&L	Accum.	P&L	Accum.
		100		100
1	50	150	50	150
2	−40	110	−60	90
3	50	160	45	135
4	−40	120	54	81

Third Sequence
(System A)

Trade No.	No Reinvestment		With Reinvestment	
	P&L	Accum.	P&L	Accum.
		100		100
1	50	150	50	150
2	50	200	75	225
3	−40	160	−90	135
4	−40	120	−54	81

Fourth Sequence
(System A)

Trade No.	No Reinvestment		With Reinvestment	
	P&L	Accum.	P&L	Accum.
		100		100
1	−40	60	−40	60
2	−40	20	−24	36
3	50	70	18	54
4	50	120	27	81

As can obviously be seen, the sequence of trades has no bearing on the final outcome, whether viewed on a reinvestment or nonreinvestment basis. What *are* affected, however, are the drawdowns. Listed next are the drawdowns to each of the sequences of trades listed above.

First Sequence

No Reinvestment	Reinvestment
100 to 60 = 40 (40%)	100 to 54 = 46 (46%)

Second Sequence

No Reinvestment	Reinvestment
150 to 110 = 40 (27%)	150 to 81 = 69 (46%)

Third Sequence

No Reinvestment	Reinvestment
200 to 120 = 80 (40%)	225 to 81 = 144 (64%)

Fourth Sequence

No Reinvestment	Reinvestment
100 to 20 = 80 (80%)	100 to 36 = 64 (64%)

Reinvestment trading is never the best based on absolute drawdown. One side benefit to trading on a reinvestment basis is that the drawdowns tend to be buffered. As a system goes into and through a drawdown period, each losing trade is followed by a trade with fewer and fewer contracts. That is why drawdowns as a percent of account equity are always less with reinvestment than with a nonreinvestment approach.

By inspection it would seem you are better off to trade on a nonreinvestment basis rather than to reinvest. This would seem so, since your probability of winning is greater. However, this is not a valid assumption, because in the real world we do not withdraw all of our profits and make up all of our losses by depositing new cash into an account. Further, the nature of investment or trading is predicated upon the effects of compounding. If we do away with compounding (as in the nonreinvestment plan), we can plan on doing little better in the future than we can today, no matter how successful our trading is between now and then. It is compounding that takes the linear function of account growth and makes it a geometric function.

Refer back to the statement that under a reinvestment plan a winning system can be turned into a losing system but not vice versa. Why, then, reinvest our profits into our trading? The sole reason is that by reinvestment, winning systems can be made to win far more than could ever be accomplished on a nonreinvestment basis.

The reader may still be inclined to prefer the nonreinvestment approach since an account that may not be profitable on a reinvestment basis may be profitable on a nonreinvestment basis. However, if a system is good enough, the profits generated on a reinvestment basis will be far greater than on a nonreinvestment basis, and that gap will widen as time goes by. If you have a system that can beat the market, it doesn't make any sense to trade it any other way than to increase your amount wagered as your stake increases.

There is another phenomenon that lures traders away from reinvestment-based trading. That phenomenon is that a losing trade, or losing streak, is inevitable after a prolonged equity run-up. This is true by inspection. The only way a streak of winning trades can end is by a losing trade. The only way a streak of profitable months can end is with a losing month. The problem with reinvestment-based trading is that when the inevitable losses come along you will have the most contracts on. Hence, the losses will be bigger. Similarly, after a losing streak, the reinvestment-basis trader will have fewer contracts on when the inevitable win comes along to break the streak.

This is not to say that there is any statistical reason to assume that winning streaks portend losing trades or vice versa. Rather, what is meant is: If you trade long enough, you will eventually encounter a loss. If you are trading on a reinvestment basis, that loss will be magnified, since, as a result of your winning streak, you will have more contracts on when the loss comes. Unfortunately, there is no way to avoid this—at least no way based on statistical fact in a stationary distribution environment, unless we are talking about a dependent trials process.

Therefore, assuming that the market system in question generates independent trades, there is no way to avoid this phenomenon. It is unavoidable when trading on a reinvestment basis, just as losses are unavoidable in trading under any basis. Losses are a part of the game. Since the goal of good money management is to exploit a profitable system to its fullest potential, the intelligent trader should realize that this phenomenon is part of the game and accept it as such to achieve the longer-term rewards that correct money-management techniques provide for.

MEASURING A GOOD SYSTEM FOR REINVESTMENT—THE GEOMETRIC MEAN

So far we have seen how a system can be sabotaged by not being consistent enough from trade to trade. Does this mean we should close up and put our money in the bank? Let's go back to System A, with its first two trades. For the sake of illustration we are going to add two winners of one point each.

System A

Trade No.	No Reinvestment		With Reinvestment	
	P&L	Accum.	P&L	Accum.
		100		100
1	50	150	50	150
2	−40	110	−60	90
3	1	111	0.9	90.9
4	1	112	0.909	91.809
Percent Wins		0.75		0.75
Avg. Trade		3		−2.04775
Profit Factor		1.3		0.86
Std. Dev.		31.88		39.00
Avg. Trade/Std. Dev.		0.09		−0.05

Now let's take System B and add two more losers of one point each.

System B

Trade No.	No Reinvestment		With Reinvestment	
	P&L	Accum.	P&L	Accum.
		100		100
1	15	115	15	115
2	−5	110	−5.75	109.25
3	−1	109	−1.0925	108.1575
4	−1	108	−1.08157	107.0759
Percent Wins		0.25		0.25
Avg. Trade		2		1.768981
Profit Factor		2.14		1.89
Std Dev.		7.68		7.87
Avg. Trade/Std. Dev.		0.26		0.22

Now, if consistency is what we're really after, let's look at a bank account, the perfectly consistent vehicle (relative to trading), paying 1 % per period. We'll call this series System C.

Notice that in reinvestment the standard deviation always goes up (and hence the Avg. Trade/Std. Dev. tends to come down). Furthermore, the Profit Factor[1] measure is never higher in reinvestment than it is in non-reinvestment trading.

---

[1]Profit Factor = Avg Win/Avg Loss × Percent Winners/(1-Percent Winners). (3.01)

System C

	No Reinvestment		With Reinvestment	
Trade No.	**P&L**	**Accum.**	**P&L**	**Accum.**
		100		100
1	1	101	1	101
2	1	102	1.01	102.01
3	1	103	1.0201	103.0301
4	1	104	1.030301	104.0604
Percent Wins		1.00		1.00
Avg. Trade		1		1.015100
Profit Factor		Infinite		Infinite
Std. Dev.		0.00		0.01
Avg. Trade/Std. Dev.		Infinite		89.89

Our aim is to maximize our profits under reinvestment trading. With that as the goal, we can see that our best reinvestment sequence came from System B. How can we have known that, given only information regarding nonreinvestment trading? By percent of winning trades? By total dollars? Average trade? The answer to these questions is no, since that would have us trading System A (but this is the solution most futures traders opt for). What if we opted for most consistency (i.e., highest ratio of Avg. Trade/Std. Dev. or lowest standard deviation). How about highest profit factor or lowest drawdown? This is not the answer, either. If it were, we should put our money in the bank and forget about trading.

System B has the right mix of profitability and consistency. Systems A and C do not. That is why System B performs the best under reinvestment trading. How best to measure this "right mix"? It turns out there is a formula that will do just that: the *geometric mean*. This is simply the Nth root of the Terminal Wealth Relative (TWR), where N is the number of periods (trades). The TWR is simply what we've been computing when we figure what the final cumulative amount is under reinvestment. In other words, the TWRs for the three systems we just saw are:

SYSTEM	TWR
System A	91.809
System B	107.0759
System C	104.0604

Since there are four trades in each of these, we take the TWRs to the fourth root to obtain the geometric mean:

SYSTEM	GEO. MEAN
System A	0.978861
System B	1.017238
System C	1.009999

$$TWR = \prod_{i=1}^{N} HPR_i \qquad (3.02)$$

$$Geometric\ Mean = TWR^{1/N} \qquad (3.03)$$

where: N = Total number of trades.
          HPR = Holding period returns (equal to 1 plus the
                rate of return).

For example, an HPR of 1.10 means a 10% return over a given period/bet/trade. TWR shows the number of dollars of value at the end of a run of periods/bets/trades per dollar of initial investment, assuming gains and losses are allowed to compound. Here is another way of expressing these variables:

TWR = Final stake / Starting stake

Geometric Mean = Your growth factor per play, or
Final stake / starting stake)$^{1/\text{number of plays}}$.

or

$$Geometric\ Mean = \exp((1/N) * \log(TWR)) \qquad (3.03a)$$

where: N = Total number of trades.
          log(TWR) = The log base 10 of the TWR.
          exp = The exponential function.

Think of the geometric mean as the "growth factor" of your stake, per play. The system or market with the highest geometric mean is the system or market with the highest utility to the trader trading on a reinvestment of returns basis. A geometric mean < 1 means that the system would have lost money if you were trading it on a reinvestment basis. Furthermore, it is vitally important that you use realistic slippage and commissions in calculating geometric means in order to have realistic results.

ESTIMATING THE GEOMETRIC MEAN

There exists a simple technique of finding the geometric mean, whereby you do not have to take the product of all HPRs to the Nth root. The geometric mean squared can be very closely approximated as the arithmetic mean of the HPRs squared minus the population standard deviation of HPRs squared. So the way to approximate the geometric mean is to square the average HPR, then subtract the squared population standard deviation of those HPRs. Now take the square root of this answer and that will be a very close approximation of the actual geometric mean. As an example, assume the following HPRs over four trades:

	1.00
	1.50
	1.00
	.60
Arithmetic Mean	1.025
Population Standard Deviation	.3191786334
Estimated Geometric Mean	.9740379869
Actual Geometric Mean	.9740037464

Here is the formula for finding the estimated geometric mean (EGM):

$$EGM = \sqrt{\text{Arithmetic Mean}^2 - \text{Pop. Std. Dev.}^2} \qquad (3.04)$$

The formula given in Chapter 1 to find the standard deviation of a Normal Probability Function is not what you use here. If you already know how to find a standard deviation, skip this section and go on to the next section entitled "How Best to Reinvest."

The standard deviation is simply the square root of the variance:

$$\text{Variance} = (1/(N-1)) \sum_{i=1}^{N} (X_i - \bar{X})^2$$

where: $\bar{X}$ = The average of the data points.
            X_i = The i'th data point.
            N = The total number of data points.

This will give you what is called the *sample* variance. To find what is called the *population* variance you simply substitute the term $(N - 1)$ with (N).

Notice that if we take the square root of the sample variance, we obtain the sample standard deviation. If we take the square root of the population variance, we will obtain the population standard deviation. Now, let's run through an example using the four data points:

$$1.00$$
$$1.50$$
$$1.00$$
$$.60$$

1. Find the average of the data points. In our example this is:

$$\bar{X} = (1.00 + 1.50 + 1.00 + .6)/4$$
$$= 4.1/4$$
$$= 1.025$$

2. For each data point, take the difference between that data point and the answer just found in step 1 (the average). For our example this would be:

$$1.00 - 1.025 = -.025$$
$$1.50 - 1.025 = .475$$
$$1.00 - 1.025 = -.025$$
$$.60 - 1.025 = -.425$$

3. Square each answer just found in step 2. Note that this will make all answers positive numbers:

$$-.025 * -.025 = .000625$$
$$.475 * .475 = .225625$$
$$-.025 * -.025 = .000625$$
$$-.425 * -.425 = .180625$$

4. Sum up all of the answers found in step 3. For our example:

$$.000625$$
$$.225625$$
$$.000625$$
$$+.180625$$
$$\overline{.4075}$$

5. Multiply the answer just found in step 4 by (1/N). If we were looking to find the sample variance, we would multiply the answer just found in step 4 by (1/(N − 1)). Since we eventually want to find the population

standard deviation of these four HPRs to find the estimated geometric mean, we will therefore multiply our answer to step 4 by the quantity (1/N).

$$\text{Population Variance} = (1/N) * (.4075)$$
$$= (1/4) * (.4075)$$
$$= .25 * .4075$$
$$= .101875$$

6. To go from variance to standard deviation, take the square root of the answer just found in step 5. For our example:

$$\text{Population Standard Deviation} = \sqrt{.101875}$$
$$= .3191786334$$

Now, let's suppose we want to figure our estimated geometric mean for our example:

$$\text{EGM} = \sqrt{\text{Arithmetic Mean}^2 - \text{Pop. Std. Dev.}^2}$$
$$= \sqrt{1.025^2 - .3191786334^2}$$
$$= \sqrt{1.050625 - .101875}$$
$$= \sqrt{.94875}$$
$$= .9740379869$$

This compares to the actual geometric mean for our example data set of:

$$\text{Geometric Mean} = \sqrt[4]{1.00 * 1.50 * 1.00 * .6}$$
$$= \sqrt[4]{.9}$$
$$= .9740037464$$

As you can see, the estimated geometric mean is very close to the actual geometric mean—so close, in fact, that we can use the two interchangeably throughout the text.

HOW BEST TO REINVEST

Thus far, we have discussed reinvestment of returns in trading whereby we reinvest 100% of our stake on all occasions. Although we know that in order to maximize a potentially profitable situation we must use reinvestment, a 100% reinvestment is rarely the wisest thing to do.

Take the case of a coin toss. Someone is willing to pay you $2 if you win the toss, but will charge you $1 if you lose. You can figure what you should make, on average, per toss by the mathematical expectation formula:

$$\text{Mathematical Expectation} = \sum_{i=1}^{N}(P_i * A_i)$$

where: P = Probability of winning or losing.
         A = Amount won or lost.
         N = Number of possible outcomes.

In the given example of the coin toss:

$$\text{Mathematical Expectation} = (2 * .5) + (1 * (-.5))$$
$$= 1 - .5$$
$$= .5$$

In other words, you would expect to make 50 cents per toss, on average. This is true of the first toss and all subsequent tosses, provided you do not step up the amount you are wagering. But in an independent trials process, that is exactly what you should do. As you win, you should commit more and more to each trade.

At this point it is important to realize the keystone rule to money-management systems, which states: *In an independent trials process, if the mathematical expectation is less than or equal to 0, no money-management technique, betting scheme, or progression can turn it into a positive expectation game.*

This rule is applicable to trading one market system only. When you begin trading more than one market system, you step into a strange environment where it is possible to include a market system with a negative mathematical expectation as one of the market being traded, and actually have a net mathematical expectation higher than the net mathematical expectation of the group before the inclusion of the negative expectation system! Further, it is possible that the net mathematical expectation for the group with the inclusion of the negative mathematical expectation market system can be higher than the mathematical expectation of any of the individual market systems!

For the time being, we will consider only one market system at a time, and therefore we must have a positive mathematical expectation in order for the money-management techniques to work.

Refer again to the two-to-one coin-toss example (which is a positive mathematical expectation game). Suppose you begin with an initial stake of $1. Now suppose you win the first toss and are paid $2. Since you had your entire stake ($1) riding on the last bet, you bet your entire stake ($3 now) on the next toss as well. However, this next toss is a loser and your entire $3 stake is gone. You have lost your original $1 plus the $2 you had won. If you had won the last toss, it would have paid you $6, since you had three full $1 bets on it. The point is that if you are betting 100% of your stake, then as soon as you encounter a losing wager (an inevitable event), you'll be wiped out.

If you were to replay the previous scenario and bet on a non-reinvestment basis (i.e., a constant bet size) you would make $2 on the first bet and lose $1 on the second. You would now be ahead $1 and have a total stake of $2. Somewhere between these two scenarios lies the optimal betting approach.

Now, consider four desirable properties of a money-management strategy. First, you want to make as much as mathematically possible, given a favorable game. Second, the trade-off between the potential rate of growth of your stake and its security should be considered as well (this may not be possible given the first property, but it should at least be considered.[2] Third, the likelihood of winning should be taken into consideration. Fourth and finally, the amounts you can win and the amounts you can lose should influence the bet size as well. If you know you have an edge over N bets, but you do not know which of those N bets will be winners, or for how much, and which will be losers, and for how much, you are best off (in the long run) treating each bet exactly the same in terms of what percentage of your total stake is at risk.

Let's go back to the coin toss. Suppose we have an initial stake of $2. We toss a coin three times; twice it comes up heads (whereby we win $1 per $1 bet) and once it comes up tails (whereby we lose $1 per every $1 bet). Also, assume this coin is flawed in that it always comes up heads two out of three times and comes up tails one out of three times. Let's further say that this flawed coin can never come up HHH or TTT on any three-toss sequence. Since we know that this coin is flawed in these ways, but do not know where that loss will come in, how can we maximize this situation? The three possible exact sequences (the sample space), because of the flaws, are:

$$\begin{array}{ccc} H & H & T \\ H & T & H \\ T & H & H \end{array}$$

---

[2] In the final sections of the text, where we look at real world implementation, this vital caveat is addressed.

Here is our dilemma: We know we will win 66% of the time, but we do not know when we will lose, and we want to maximize what we make out of this situation.

Suppose now that rather than bet an equal fraction of our stake—which optimally is one-third of our stake on each bet (more on how to calculate this later)—we arbitrarily bet $2 on the first bet and $1 on each bet thereafter. Our $2 stake would grow to $4 at the end of both the HHT and the HTH sequences. However, for the THH sequence we would have been tapped out on the first bet. Since there are three exact sequences, and two of them resulted in profits of $2 and one resulted in a complete loss, we can say that the sum of all sequences was $4 gained (2 + 2 + 0). The average sequence was therefore a gain of $1.33 (4/3).

You can try any other combination like this for yourself. Ultimately, you will find that, since you do not know where the loss is due to crop up, you are best to bet the same fraction of your stake on each bet. Optimally, this fraction is one-third, or 33%, whereby you would make a profit of about $1.41 on each sequence, regardless of sequence(!), for a sum of all sequences of $4.23 gained (1.41 + 1.41 + 1.41). The average sequence was therefore a gain of $1.41 (4.23/3).

Many "staking"systems have been created by gamblers throughout history. One, the martingale, has you double your bet after each loss until ultimately, when a win does occur, you are ahead by one unit. However, the martingale strategy can have you making enormous bets during a losing streak. On the surface, this would appear to be the ultimate betting progression, as you will always come out ahead by one unit if you can follow the progression to termination. Of course, if you have a positive mathematical expectation, there is no need to use a scheme such as this. Yet it seems this should work for an even-money game as well as for a game where you have a small negative expectancy.

Yet, as we saw in Chapter 1, the sum of a series of negative expectancy bets must be a negative expectation. Suppose you are betting à la martingale. You lose the first 10 bets in succession. Going into the eleventh bet, you are now betting 1,024 units. The probabilities of winning are again the same as if you were betting one unit (assuming an independent trials process). Your mathematical expectation therefore, as a percentage, is the same as in the first bet, but in terms of units it is 1,024 times greater than the first bet. If you are betting with any kind of a negative expectation, it is now multiplied 1,024 times over.

"It doesn't matter," you, the martingale bettor, reply, "since I'll just double up for the twelfth bet if I lose the eleventh, and eventually I will come out ahead one unit." What eventually stymies the martingale bettor is a ceiling on the amount that may be bet, either by a house limit or inadequate capital to continue the progression on the bettor's part.

Theoretically, if you are gambling in a situation with no house limit, it would seem you could work this progression successfully if you had unlimited capital. Yet who has unlimited capital?

Ultimately, the martingale bettor has a maximum bet size, imposed by either the house (as in casino gambling) or his capitalization (as in the markets). Eventually, the bettor will bet and lose this maximum bet size and thus go bust. Furthermore, this will happen regardless of mathematical expectation—that is why the martingale is completely foolish if you have a positive mathematical expectation, and just futile if you have an even-money game or a negative expectation. True, betting à la martingale you will most often walk away from the tables a winner. However, when you lose, it will be for an amount that will more than compensate the casino for letting you walk away a winner the vast majority of the time.

It is not the maximum bet size that stymies the martingale as much as it is the number of bets required to reach the maximum bet size (this is also one of the reasons why there are house minimums). To overcome this, gamblers have tried what is known as the small martingale—a somewhat watered-down version of the martingale.

The small martingale tries to provide survival for the bettor by increasing the number of bets required to reach the maximum bet size. Ultimately, the small martingale tries to win one unit per cycle. Since the system rules are easier to demonstrate than to describe, I will show this system through the use of examples. In the small martingale you keep track of a "progression list," and bet the amount that is the sum of the first and last values on the list. When a win is encountered, you cross off the first and last values on the list, thus obtaining new first and last values, giving you a new amount to wager on the next bet. The list starts at simply the number 1. When a loss is encountered, the next number is added on to the end of the list (i.e., 2, 3, 4, etc.). A cycle ends when one unit is won. If a list is ever composed of just the number 2, then convert it to a list of 1, 1. The following examples of four different cycles should make the progression clear:

Bet Number	List	Bet Size	Win/Loss
1	1	1	W

Bet Number	List	Bet Size	Win/Loss
1	1	1	L
2	1, 1	2	W

Bet Number	List	Bet Size	Win/Loss
1	1	1	L
2	1, 1	2	L
3	1, 1, 2	3	W
4	1	1	W

Bet Number	List	Bet Size	Win/Loss
1	1	1	L
2	1, 1	2	L
3	1, 1, 2	3	L
4	1, 1, 2, 3	4	W
5	1, 2	3	L
6	1, 2, 3	4	W
7	1, 1	2	L
8	1, 1, 2	... and continuing until the bettor is ahead by 1 unit.	

The small martingale is ultimately a loser, too, for the same reasons that the martingale system is a loser. A sum of negative expectancy bets must have a negative expectancy.

Another system, the antimartingale, is just the opposite of the martingale (as its name implies). Here, you increase your bet size after each win. The idea is to hit a streak of winning bets and maximize the gains from that streak. Of course, just as the martingale ultimately makes one unit, the antimartingale will ultimately lose all of its units (even in a positive mathematical expectation game) once a loss is incurred, if 100% of the stake is placed on each bet.

Notice, however, that fixed fractional trading is actually a small antimartingale! Recall our flawed-coin example earlier in this chapter. In that example we saw how our "best" strategy was the small antimartingale. In the final analysis, fixed fractional trading, the small antimartingale, is the optimal betting system—provided you have a positive mathematical expectation.[3]

Another famous system is the reserve strategy. Here, you trade a base bet plus a fraction of your winnings. However, in the reserve strategy, if the last bet was a winner, then you bet the same amount on the next bet as you did the last. Suppose you encounter the sequence of win $1, win $1, lose $1, then win $1 for every $1 bet. If you are betting $1 plus 50% of winnings (in the reserve strategy), you would bet $1 on the first bet. Since it was a winner, you would still bet $1 on the second bet—which was also a winner, boosting your total winnings to $2. Since the second bet was also a winner, you would not increase your third bet; rather, you would still bet $1. The

---

[3]This is critical. Optimal fixed fractional trading therefore possesses those characteristics, pro and con, of the small antimartingale. It *will* maximize growth, *but* it will cause you to endure severe and protracted drawdowns. Just as the martingale strategy has you leave the tables a winner most of the time, the antimartingale, and to a lesser extent, the small antimartingale (i.e., "fixed fractional trading") has you leave the tables—or a performance period—a loser more frequently than you would have betting on a constant-bet-size or martingale basis.

third bet, being a loss of $1, lowers your total winnings to $1. Since you encountered a loss, however, you recapitalize your bet size to the base bet ($1) plus 50% of your winnings (.5 * $1) and hence bet $1.50 on the fourth bet. The fourth bet was a winner, paying 1 to 1, so you made $1.50 on the fourth bet, bringing your total winnings to $2.50. Since this last bet was a winner, you will not recapitalize (step up or down) your bet size into the fifth bet; instead, you stay with a bet size of $1.50 into the fifth bet.

On the surface, the reserve strategy seems like an ideal staking system. However, like all staking systems, its long-term performance falls short of the simple fixed fraction (small antimartingale) approach. Another popular idea of gamblers/traders has been the base bet plus square root strategy, whereby you essentially are always betting the same amount you started with plus the square root of any winnings. As you can see, the possibilities of staking systems are endless.

Many people seem to be partial, for whatever reason, to adding contracts after a losing trade, a streak of losing trades, or a drawdown. Over and over again in computer simulations (by myself and others) this turns out to be a very poor method of money management. It is akin to the martingale and small martingale. Since we have determined that trading is largely an independent trials process, the past trades have no effect on the present trade. It doesn't matter whether the last 20 trades were all winners or all losers.

It is interesting to note that those computer tests that have been performed all bear out the same thing. In an independent trials process where you have an edge, you are better off to increase your bet size as your stake increases, and the bet size optimally is a fixed fraction of your entire stake. Time and again authors have performed studies that take a very long stream of independent outcomes with a net positive result, and have applied various staking systems to betting/trading on these outcomes. In a nutshell, the outcomes of every study of this type reach the same conclusion: that you are better off using a staking system that increases the size of the bet in direct proportion to the size of the total stake.

In another study, William T. Ziemba demonstrated in the June 1987 issue of *Gambling Times* magazine that proportional betting was superior to any other staking strategy.[4] Furthermore, Ziemba's article demonstrated how the optimal proportion (determined correctly by the Kelly formula in this study) far outperforms any other proportion. The study simulated 1,000 seasons of betting on 700 horse races, starting you out with an initial stake of $1,000. The test looked at such outcomes as how many seasons would

---

[4]Ziemba, William T., "A Betting Simulation, The Mathematics of Gambling and Investment," *Gambling Times*, pp. 46–47, 80, June 1987.

have tapped you out, how many seasons were profitable, how many made more than $5,000, $10,000, $100,000, and so on, as well as what the minimum, maximum, mean, and median terminal amounts were. The results of this test, too, were quite clear—betting a fixed fraction of your bankroll is far and away the best staking system.

"Wait," you say. "Aren't staking systems foolish to begin with? Didn't we see in Chapter 1 that they do not overcome the house advantage; rather, all they do is increase our total action?"

This is absolutely true for a situation with a negative mathematical expectation. For a positive mathematical expectation it is a different story altogether. In a positive expectancy situation the trader/gambler is posed with the question of how best to exploit the positive expectation.

CHAPTER 4

Optimal *f*

OPTIMAL FIXED FRACTION

We have seen that in order to consider betting/trading a given situation/ system you must first determine if a positive mathematical expectation exists. We have seen in the previous chapter that what is seemingly a "good bet" on a mathematical expectation basis (i.e., the mathematical expectation is positive) may in fact not be such a good bet when you consider reinvestment of returns.[1] Reinvesting returns never raises the mathematical expectation (as a percentage—although it can raise the mathematical expectation in terms of dollars, which it does geometrically, which is why we want to reinvest). If there is in fact a positive mathematical expectation, however small, the next step is to exploit this positive expectation to its fullest potential. This has been shown, for an independent trials process, to be by reinvesting a fixed fraction of your total stake,[2] which leads to the following axiom: *For any given independent trials situation where you*

---

[1] If you are reinvesting too high a percentage of your winnings relative to the dispersion of outcomes of the system.

[2] For a dependent trials process the idea of betting a proportion of your total stake also yields the greatest exploitation of a positive mathematical expectation, just like an independent trials process. However, in a dependent trials process you optimally bet a variable fraction of your total stake, the exact fraction for each individual bet determined by the probabilities and payoffs involved for each individual bet.

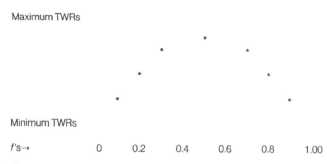

FIGURE 4.1   The curve of optimal *f*

have an edge (i.e., a positive mathematical expectation), there exists an optimal fixed fraction (f) *between 0 and 1 as a divisor of your biggest loss to bet on each and every event.*

Most people think that the optimal fixed fraction is the percentage of your total stake to bet. Optimal *f* is not in itself the percentage of our total stake to bet; it is the divisor of our biggest loss, the result of which we divide our total stake by to know how many bets to make or contracts to have on.

You will also notice that *margin has nothing to do whatsoever with what is the mathematically optimal number of contracts to have on.*

As you can see in Figure 4.1, *f* is a curve cupped downward from 0 to 1. The highest point for *f* is that fraction of your stake to bet on each and every event (bet) to maximize your winnings.

Most people incorrectly believe that *f* is a straight-line function rising up and to the right. They believe this because they think it would mean that the more you are willing to risk, the more you stand to make. People reason this way because they think that a positive mathematical expectation is just the mirror image of a negative expectancy. They mistakenly believe that if increasing your total action in a negative expectancy game results in losing faster, then increasing your total action in a positive expectancy game will result in winning faster. This is not true. At some point in a positive expectancy situation, to increase your total action further works against you. That point is a function of both the system's profitability and its consistency (i.e., its geometric mean), since you are reinvesting the returns.

ASYMMETRICAL LEVERAGE

Recall that the amount required to recoup a loss increases geometrically with the loss. We can show that the percentage gain to recoup a loss is:

$$\text{Required Gain} = (1/(1 - \text{loss in percent})) - 1 \qquad (4.01)$$

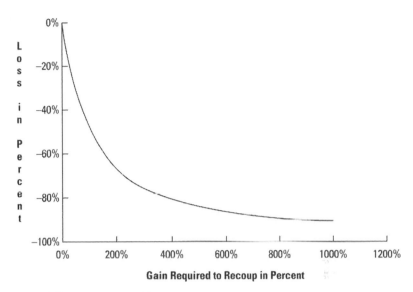

FIGURE 4.2 Asymmetrical leverage

A 20% loss requires a 25% gain afterwards to recoup. A 30% loss requires a 42% gain afterwards to recoup. This is asymmetrical leverage. In fixed fractional trading we have seen that the trader will tend to have on more contracts when she takes a loss than when she has a win. This is what amplifies the asymmetrical leverage. It is also what curves the f function, since the peak of the f function represents that point where the trader has the right amount of contracts on to go into the losses and come out of the losses (with asymmetrical leverage) and achieve the maximum growth on her money at the end of a sequence of trades (see Figure 4.2). The f value (X axis) that corresponds to the peak of this f curve will be known as the optimal f (f is always in lowercase).

So f is a curved-line function, and this is due, in part, to the fact that asymmetrical leverage is amplified when reinvesting profits.

And how do we find this optimal f? Much work has been done in recent decades on this topic in the gambling community, the most famous and accurate of which is known as the Kelly Betting System. This is actually an application of a mathematical idea developed in early 1956 by John L. Kelly, Jr., and published in the July 1956 *Bell System Technical Journal.*[3]

[3]Kelly, J. L., Jr., "A New Interpretation of Information Rate," *Bell System Technical Journal*, pp. 917–926, July 1956.

The Kelly criterion states that we should bet that fixed fraction of our stake (f) which maximizes the growth function $G(f)$:

$$G(f) = P * \ln(1 + B * f) + (1 - P) * \ln(1 - f) \qquad (4.02)$$

where: f = The optimal fixed fraction.
  P = The probability of a winning bet/trade.
  B = The ratio of amount won on a winning bet to amount lost on a losing bet.
  $\ln() $ = The natural logarithm function to the base $e = 2.71828....$

As it turns out, for an event with two possible outcomes, this optimal f can be found quite easily with the Kelly formulas.

KELLY

Beginning around the late 1940s, Bell System engineers were working on the problem of data transmission over long distance lines. The problem facing them was that the lines were subject to seemingly random, unavoidable "noise" that would interfere with the transmission. Some rather ingenious solutions were proposed by engineers at Bell Labs. Oddly enough, there are great similarities between this data communications problem and the problem of geometric growth as it pertains to gambling money management (as both problems are the product of an environment of favorable uncertainty). The Kelly formula is one of the outgrowths of these solutions.

The first equation here is:

$$f = 2 * P - 1 \qquad (4.03)$$

where: f = The optimal fixed fraction.
  P = The probability of a winning bet/trade.

This formula will yield the correct answer for optimal f provided the sizes of wins and losses are the same. As an example, consider the following stream of bets:

$$-1, +1, +1, -1, -1, +1, +1, +1, +1, -1$$

There are 10 bets, 6 winners, hence:

$$f = (.6 * 2) - 1$$
$$= 1.2 - 1$$
$$= .2$$

If the winners and losers were not all the same size, this formula would not yield the correct answer. Such a case would be our two-to-one coin-toss example, where all of the winners were for 2 units and all of the losers for 1 unit. For this situation the Kelly formula is:

$$f = ((B + 1) * P - 1)/B \qquad (4.04)$$

where: f = The optimal fixed fraction.
   P = The probability of a winning bet/trade.
   B = The ratio of amount won on a winning bet to amount lost on a losing bet.

For the two-to-one coin toss:

$$
\begin{aligned}
f &= ((2 + 1) * .5 - 1)/2 \\
&= (3 * .5 - 1)/2 \\
&= (1.5 - 1)/2 \\
&- .5/2 \\
&= .25
\end{aligned}
$$

This formula will yield the correct answer for optimal f, provided all wins are always for the same amount and all losses are always for the same amount. If this condition is not met, the formula will not yield the correct answer.

Consider the following sequence of bets/trades:

$$+9, +18, +7, +1, +10, -5, -3, -17, -7$$

Since all wins and all losses are of different amounts, the previous formula does not apply. However, let's try it anyway and see what we get.

Since five of the nine events are profitable, $P = .555$. Now let's take averages of the wins and losses to calculate B (here is where so many traders go wrong). The average win is 9 and the average loss is 8. Therefore, we will say that $B = 1.125$. Plugging in the values we obtain:

$$
\begin{aligned}
f &= ((1.125 + 1) * .555 - 1)/1.125 \\
&= (2.125 * .555 - 1)/1.125 \\
&= (1.179375 - 1)/1.125 \\
&= .179375/1.125 \\
&= .159444444
\end{aligned}
$$

So we say $f = .16$. We will see later in this chapter that this is not the optimal f. The optimal f for this sequence of trades is .24. Applying Kelly when wins are not all for the same amount and/or losses are not all for the same amount is a mistake. It will not yield the optimal f.

Notice that the numerator in this formula equals the mathematical expectation for an event with two possible outcomes as defined in Chapter 1. Therefore, we can say that as long as all wins are for the same amount and all losses are for the same amount (regardless of whether the amount that can be won equals the amount that can be lost), the optimal f is:

$$f = \text{Mathematical Expectation/B} \qquad (4.05)$$

where: f = The optimal fixed fraction.
         B = The ratio of amount won on a winning bet to amount lost on a losing bet.

FINDING THE OPTIMAL f BY THE GEOMETRIC MEAN

In trading we can count on our wins being for various amounts and our losses being for various amounts. Therefore, the Kelly formula cannot give us the correct optimal f. How then can we find the optimal f to tell us how many contracts to have on and have it be mathematically correct?

As you will see later in this chapter, trading the correct quantities of contracts/shares is a far bigger problem than was previously thought. Quantity can mean the difference between winning and losing. All systems experience losing trades. All systems experience drawdown. These are givens, facts of life. Yet if you can always have the right amount of contracts on (i.e., the mathematically correct amount), then there is consolation in the losses.

Now here is the solution. To begin with, we must amend our formula for finding HPRs to incorporate f.

$$\text{HPR} = 1 + f * (-\text{trade/biggest loss}) \qquad (4.06)$$

And again, TWR is simply the geometric product of the HPRs and working from (3.03), geometric mean is simply the Nth root of the TWR.

$$\text{TWR} = \prod_{i=1}^{N}(1 + f * (-\text{trade}_i/\text{biggest loss})) \qquad (4.07)$$

$$\text{Geo. Mean} = \left(\prod_{i=1}^{N}(1 + f * (-\text{trade}_i/\text{biggest loss}))\right)^{1/N} \qquad (4.08)$$

The geometric mean can also be calculated here by the procedure for finding the estimated geometric mean by using the HPRs as formulated above, or by taking the TWR, as formulated above, as an input to the equation:

$$\text{Geo. Mean} = \exp((1/N) * \log(\text{TWR}))$$

where: N = Total number of trades.
log(TWR) = The log base 10 of the TWR.
exp = The exponential function.

By looping through all values for f *between .01 and 1, we can find that value for* f *which results in the highest TWR.* This is the value for f that would provide us with the maximum return on our money using fixed fraction. We can also state that the optimal f is the f that yields the highest geometric mean. It matters not whether we look for highest TWR or geometric mean, as both are maximized at the same value for f.

Doing this with a computer is easy. Simply loop from f = .01 to f = 1.0 by .01. As soon as you get a TWR that is less than the previous TWR, you know that the f corresponding to the previous TWR is the optimal f. You can also calculate this by hand, but that is a lot more tedious, especially as the number of trades increases. A quicker way to do it is to use iteration to solve for the optimal f (you can use the iterative approach whether you are doing it by hand or by computer). Here, you are initially bounded on f at f = 0 and f = 1.00. Pick a start value, say f = .10, and find the corresponding TWR. Now step the f value up an arbitrary amount. The example that follows steps it up by .10, but you can use any amount you want to (so long as you do not have an f value greater than 1.00, the upper bound). As long as your TWRs keep increasing, you must step up the f value you are testing. Do this until f = .30, where your TWR is less than at f = .20. Now, your f bounds are .20 and .30. Keep on repeating the process until you zero in on the optimal f. The following illustration demonstrates the iterative process as well as the calculations:

At f = .10

TRADE	HPR	
9	1.052941	
18	1.105882	The HPRs are equal to 1 + (f * (−trade/biggest loss))
7	1.041176	
1	1.005882	
10	1.058823	
−5	0.970588	
−3	0.982352	
−17	0.9	
−7	0.958823	
	TWR = 1.062409	The TWR is all of the HPRs multiplied together

At $f = .20$

TRADE	HPR
9	1.105882
18	1.211764
7	1.082352
1	1.011764
10	1.117647
−5	0.941176
−3	0.964705
−17	0.8
−7	0.917647

TWR = 1.093231

At $f = .30$

TRADE	HPR
9	1.158823
18	1.317647
7	1.123529
1	1.017647
10	1.176470
−5	0.911764
−3	0.947058
−17	0.7
−7	0.876470

TWR = 1.088113

At $f = .25$

TRADE	HPR
9	1.132352
18	1.264705
7	1.102941
1	1.014705
10	1.147058
−5	0.926470
−3	0.955882
−17	0.75
−7	0.897058

TWR = 1.095387

At $f = .23$

TRADE	HPR
9	1.121764
18	1.243529
7	1.094705
1	1.013529
10	1.135294
−5	0.932352
−3	0.959411
−17	0.77
−7	0.905294

TWR = 1.095634

At $f = .24$

TRADE	HPR
9	1.127058
18	1.254117
7	1.098823
1	1.014117
10	1.141176
−5	0.929411
−3	0.957647
−17	0.76
7	0.901176

TWR = 1.095698

TO SUMMARIZE THUS FAR

In the previous chapter we demonstrated that a good system is the one with the highest geometric mean. Yet, to find the geometric mean you must know f. Understandably, the reader must be confused. Here now is a summary and clarification of the process:

1. Take the trade listing of a given market system.
2. Find the optimal f, either by testing various f values from 0 to 1 or through iteration. The optimal f is that which yields the highest TWR.
3. Once you have found f you can take the Nth root of that TWR corresponding to your f, where N is the total number of trades. This is your geometric mean for this market system. You can now use this geometric

mean to make apples-to-apples comparisons with other market systems, as well as use the f to know how many contracts to trade for that particular market system.

Once the highest f *is found, it can readily be turned into a dollar amount by dividing the biggest loss by the negative optimal* f. For example, if our biggest loss is \$100 and our optimal f is .25, then $-\$100/-.25 = \$400$. In other words, we should bet one unit for every \$400 we have in our stake.

In the sequence of bets that our coin-toss example would generate, we find that the optimal f value for the sequence $+2, -1$ is .25. Since our biggest loss is \$1, $1/.25 = \$4$. In other words, we should bet \$1 for every \$4 we have in our stake in order to make the most money out of this game. To bet a higher number or a lower number will not result in a greater gain! After 10 bets, for every \$4 we started out with in our stake, we will have \$9.

This approach to finding the optimal f will yield the same result as:

$$f = ((B + 1) * P - 1)/B$$

You obtain the same result, of course, when losses are all for the same amount and wins are all for the same amount. In such a case, either technique is correct. When both wins and losses are for the same amount, you can use any of the three methods—the Kelly formula just shown, the f that corresponds to the highest TWR, or:

$$f = 2 * P - 1$$

Any of the three methods will give you the same answer when all wins and losses are for the same amount.

All three methods for finding the optimal f meet the four desirable properties of a money-management strategy outlined earlier, given the constraints of the two formulas (i.e., all wins being for the same amount and all losses being for the same amount, or all wins and losses being for the same amount). Regardless of constraints, the optimal f via the highest TWR will always meet the four desirable properties of a money-management strategy.

If you're having trouble with some of these concepts, try thinking in terms of betting in units, not dollars (e.g., one \$5 chip or one futures contract or one 100-share unit of stock). The amount of dollars you allocate to each unit is calculated by figuring your largest loss divided by the negative optimal f.

The optimal f is a result of the balance between a system's profit-making ability (on a constant one-unit basis) and its risk (on a constant one-unit basis). Notice that margin doesn't matter, because the size of individual profits and losses are not the product of the amount of money put up as margin (they would be the same whatever the size of the margin). Rather, the profits and losses are the product of the exposure of one unit (one futures contract). The amount put up as margin is further made meaningless in a money-management sense, since the size of the loss is not limited to the margin.

HOW TO FIGURE THE GEOMETRIC MEAN USING SPREADSHEET LOGIC

Here is an example of how to use a spreadsheet like to calculate the geometric mean and TWR when you know the optimal f or want to test a value for f.

(Assume $f = .5$, biggest loss $= -50$)

	col A	col B	col C	col D	col E
row 1					1
row 2	15	0.3	0.15	1.15	1.15
row 3	−5	−0.1	−0.05	0.95	1.0925

cell(s) explanation

A1 through D1 are blank.
E1   Set equal to 1 to begin with.
A2 down These are the individual trade P&Ls.
B2 down = A2/abs value of (biggest loss)
C2 down = B2/f
D2 down = C2 + 1
E2 down − E1 * D2

When you get to the end of the trades (the last row), your last value in column E is your TWR. Now take the Nth root of this TWR (N is the total number of trades); that is your geometric mean. In the above example, your TWR (cell E3) raised to the power 1/2 (there are a total of two trades here) = 1.045227. That is your geometric mean.

GEOMETRIC AVERAGE TRADE

At this point you may be interested in figuring your geometric average trade. That is, what is the average garnered per contract per trade, assuming profits are always reinvested and fractional contracts can be purchased. In effect, this is the mathematical expectation when you are trading on a fixed fractional basis. *This figure shows you what effect there is from losers occurring when you have many contracts on and winners occurring when you have fewer contracts on. In effect, this approximates how a system would have fared per contract per trade doing fixed fraction.* (Actually, the geometric average trade is your mathematical expectation in dollars per contract per trade. The geometric mean minus 1 is your percent mathematical expectation per trade—e.g., a geometric mean of, say, 1.025 represents a mathematical expectation of 2.5% per trade, irrespective of size.) So many traders look simply at the average trade of a market system to see if it is high enough to justify trading the system. However, in making their decision, they should be looking at the geometric average trade (which is never greater than the average trade) as well as at the PRR.

$$\text{Geo. Avg. Trade} = G * (\text{biggest loss}/- f) \qquad (4.09)$$

where: G = Geometric mean -1.
f = Optimal fixed fraction.
(And, of course, our biggest loss is always a negative number.)

For example, suppose a system has a geometric mean of 1.017238, the biggest loss is $8,000, and the optimal f is .31. Our geometric average trade would equal:

$$
\begin{aligned}
\text{Geo. Avg. Trade} &= (1.017238 - 1) * (-8,000/-.31) \\
&= .017238 * 25,806.45 \\
&= \$444.85
\end{aligned}
$$

A SIMPLER METHOD FOR FINDING THE OPTIMAL *f*

There are numerous ways to arrive at the optimal value for f. The technique for finding the optimal f that has been presented thus far in this chapter is perhaps the most mathematically logical. That is to say, it is obvious upon inspection that this technique will yield the optimal f. It makes more intuitive sense when you can see the HPRs laid out than does the next and somewhat easier method. So here is another way for calculating the optimal f, one that some readers may find simpler and more to their liking. It will give the same answer for optimal f as the technique previously described.

Under this method we still need to loop through different test values for f to see which value for f results in the highest TWR. However, we calculate our TWR without having to calculate the HPRs. Let's assume the following stream of profits and losses from a given system:

$$+\$100$$
$$-\$500$$
$$+\$1500$$
$$-\$600$$

Again, we must isolate the largest losing trade. This is $-\$600$.

Now we want to obtain the TWR for a given test value for f. Our first step is to calculate what we'll call the *starting value*. To begin with, take the largest loss and divide it by the test value for f. Let's start out by testing a value of .01 for f. So we will divide the largest loss, $-\$600$, by .01. This yields an answer of $-\$60,000$. Now we make it a positive value. Therefore, our starting value for this example sequence of a .01 test value for f is $\$60,000$.

For each trade we must now calculate a *working value*. To do this, for each trade we must take the previous working value and divide it by the starting value. (For the first trade, the answer will be 1, since the previous working value is the same as the starting value.) Next, we multiply the answer by the current trade amount. Finally, we add this answer and the previous working value to obtain the current working value.

P&L	WORKING VALUE	
	60000 ⟵	This is the starting value
+100	60100	
−500	59599.166667	
+1500	61089.14583	
−600	60478.25437	

Our TWR is obtained simply by taking the last entry in the working value column and dividing it by our starting value. In this instance:

$$\text{TWR} = 60478.25437/60000$$
$$= 1.007970906$$

Now we repeat the process, only we must increment our test value for f. This time through, rather than dividing the absolute value of the largest loss of $-\$600$ by .01 we will divide it by .02. Therefore, we will begin this next pass through with a starting value of $600/.02 = 30000$.

P&L	WORKING VALUE	
	30000	
+100	30100	((30000/30000) * 100) + 30000
−500	29598.33	((30100/30000) * −500) + 30100
+1500	31078.2465	((29598.33/30000) * 1500) + 29598.33
−600	30456.68157	((31078.2465/30000) * −600) + 31078.2465

Here the TWR = 30456.68157/30000 = 1.015222719

We keep on repeating the procedure until we obtain the value for f that results in the highest TWR. The answers we obtain for TWRs, as well as for the optimal f, will be the same with this technique as with the previous technique using the HPRs.

THE VIRTUES OF THE OPTIMAL f

It is a mathematical fact that when two people face the same sequence of favorable betting/trading opportunities, if one uses the optimal f and the other uses any different money-management system, then the ratio of the optimal f bettor's stake to the other person's stake will increase as time goes on, with higher and higher probability. In the long run, the optimal f bettor will have infinitely greater wealth than any other money-management-system bettor with a probability approaching one.

Furthermore, if a bettor has the goal of reaching a prespecified fortune, and is facing a series of favorable betting/trading opportunities, the expected time needed to reach the fortune will be less with optimal f than with any other betting system.

Obviously, the optimal f strategy satisfies desirable property number 1 for money management, as it makes the most amount of money that is mathematically possible using a fixed fractional betting strategy on a game where you have the edge. Since optimal f incorporates the probability of winning as well as the amounts won and lost, it also satisfies desirable property numbers 3 and 4. Not much has been discussed about desirable property number 2, the security aspect, but this will be treated in the closing chapters.

Let's go back and reconsider the following sequence of bets/trades:

$$+9, +18, +7, +1, +10, -5, -3, -17, -7$$

Recall that we determined earlier in this chapter that the Kelly formula did not apply to this sequence, since wins were not all for the same amount and losses were not all for the same amount. We also decided to average the wins and average the losses and take these averages as our values into the Kelly formula (as many traders mistakenly do). Doing this we arrived at an f value of .16. It was stated that this is an incorrect application of Kelly, that it would not yield the optimal f. The Kelly formula must be specific to a single bet. We cannot average our wins and losses from trading and obtain the true optimal f using the Kelly formula.

Our highest TWR on this sequence of bets/trades is obtained at .24, or betting $1 for every $71 in our stake. That is the optimal geometric growth we can squeeze out of this sequence of bets/trades trading fixed fraction. Let's look at the TWRs at different points along 100 loops through this sequence of bets.

At one loop through, nine bets/trades, the TWR for $f = .16$ is 1.085; for $f = .24$ it is 1.096. This means that for one pass through this sequence of bets an $f = .16$ made 99% of what an $f = .24$ would have made. To continue:

Passes Through	Total Bets/Trades	TWR for $f = .24$	TWR for $f = .16$	Percentage Difference
1	9	1.096	1.085	1%
10	90	2.494	2.261	9.4%
40	360	38.694	26.132	32.5%
100	900	9313.312	3490.761	62.5%

As can be seen, using an f value that we mistakenly figured from Kelly made only 37.5% as much as our optimal f of .24 after 900 bets/trades (100 cycles through the series of nine outcomes). In other words, our optimal f of .24 (which is only .08 more than .16) made almost 267% the profit that $f = .16$ did after 900 bets!

Let's go another 11 cycles through this sequence of trades, so we have a total of 999 trades. Now our TWR for $f = .16$ is 8563.302 (not even what it was for $f = .24$ at 900 trades) and our TWR for $f = .24$ is 25,451.045. At 999 trades $f = .16$ is only 33.6% of $f = .24$, or $f = .24$ is 297% of $f = .16$! Here you can see that using the Kelly formula does not yield the true optimal f for trading.

As can be seen from the above, using the optimal f *does not appear to offer much advantage over the short run, but over the long run it becomes more and more important to use the optimal* f. *The point is you must give the program time when trading at the optimal* f *and not expect miracles in the short run. The more time (i.e., bets/trades) that elapses, the greater the difference between using the optimal* f *and any other money-management strategy.*

WHY YOU MUST KNOW YOUR OPTIMAL *f*

Figures 4.3 through 4.6 demonstrate the importance of using optimal f in fixed fractional trading. The graphs are constructed by plotting the f values from 0 to 1.0 along the X axis and the respective TWRs along the Y axis. Values are plotted at intervals of .05.

Each graph has a corresponding spreadsheet. Each column heading in the spreadsheet has a different f value. Under each f value is the corresponding start value, figured as the biggest loss divided by the negative f value. For every unit of start value you have in your stake, you bet one unit. Along the far left is the sequence of 40 bets. This sequence is the only difference between the various spreadsheets and graphs.

As you go down through the sequence of trades you will notice that each cell equals the previous cell divided by that cell's starting value. This result is then multiplied by the outcome of the current bet, and the product added to the original value of the previous cell, to obtain the value of the current cell. When you reach the end of the column you can figure your TWR as the last value of the column divided by the start value of the column (i.e., the biggest loss divided by negative f). This is the alternative and somewhat easier way to figure your TWRs. Both methods shown thus far make the calculations non-quantum. In other words, you do not need an integer amount to multiply the current bet result by; you can use a decimal amount of the starting value as well. An example may help clarify.

In the $+1.2$, -1 sequence (Figure 4.3), for an f value of .05, we have a starting value of 20:

$$-1/-.05 = 20$$

In other words, we will bet one unit for every 20 units in our stake. With the first bet, a gain of 1.2, we now have 21.2 units in our stake. (Since we had 20 units in our stake prior to this bet and we bet one unit for every 20 in our stake, we bet only one unit on this bet.) Now the next bet is a loss of one unit. The question now is, "How many units were we betting on this one?"

We could argue that we were betting only one unit, since 21.20 (our stake prior to the bet) divided by 20 (the starting value) = 1.06. Since most bets must be in integer form—that is, no fractional bets (chips are not divisible and neither are futures contracts)—we could bet only one unit in real life in this situation. However, in these simulations the fractional bet is allowed. The reasoning here behind allowing the fractional bet is to keep the outcome consistent regardless of the starting stake. Notice that each simulation starts with only enough stake to make one full bet. What if each simulation started with more than that? Say each simulation started with enough to make 1.99 bets. If we were only allowing integer bets, our outcomes (TWRs) would be altogether different.

Further, the larger the amount we begin trading with is, relative to the starting value (biggest loss/ − optimal f), the closer the integer bet will be to the fractional bet. Again, clarity is provided by way of an example. What if we began trading with 400 units in the previous example? After the first bet our stake would have been:

$$Stakc = 400 + ((400/20) * 1.2)$$
$$= 400 + (20 * 1.2)$$
$$= 400 + 24$$
$$= 424$$

For the next bet, we would wager 21.2 units (424/20), or the integer amount of 21 units. Note that the percentage difference between the fractional and the integer bet here is only .952381% versus a 6.0% difference, had the amount we began trading with been only one starting value, 20 units. The following axiom can now be drawn: *The greater the ratio of the amount you have as a stake to begin trading relative to the starting value (biggest loss/ − optimal* f *), the more the percentage difference will tend to zero between integer and fractional betting.*

By allowing fractional bets, making the process nonquantum, we obtain a more realistic assessment of the relationship of f values to TWRs. *The fractional bets represent the average (of all possible values of the size of initial bankrolls) of the integer bets.* So the argument that we cannot make fractional bets in real life does not apply, since the fractional bet represents the average integer bet. If we made graphs of the TWRs at each f value for the +2, −1 coin toss, and used integer bets, we would have to make a different graph for each different initial bankroll. If we did this and then averaged the graphs to create a composite graph of the TWRs at each f value, we would have a graph of the fractional bet situation exactly as shown.

20 TRIALS

f VALUES ⟶

EVENT	0.05	0.1	0.15	0.2	0.25	0.3	0.35	0.4	0.45
START VALUES ⟶	20.00	10.00	6.67	5.00	4.00	3.33	2.86	2.50	2.22
1.2	21.20	11.20	7.87	6.20	5.20	4.53	4.06	3.70	3.42
−1	20.14	10.08	6.69	4.96	3.90	3.17	2.64	2.22	1.88
1.2	21.35	11.29	7.89	6.15	5.07	4.32	3.74	3.29	2.90
−1	20.28	10.16	6.71	4.92	3.80	3.02	2.43	1.97	1.59
1.2	21.50	11.38	7.91	6.10	4.94	4.11	3.46	2.92	2.46
−1	20.42	10.24	6.73	4.88	3.71	2.88	2.25	1.75	1.35
1.2	21.65	11.47	7.94	6.05	4.82	3.91	3.19	2.59	2.08
−1	20.57	10.32	6.75	4.84	3.61	2.74	2.07	1.55	1.14
1.2	21.80	11.56	7.96	6.00	4.70	3.72	2.94	2.30	1.76
−1	20.71	10.41	6.77	4.80	3.52	2.61	1.91	1.38	0.97
1.2	21.95	11.66	7.99	5.96	4.58	3.54	2.72	2.04	1.49
−1	20.85	10.49	6.79	4.76	3.44	2.48	1.77	1.23	0.82
1.2	22.11	11.75	8.01	5.91	4.47	3.37	2.51	1.81	1.26
−1	21.00	10.57	6.81	4.73	3.35	2.36	1.63	1.09	0.69
1.2	22.26	11.84	8.03	5.86	4.36	3.21	2.32	1.61	1.07
−1	21.15	10.66	6.83	4.69	3.27	2.25	1.51	0.97	0.59
1.2	22.42	11.94	8.06	5.81	4.25	3.06	2.14	1.43	0.91
−1	21.30	10.74	6.85	4.65	3.18	2.14	1.39	0.86	0.50
1.2	22.57	12.03	8.08	5.77	4.14	2.91	1.97	1.27	0.77
−1	21.44	10.83	6.87	4.61	3.11	2.04	1.28	0.76	0.42
1.2	22.73	12.13	8.11	5.72	4.04	2.77	1.82	1.13	0.65
−1	21.60	10.92	6.89	4.58	3.03	1.94	1.18	0.68	0.36
1.2	22.89	12.23	8.13	5.68	3.94	2.64	1.68	1.00	0.55
−1	21.75	11.00	6.91	4.54	2.95	1.85	1.09	0.60	0.30
1.2	23.05	12.32	8.15	5.63	3.84	2.51	1.55	0.89	0.47
−1	21.90	11.09	6.93	4.50	2.88	1.76	1.01	0.53	0.26
1.2	23.21	12.42	8.18	5.59	3.74	2.39	1.43	0.79	0.40
−1	22.05	11.18	6.95	4.47	2.81	1.67	0.93	0.47	0.22
1.2	23.37	12.52	8.20	5.54	3.65	2.28	1.32	0.70	0.33
−1	22.21	11.27	6.97	4.43	2.74	1.59	0.86	0.42	0.18
1.2	23.54	12.62	8.23	5.50	3.56	2.17	1.22	0.62	0.28
−1	22.36	11.36	6.99	4.40	2.67	1.52	0.79	0.37	0.16
1.2	23.70	12.72	8.25	5.43	3.47	2.06	1.13	0.55	0.24
−1	22.52	11.45	7.01	4.36	2.60	1.44	0.73	0.33	0.13
1.2	23.87	12.82	8.28	5.41	3.38	1.96	1.04	0.49	0.20
−1	22.68	11.54	7.04	4.33	2.54	1.38	0.68	0.29	0.11
1.2	24.04	12.93	8.30	5.37	3.30	1.87	0.96	0.44	0.17
−1	22.83	11.63	7.06	4.29	2.47	1.31	0.62	0.26	0.09
1.2	24.20	13.03	8.33	5.32	3.21	1.78	0.89	0.39	0.15
−1	22.99	11.73	7.08	4.26	2.41	1.25	0.58	0.23	0.08
TWR ⟶	1.15	1.17	1.06	0.85	0.60	0.37	0.20	0.09	0.04

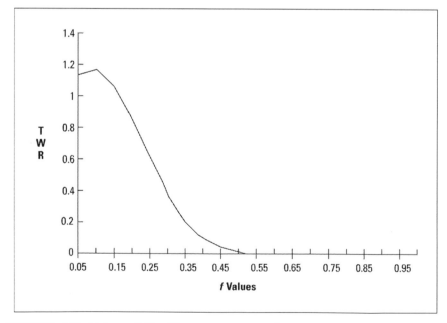

FIGURE 4.3 Values of f for 20 sequences at +1.2, −1

0.5	0.55	0.6	0.65	0.7	0.75	0.8	0.85	0.9	0.95	1
2.00	1.82	1.67	1.54	1.43	1.33	1.25	1.18	1.11	1.05	1.00
3.20	3.02	2.87	2.74	2.63	2.53	2.45	2.38	2.31	2.25	2.20
1.60	1.36	1.15	0.96	0.79	0.63	0.49	0.36	0.23	0.11	0.00
2.56	2.25	1.97	1.71	1.45	1.20	0.96	0.72	0.48	0.24	0.00
1.28	1.01	0.79	0.60	0.44	0.30	0.19	0.11	0.05	0.01	0.00
2.05	1.68	1.36	1.06	0.80	0.57	0.38	0.22	0.10	0.03	0.00
1.02	0.76	0.54	0.37	0.24	0.14	0.08	0.03	0.01	.00	0.00
1.64	1.26	0.93	0.66	0.44	0.27	0.15	0.07	0.02	.00	0.00
0.82	0.57	0.37	0.23	0.13	0.07	0.03	0.01	.00	.00	0.00
1.31	0.94	0.64	0.41	0.24	0.13	0.06	0.02	.00	.00	0.00
0.66	0.42	0.26	0.14	0.07	0.03	0.01	.00	.00	.00	0.00
1.05	0.70	0.44	0.26	0.13	0.06	0.02	0.01	.00	.00	0.00
0.52	0.32	0.18	0.09	0.04	0.02	.00	.00	.00	.00	0.00
0.84	0.52	0.30	0.16	0.07	0.03	0.01	.00	.00	.00	0.00
0.42	0.24	0.12	0.06	0.02	0.01	.00	.00	.00	.00	0.00
0.67	0.39	0.21	0.10	0.04	0.01	.00	.00	.00	.00	0.00
0.34	0.18	0.08	0.03	0.01	.00	.00	.00	.00	.00	0.00
0.54	0.29	0.14	0.06	0.02	0.01	.00	.00	.00	.00	0.00
0.27	0.13	0.06	0.02	0.01	.00	.00	.00	.00	.00	0.00
0.43	0.22	0.10	0.04	0.01	.00	.00	.00	.00	.00	0.00
0.21	0.10	0.04	0.01	.00	.00	.00	.00	.00	.00	0.00
0.34	0.16	0.07	0.02	0.01	.00	.00	.00	.00	.00	0.00
0.17	0.07	0.03	0.01	.00	.00	.00	.00	.00	.00	0.00
0.27	0.12	0.05	0.02	.00	.00	.00	.00	.00	.00	0.00
0.14	0.05	0.02	0.01	.00	.00	.00	.00	.00	.00	0.00
0.22	0.09	0.03	0.01	.00	.00	.00	.00	.00	.00	0.00
0.11	0.04	0.01	.00	.00	.00	.00	.00	.00	.00	0.00
0.18	0.07	0.02	0.01	.00	.00	.00	.00	.00	.00	0.00
0.09	0.03	0.01	.00	.00	.00	.00	.00	.00	.00	0.00
0.14	0.05	0.02	.00	.00	.00	.00	.00	.00	.00	0.00
0.07	0.02	0.01	.00	.00	.00	.00	.00	.00	.00	0.00
0.11	0.04	0.01	.00	.00	.00	.00	.00	.00	.00	0.00
0.06	0.02	.00	.00	.00	.00	.00	.00	.00	.00	0.00
0.09	0.03	0.01	.00	.00	.00	.00	.00	.00	.00	0.00
0.05	0.01	.00	.00	.00	.00	.00	.00	.00	.00	0.00
0.07	0.02	.00	.00	.00	.00	.00	.00	.00	.00	0.00
0.04	0.01	.00	.00	.00	.00	.00	.00	.00	.00	0.00
0.06	0.02	.00	.00	.00	.00	.00	.00	.00	.00	0.00
0.03	0.01	.00	.00	.00	.00	.00	.00	.00	.00	0.00
0.05	0.01	.00	.00	.00	.00	.00	.00	.00	.00	0.00
0.02	0.01	.00	.00	.00	.00	.00	.00	.00	.00	0.00
0.01	.00	.00	.00	.00	.00	.00	.00	.00	.00	0.00

This is not a contention that the fractional bet situation is the same as the real-life integer-bet situation. Rather, the contention is that for the purposes of studying these functions we are better off considering the fractional bet, since it represents the universe of integer bets. The fractional bet situation is what we can expect in real life in an asymptotic sense (i.e., in the long run).

This discussion leads to another interesting point that is true in a fixed fractional betting situation where fractional bets are allowed (think of fractional bets as the average outcome of all integer bets at different initial bankroll values, since that is what fractional betting represents here). This point is that *the TWR is the same regardless of the starting value.* In the examples just cited, if we have an initial stake of one starting value, 20 units, our TWR (ending stake divided by initial stake) is 1.15. If we have an initial stake of 400 units, 20 starting values, our TWR is still 1.15.

Figure 4.4 shows the f curve for 20 sequences of the $+1.5, -1$.

Refer now to the $+2, -1$ graph (Figure 4.5). Notice that here the optimal f is .25 where the TWR is 10.55 after 40 bets (20 sequences of $+2, -1$). Now look what happens if you bet only 15% away from the optimal .25 f. At an f

20 TRIALS

EVENT	f VALVES → 0.05	0.1	0.15	0.2	0.25	0.3	0.35	0.4	0.45
START VALUES →	20.00	10.00	6.67	5.00	4.00	3.33	2.86	2.50	2.22
1.5	21.50	11.50	8.17	6.50	5.50	4.83	4.36	4.00	3.72
−1	20.43	10.35	6.94	5.20	4.13	3.38	2.83	2.40	2.05
1.5	21.96	11.90	8.50	6.76	5.67	4.91	4.32	3.84	3.43
−1	20.86	10.71	7.23	5.41	4.25	3.43	2.81	2.30	1.89
1.5	22.42	12.32	8.85	7.03	5.85	4.98	4.28	3.69	3.16
−1	21.30	11.09	7.53	5.62	4.39	3.49	2.78	2.21	1.74
1.5	22.90	12.75	9.22	7.31	6.03	5.05	4.24	3.54	2.91
−1	21.75	11.48	7.84	5.85	4.52	3.54	2.76	2.12	1.60
1.5	23.39	13.20	9.60	7.60	6.22	5.13	4.21	3.40	2.68
−1	22.22	11.88	8.16	6.08	4.67	3.59	2.73	2.04	1.47
1.5	23.88	13.66	10.00	7.91	6.41	5.21	4.17	3.26	2.47
−1	22.69	12.29	8.50	6.33	4.81	3.64	2.71	1.96	1.36
1.5	24.39	14.14	10.41	8.22	6.62	5.28	4.13	3.13	2.28
−1	23.17	12.72	8.85	6.58	4.96	3.70	2.69	1.88	1.25
1.5	24.91	14.63	10.84	8.55	6.82	5.36	4.10	3.01	2.10
−1	23.66	13.17	9.21	6.84	5.12	3.75	2.66	1.80	1.15
1.5	25.44	15.14	11.28	8.90	7.04	5.44	4.06	2.89	1.93
−1	24.17	13.63	9.59	7.12	5.28	3.81	2.64	1.73	1.06
1.5	25.98	15.67	11.75	9.25	7.26	5.53	4.03	2.77	1.78
−1	24.68	14.11	9.99	7.40	5.44	3.87	2.62	1.66	0.98
1.5	26.53	16.22	12.23	9.62	7.48	5.61	3.99	2.66	1.64
−1	25.20	14.60	10.40	7.70	5.61	3.93	2.59	1.60	0.90
1.5	27.10	16.79	12.74	10.01	7.72	5.69	3.96	2.55	1.51
−1	25.74	15.11	10.83	8.01	5.79	3.99	2.57	1.53	0.83
1.5	27.67	17.38	13.26	10.41	7.96	5.78	3.92	2.45	1.39
−1	26.29	15.64	11.28	8.33	5.97	4.05	2.55	1.47	0.77
1.5	28.26	17.99	13.81	10.82	8.21	5.87	3.89	2.35	1.28
−1	26.85	16.19	11.74	8.66	6.15	4.11	2.53	1.41	0.70
1.5	28.86	18.61	14.38	11.26	8.46	5.95	3.85	2.26	1.18
−1	27.42	16.75	12.22	9.00	6.35	4.17	2.50	1.36	0.65
1.5	29.47	19.27	14.98	11.71	8.73	6.04	3.82	2.17	1.09
−1	28.00	17.34	12.73	9.36	6.54	4.23	2.48	1.30	0.60
1.5	30.10	19.94	15.59	12.17	9.00	6.13	3.79	2.08	1.00
−1	28.59	17.95	13.25	9.74	6.75	4.29	2.46	1.25	0.55
1.5	30.74	20.64	16.24	12.66	9.28	6.23	3.75	2.00	0.92
−1	29.20	18.57	13.80	10.13	6.96	4.36	2.44	1.20	0.51
1.5	31.39	21.36	16.91	13.17	9.57	6.32	3.72	1.92	0.85
−1	29.82	19.23	14.37	10.53	7.18	4.42	2.42	1.15	0.47
1.5	32.06	22.11	17.60	13.69	9.87	6.41	3.69	1.84	0.78
−1	30.46	19.90	14.96	10.96	7.40	4.49	2.40	1.11	0.43
TWR →	1.52	1.99	2.24	2.19	1.85	1.35	0.84	0.44	0.19

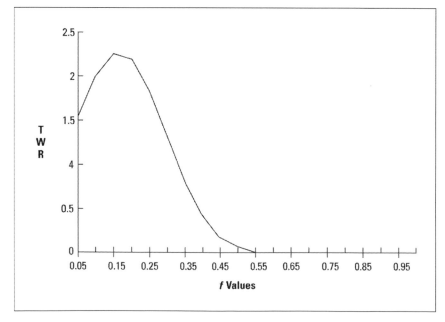

FIGURE 4.4   Values of *f* for 20 sequences at +1.5, −1

0.5	0.55	0.6	0.65	0.7	0.75	0.8	0.85	0.9	0.95	1
2.00	1.82	1.67	1.54	1.43	1.33	1.25	1.18	1.11	1.05	1.00
3.50	3.32	3.17	3.04	2.93	2.83	2.75	2.68	2.61	2.55	2.50
1.75	1.49	1.27	1.06	0.88	0.71	0.55	0.40	0.26	0.13	0.00
3.06	2.73	2.41	2.10	1.80	1.51	1.21	0.91	0.61	0.31	0.00
1.53	1.23	0.96	0.74	0.54	0.38	0.24	0.14	0.06	0.02	0.00
2.68	2.24	1.83	1.45	1.11	0.80	0.53	0.31	0.14	0.04	0.00
1.34	1.01	0.73	0.51	0.33	0.20	0.11	0.05	0.01	.00	0.00
2.34	1.84	1.39	1.00	0.68	0.42	0.23	0.11	0.03	.00	0.00
1.17	0.83	0.56	0.35	0.20	0.11	0.05	0.02	.00	.00	0.00
2.05	1.51	1.06	0.69	0.42	0.23	0.10	0.04	0.01	.00	0.00
1.03	0.68	0.42	0.24	0.13	0.06	0.02	0.01	.00	.00	0.00
1.80	1.24	0.80	0.48	0.26	0.12	0.05	0.01	.00	.00	0.00
0.90	0.56	0.32	0.17	0.08	0 03	0.01	.00	.00	.00	0.00
1.57	1.02	0.61	0.33	0.16	0.06	0.02	.00	.00	.00	0.00
0.79	0.46	0.24	0.12	0.05	0.02	.00	.00	.00	.00	0.00
0.69	0.38	0.19	0.08	0.03	0.01	.00	.00	.00	.00	0.00
1.20	0.69	0.35	0.16	0.06	0.02	.00	.00	.00	.00	0.00
0.60	0.31	0.14	0.06	0.02	.00	.00	.00	.00	.00	0.00
1.05	0.56	0.27	0.11	0.04	0.01	.00	.00	.00	.00	0.00
0.53	0.25	0.11	0.04	0 01	.00	.00	.00	.00	.00	0.00
0.92	0.46	0.20	0.08	0.02	0.01	.00	.00	.00	.00	0.00
0.46	0.21	0.08	0.03	0.01	.00	.00	.00	.00	.00	0.00
0.81	0.38	0.15	0.05	0.01	.00	.00	.00	.00	.00	0.00
0.40	0.17	0.06	0.02	.00	.00	.00	.00	.00	.00	0.00
0.70	0.31	0.12	0.04	0.01	.00	.00	.00	.00	.00	0.00
0.35	0.14	0.05	0.01	.00	.00	.00	.00	.00	.00	0.00
0.62	0.26	0.09	0.02	0.01	.00	.00	.00	.00	.00	0.00
0.31	0.12	0.04	0.01	.00	.00	.00	.00	.00	.00	0.00
0.54	0.21	0.07	0.02	.00	.00	.00	.00	.00	.00	0.00
0.27	0.09	0.03	0.01	.00	.00	.00	.00	.00	.00	0.00
0 47	0.17	0.05	0.01	.00	.00	.00	.00	.00	.00	0.00
0.24	0.08	0.02	.00	.00	.00	.00	.00	.00	.00	0.00
0.41	0.14	0.04	0.01	.00	.00	.00	.00	.00	.00	0.00
U 21	U.U6	0.02	.00	.00	.00	.00	.00	.00	.00	0.00
0 36	0 12	0 03	0 01	.00	.00	.00	.00	.00	.00	0.00
0.18	0.05	0.01	.00	.00	.00	.00	.00	.00	.00	0.00
0.32	0.10	0.02	.00	.00	.00	.00	.00	.00	.00	0.00
0.16	0.04	0.01	.00	.00	.00	.00	.00	.00	.00	0.00
0.28	0.08	0.02	.00	.00	.00	.00	.00	.00	.00	0.00
0.14	0.04	0.01	.00	.00	.00	.00	.00	.00	.00	0.00
0 07	0 02	00	00	.00	.00	.00	.00	.00	.00	0.00

of .1 or .4 your TWR is 4.66. This is not even half of what it is at .25, yet you are only 15% away from the optimal and only 40 bets have elapsed! What does this mean in terms of dollars? At $f = .1$, you would be making one bet for every \$10 in your stake. At $f = .4$ you would be making one bet for every \$2.50 in your stake. Both make the same amount, with a TWR of 4.66. At $f = .25$, you are making one bet for every \$4 in your stake. Notice that if you make one bet for every \$4 in your stake, you will make more than twice as much as you would if you were making one bet for every \$2.50 in your stake! Clearly, it does not pay to overbet. At one bet for every \$10 in your stake you make the same amount as if you had bet four times that amount, one bet for every \$2.50 in your stake! Notice that in a 50/50 game where you win twice the amount that you lose, at an f of .5 you are only breaking even! That means you are only breaking even if you made one bet for every \$2 in your stake. At an f greater than .5 you are losing in this game, and it is simply a matter of time until you are completely tapped out!

Now let's increase the winning payout from two units to five units, as is demonstrated in the data in Figure 4.6. Here your optimal f is .4, or bet \$1 for every \$2.50 in your stake. After 20 sequences of $+5, -1$, 40 bets, your

20 TRIALS

EVENT	f VALUES → 0.05	0.1	0.15	0.2	0.25	0.3	0.35	0.4	0.45
START VALUES →	20.00	10.00	6.67	5.00	4.00	3.33	2.86	2.50	2.22
2	22.00	12.00	8.67	7.00	6.00	5.33	4.86	4.50	4.22
−1	20.90	10.80	7.37	5.60	4.50	3.73	3.16	2.70	2.32
2	22.99	12.96	9.58	7.84	6.75	5.97	5.37	4.86	4.41
−1	21.84	11.66	8.14	6.27	5.06	4.18	3.49	2.92	2.43
2	24.02	14.00	10.59	8.78	7.59	6.69	5.93	5.25	4.61
−1	22.82	12.60	8.99	7.02	5.70	4.68	3.85	3.15	2.54
2	25.11	15.12	11.69	9.83	8.54	7.49	6.55	5.67	4.82
−1	23 85	13.60	9.94	7.87	6.41	5.25	4.26	3.40	2.65
2	26.24	16.33	12.92	11.01	9.61	8.39	7.24	6.12	5.04
−1	24.92	14.69	10.98	8.81	7.21	5.87	4.71	3.67	2.77
2	27.42	17.63	14.28	12.34	10.81	9.40	8.00	6.61	5.26
−1	26.05	15.87	12.14	9.87	8.11	6.58	5.20	3.97	2.89
2	28.65	19.04	15.78	13.82	12.16	10.53	8.84	7.14	5.50
−1	27.22	17.14	13.41	1.05	9.12	7.37	5.75	4.28	3.02
2	29.94	20.57	17.43	15.47	13.68	11.79	9.77	7.71	5.75
−1	28.44	18.51	14.82	12.38	10.26	8.25	6.35	4.63	3.16
2	31.29	22.21	19.26	17.33	15.39	13.21	10.80	8.33	6.00
−1	29.72	19.99	16.37	13.87	11.55	9.24	7.02	5.00	3.30
2	32.69	23.99	21.29	19.41	17.32	14.79	11.93	9.00	6.27
−1	31.06	21.59	18.09	15.53	12.99	10.35	7.75	5.40	3.45
2	34.17	25.91	23.52	21.74	19.48	16.56	13.18	9.72	6.56
−1	32.46	23.32	19.99	17.39	14.61	11.60	8.57	5.83	3.61
2	35.70	27.98	25.99	24.35	21.92	18.55	14.57	10.49	6.85
−1	33.92	25.18	22.09	19.48	16.44	12.99	9.47	6.30	3.77
2	37.31	30.22	28.72	27.27	24.66	20.78	16.10	11.33	7.16
−1	35.44	27.20	24.41	21.82	18.49	14.54	10.46	6.80	3.94
2	38.99	32.64	31.74	30.54	27.74	23.27	17.79	12.24	7.48
−1	37.04	29.37	26.98	24.44	20.81	16.29	11.56	7.34	4.12
2	40.74	35.25	35.07	34.21	31.21	26.06	19.65	13.22	7.82
−1	38.71	31.72	29.81	27.37	23.41	18.25	12.78	7.93	4.30
2	42.58	38.07	38.75	38.31	35.11	29.19	21.72	14.27	8.17
−1	40.45	34.26	32.94	30.65	26.33	20.43	14.12	0.56	4.49
2	44.49	41.11	42.82	42.91	39.50	32.70	24.00	15.42	8.54
−1	42.27	37.00	36.40	34.33	29.62	22.89	15.60	9.25	4.70
2	46.49	44.40	47.32	40.06	44.44	36.62	26.52	16.65	8.92
−1	44.17	39.96	40.22	38.45	33.33	25.63	17.24	9.99	4.91
2	46.59	47.95	52.28	53.83	49.99	41.01	29.30	17.98	9.32
−1	46.16	43.16	44.44	43.06	37.49	21.71	19.05	10.79	5.13
2	50.77	51.79	57.77	60.29	56.24	45.93	32.38	19.42	9.74
−1	48.23	46.61	49.11	48.23	42.18	32.15	21.05	11.65	5.36
TUR →	2.41	4.66	7.37	9.65	10.55	9.65	7.37	4.66	2.41

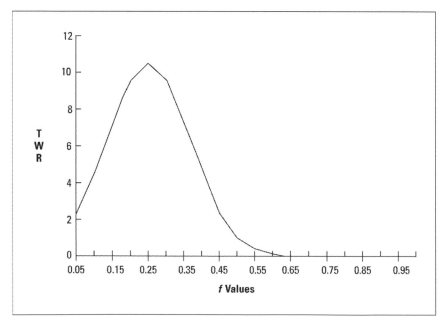

FIGURE 4.5 Values of f for 20 sequences at +2, −1

0.5	0.55	0.6	0.65	0.7	0.75	0.8	0.85	0.9	0.95	1
2.00	1.82	1.67	1.54	1.43	1.33	1.25	1.18	1.11	1.05	1.00
4.00	3.82	3.67	3.54	3.43	3.33	3.25	3.18	3.11	3.05	3.00
2.00	1.72	1.47	1.24	1.03	0.83	0.65	0.48	0.31	0.15	0.00
4.00	3.61	3.23	2.85	2.47	2.08	1.69	1.29	0.87	0.44	0.00
2.00	1.62	1.29	1.00	0.74	0.52	0.34	0.19	0.09	0.02	0.00
4.00	3.41	2.84	2.29	1.78	1.30	0.88	0.52	0.24	0.06	0.00
2.00	1.53	1.14	0.80	0.53	0.33	0.18	0.08	0.02	.00	0.00
4.00	3.22	2.50	1.85	1.28	0.81	0.46	0.21	0.07	0.01	0.00
2.00	1.45	1.00	0.65	0.38	0.20	0.09	0.03	0.01	.00	0.00
4.00	3.04	.20	1.49	0.92	0.51	0.24	0.09	0.02	.00	0.00
2.00	1.37	0.88	0.52	0.28	0.13	0.05	0.01	.00	.00	0.00
4.00	2.88	1.94	1.20	0.66	0.32	0.12	0.03	0.01	.00	0.00
2.00	1.29	0.77	0.42	0.20	0.08	0.02	0.01	.00	.00	0.00
4.00	2.72	1.70	0.96	0.48	0.20	0.06	0.01	.00	.00	0.00
2.00	1.22	0.68	0.34	0.14	0.05	0.01	.00	.00	.00	0.00
4.00	2.57	1.50	0.78	0.34	0.12	0.03	0.01	.00	.00	0.00
2.00	1.16	0.60	0.25	0.08	0.03	0.01	.00	.00	.00	0.00
4.00	2.43	1.32	0.62	0.25	0.08	0.02	.00	.00	.00	0.00
2.00	1.09	0.53	0.22	0.07	0.02	0.02	.00	.00	.00	0.00
4.00	2.29	1.16	0.50	0.18	0.05	0.01	.00	.00	.00	0.00
2.00	1.03	0.46	0.18	0.05	0.01	.00	.00	.00	.00	0.00
4.00	2.17	1.02	0.40	0.13	0.03	.00	.00	.00	.00	0.00
2.00	0.98	0.41	0.14	0.04	0.01	.00	.00	.00	.00	0.00
4.00	2.05	0.90	0.33	0.09	0.02	.00	.00	.00	.00	0.00
2.00	0.92	0.36	0.11	0.03	.00	.00	.00	.00	.00	0.00
4.00	1.94	0.79	0.26	0.07	0.01	.00	.00	.00	.00	0.00
2.00	0.87	0.32	0.09	0.02	.00	.00	.00	.00	.00	0.00
4.00	1.83	0.70	0.21	0.05	0.01	.00	.00	.00	.00	0.00
2.00	0.82	0.21	0.07	0.01	.00	.00	.00	.00	.00	0.00
4.00	1.73	0.61	0.17	0.03	.00	.00	.00	.00	.00	0.00
2.00	0.78	0.24	0.06	0.01	.00	.00	.00	.00	.00	0.00
4.00	1.63	0.54	0.14	0.02	.00	.00	.00	.00	.00	0.00
2.00	0.74	0.22	0.05	0.01	.00	.00	.00	.00	.00	0.00
4.00	1.54	0.47	0.11	0.02	.00	.00	.00	.00	.00	0.00
2.00	0.69	0.19	0.04	0.01	.00	.00	nn	nn	.00	0.00
4.00	1.46	0.42	0.09	0.01	.00	.00	.00	.00	.00	0.00
2.00	0.66	0.17	0.03	.00	.00	.00	.00	.00	.00	0.00
4.00	1.38	0.37	0.07	0.01	.00	.00	.00	.00	.00	0.00
2.00	0.62	0.19	0.02	.00	.00	.00	.00	.00	.00	0.00
4.00	1.30	0.32	0.06	0.01	.00	.00	.00	.00	.00	0.00
2.00	0.59	0.13	0.02	.00	.00	.00	.00	.00	.00	0.00
1.00	0.32	0.08	0.01	.00	.00	.00	.00	.00	.00	0.00

$2.50 stake has grown to $127,482, thanks to optimal f. Now look what happens in this extremely favorable situation if you miss the optimal f by 20%. At f values of .6 and .2 you don't make one-tenth as much as you do at .4 in this case! This particular situation, a 50/50 bet paying 5 to 1, has a mathematical expectation of $(5 * .5) + (1 * (-.5)) = 2$. Yet if you bet using an f value greater than .8, you lose money in this situation. Clearly, the question of what is the correct quantity to bet or trade has been terribly underrated.

The graphs bear out a few more interesting points. The first is that *at no other fixed fraction will you make more money than optimal* f. In other words, it does not pay to bet $1 for every $2 in your stake in the above example of $+5$, -1. In such a case, you would make less money than if you bet $1 for every $2.50 in your stake. *It does not pay to risk more than the optimal* f—*in fact, you pay a price to do so!* Notice in Figure 4.7 that you make less at $f = .55$ than at $f = .5$. The second interesting point to notice is how important the biggest loss is in the calculations. Traders may be incorrectly inclined to use maximum drawdown rather than biggest loss.

20 TRIALS

EVENT	f VALUES → 0.06	0.1	0.15	0.2	0.29	0.3	0.35	0.4	0.45
START VALUES →	20.00	10.00	6.67	5.00	4.00	3.33	2.86	2.90	2.22
5	25.00	15.00	11.67	10.00	9.00	8.33	7.86	7.50	7.22
−1	23.75	13.50	3.92	8.00	6.75	5.83	5.11	4.50	3.97
5	29.89	20.25	17.35	16.00	15.19	14.58	14.04	13.50	12.91
−1	28.20	18.23	14.75	12.80	11.39	10.21	9.13	8.10	7.10
5	35.25	27.34	25.81	25.60	25.63	25.52	25.10	24.30	23.08
−1	33.49	24.60	21.94	20.48	19.22	17.86	16.32	14.50	12.69
5	41.66	36.91	38.40	40.96	43.25	44.66	44.87	43.74	41.25
−1	39.77	33.22	32.64	32.77	32.44	31.26	29.17	26.24	22.69
5	49.71	49.82	57.12	65.54	72.98	78.16	80.21	78.73	73.73
−1	47.23	44.84	48.55	52.43	54.74	54.71	52.14	47.24	40.55
5	59.03	67.26	84.96	104.86	123.16	136.78	143.38	141.72	131.80
−1	56.08	60.53	72.22	83.89	92.37	96.74	93.20	85.03	72.49
5	70.10	90.80	126.38	167.77	207.83	239.36	256.30	255.09	230.58
−1	66.60	81.72	107.43	134.22	155.87	167.55	166.59	153.06	129.57
5	83.25	122.58	187.99	268.44	350.71	418.88	458.13	459.17	421.11
−1	79.09	110.32	199.80	214.75	263.03	293.21	297.78	275.50	231.61
5	98.86	165.49	279.64	429.50	591.82	733.03	818.90	826.50	752.73
−1	93.91	148.94	237.70	343.60	443.87	513.12	532.29	496.90	414.00
5	117.39	223.41	415.97	687.19	998.70	1282.81	1463.79	1487.69	1345.50
−1	111.52	201.07	353.57	549.76	749.03	897.96	951.46	892.62	740.03
5	139.40	301.60	618.75	1099.51	1685.31	2244.91	2616.53	2677.85	2405.09
−1	132.43	271.44	525.94	879.61	1263.98	1571.44	1700.74	1606.71	1322.60
5	165.54	407.16	920.39	1759.22	2843.96	3928.60	4677.04	4820.13	4299.10
−1	157.27	366.44	782.33	1407.37	2132.97	2750.02	3040.08	2892.08	2364.50
5	196.58	549.66	1369.09	2814.75	4799.19	6875.04	8360.21	8676.24	7684.64
−1	186.75	494.70	1163.72	2251.80	3599.39	4812.53	5434.14	5205.74	4226.55
5	233.44	742.05	2036.51	4503.60	8098.63	12031.32	14943.88	15617.22	13736.29
−1	221.77	667.84	1731.04	3602.88	6073.97	9421.93	9713.52	9370.33	7554.96
5	277.21	1001.76	3029.31	7205.76	13666.44	21054.82	26712.18	2811.00	24553.62
−1	263.35	901.58	2574.92	5764.61	10249.83	14738.37	17362.92	16866.60	13504.49
5	329.19	1352.38	4506.11	11529.22	23062.12	36845.93	47748.02	50599.80	43889.60
−1	312.73	1217.14	3830.19	9223.37	17296.59	25792.15	31036.22	30359.88	24139.28
5	390.91	1825.71	6702.83	18446.74	38917.33	64480.37	85349.59	91079.65	78452.66
−1	371.37	1643.14	5697.41	14757.40	29188.00	45136.26	55477.24	54647.79	43148.96
5	464.21	2464.71	9970.46	29514.79	65672.99	112840.65	152562.40	163943.37	140234.12
−1	441.00	2218.24	8474.89	23611.83	49254.74	78988.46	99165.56	98366.02	77128.77
5	551.25	3327.35	14831.06	47723.66	110823.17	197471.14	272705.29	295098.06	250668.50
−1	523.68	2994.62	12606.40	37778.93	83117.38	138229.80	177258.44	177058.84	137867.67
5	654.60	4491.93	22061.21	75557.86	187014.10	345574.49	487460.70	531176.51	448069.94
−1	621.87	4042.74	18752.03	60446.29	140260.58	241902.14	316849.46	318705.91	246438.47
TWR →	31.09	404.27	2812.80	12089.26	35065.14	72570.64	110897.31	127482.36	110897.31

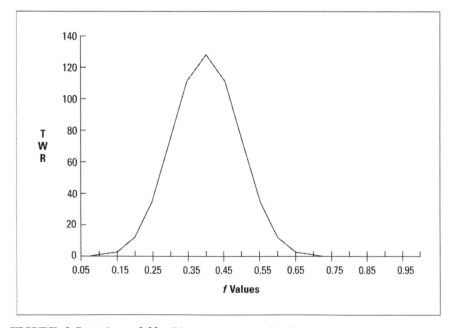

FIGURE 4.6   Values of f for 20 sequences at +5, −1

0.5	0.55	0.6	0.65	0.7	0.75	0.8	0.85	0.9	0.95	1
2.00	1.82	1.67	1.54	1.43	1.33	1.25	1.18	1.11	1.05	1.00
7.00	6.82	6.67	6.54	6.43	6.33	6.25	6.18	6.11	6.05	6.00
3.50	3.07	2.67	2.29	1.93	1.59	1.25	0.93	0.61	0.30	0.00
12.25	11.51	10.67	9.73	8.68	7.52	6.25	4.86	3.36	1.74	0.00
6.13	5.18	4.27	3.40	2.60	1.88	1.25	0.73	0.34	0.09	0.00
21.44	19.42	17.07	14.47	11.72	8.93	6.25	3.83	1.85	0.50	0.00
10.72	8.74	6.83	5.06	3.51	2.23	1.25	0.57	0.18	0.03	0.00
37.52	32.76	27.31	21.52	15.82	10.61	6.25	3.02	1.02	0.14	0.00
18.76	14.74	10.92	7.53	4.75	2.65	1.25	0.45	0.10	0.01	0.00
65.65	55.29	43.69	32.01	21.35	12.59	6.25	2.38	0.56	0.04	0.00
32.83	24.88	17.48	11.20	6.41	3.15	1.25	0.36	0.06	.00	0.00
114.89	93.30	69.91	47.62	28.83	14.96	6.25	1.87	0.31	0.01	0.00
57.45	41.99	27.96	16.67	8.65	3.74	1.25	0.28	0.03	.00	0.00
201.06	157.45	111.85	70.83	38.92	17.76	6.25	1.47	0.17	.00	0.00
100.53	70.85	44.74	24.79	11.67	4.44	1.25	0.22	0.02	.00	0.00
351.86	265.69	178.96	105.36	52.54	21.09	6.25	1.16	0.09	.00	0.00
175.93	119.56	71.58	36.88	15.76	5.27	1.25	0.17	0.01	.00	0.00
615.75	448.35	286.33	156.72	70.92	25.04	6.25	0.91	0.05	.00	0.00
307.87	201.76	114.53	54.85	21.28	6.26	1.25	0.14	0.01	.00	0.00
1077.56	756.59	458.13	233.12 ·	95.75	29.74	6.25	0.72	0.03	.00	0.00
538.78	340.47	183.25	81.59	28.72	7.43	1.25	0.11	.00	.00	0.00
1885.73	1276.75	733.01	346.77	129.26	35.32	6.25	0.57	0.02	.00	0.00
942.86	574.54	293.20	121.37	38.78	8.83	1.25	0.08	.00	.00	0.00
3300.02	2154.52	1172.81	515.82	174.50	41.94	6.25	0.45	0.01	.00	0.00
1650.01	969.53	469.12	180.54	52.35	10.48	1.25	0.07	.00	.00	0.00
5775.04	3635.75	1876.50	767.29	235.57	49.80	6.25	0.35	.00	.00	0.00
2887.52	1636.09	750.60	268.55	70.67	12.45	1.25	0.05	.00	.00	0.00
10106.31	6135.33	3002.40	1141.34	318.02	59.14	6.25	0.28	.00	.00	0.00
5053.16	2760.90	1200.96	399.47	95.41	14.78	1.25	0.04	.00	.00	0.00
17686.04	10353.36	4803.84	1697.75	429.33	70.23	6.25	0.22	.00	.00	0.00
8843.02	4659.01	1921.54	594.21	128.80	17.56	1.25	0.03	.00	.00	0.00
30950.58	17471.30	7686.14	2525.40	579.59	83.39	6.25	0.17	.00	.00	0.00
15475.29	7862.09	3074.46	883.89	173.88	20.85	1.25	0.03	.00	.00	0.00
54163.51	29482.82	12297.83	3756.53	782.45	99.03	6.25	0.14	.00	.00	0.00
27081.76	13267.27	4919.13	1314.79	234.73	24.76	1.25	0.02	.00	.00	0.00
94786.15	49752.27	19676.53	5587.84	1056.30	117.60	6.25	0.11	.00	.00	0.00
47393.07	22388.52	7870.61	1955.74	316.89	29.40	1.25	0.02	.00	.00	0.00
165875.76	83956.95	31482.44	8311.92	1426.01	139.65	6.25	0.08	.00	.00	0.00
82937.88	37780.63	12592.98	2909.17	427.80	34.91	1.25	0.01	.00	.00	0.00
290282.57	141677.35	50371.91	12363.97	1925.11	165.83	6.25	0.07	.00	.00	0.00
145141.29	63754.81	20148.76	4327.39	577.53	41.46	1.25	0.01	.00	.00	0.00
72570.64	35065.14	12089.26	2812.80	404.27	31.09	1.00	0.01	.00	.00	0.00

DRAWDOWN AND LARGEST LOSS WITH *f*

First, if you have $f = 1.00$, then as soon as the biggest loss is encountered, you would be tapped out. This is as it should be. You want f to be bounded at 0 (nothing at stake) and 1 (the lowest amount at stake where you would lose 100%).

Second, in an independent trials process the sequence of trades that results in the drawdown is, in effect, arbitrary (as a result of the independence). Suppose we toss a coin six times, and we get heads three times and tails three times. Suppose that we win $1 every time heads comes up and lose $1 every time tails comes up. Considering all possible sequences here our drawdown could be $1, $2, or $3, the extreme case where all losses bunch together. If we went through this exercise once and came up with a $2 drawdown, it wouldn't mean anything. Since drawdown is an *extreme* case situation, and we are speaking of exact sequences of trades that are independent, we have to assume that the extreme case can be all losses bunching together in a row (the extreme worst case in the sample space).

20 TRIALS

EVENT	f VALUES → 0.05	0.1	0.15	0.2	0.25	0.3	0.35	0.4	0.45
START VALUES →	20.00	10.00	6.67	5.00	4.00	3.33	2.86	2.50	2.22
-1	19.00	9.00	5.67	4.00	3.00	2.33	1.86	1.50	1.22
-1	18.05	8.10	4.82	3.20	2.25	1.63	1.21	0.90	1.67
-1	17.15	7.29	4.09	2.56	1.69	1.14	0.78	0.54	0.37
-1	16.29	6.56	3.48	2.05	1.27	0.80	0.51	0.32	0.20
-1	15.48	5.90	2.96	1.64	0.95	0.56	0.33	0.19	0.11
-1	14.70	5.31	2.51	1.31	0.71	0.39	0.22	0.12	0.06
-1	13.97	4.78	2.14	1.05	0.53	0.27	0.14	0.07	0.03
-1	13.27	4.30	1.82	0.84	0.40	0.19	0.09	0.04	0.02
-1	12.60	3.87	1.54	0.67	0.30	0.13	0.06	0.03	0.01
-1	11.97	3.49	1.31	0.54	0.23	0.09	0.04	0.02	0.01
-1	12.57	3.84	1.51	0.64	0.28	0.12	0.05	0.02	0.01
-1	13.20	4.22	1.74	0.77	0.35	0.16	0.07	0.03	0.01
1	13.86	4.64	2.00	0.93	0.44	0.21	0.09	0.04	0.02
1	14.56	5.11	2.30	1.11	0.55	0.27	0.13	0.06	0.02
1	15.28	5.62	2.64	1.34	0.69	0.35	0.17	0.08	0.04
1	16.05	6.18	3.04	1.60	0.86	0.45	0.23	0.11	0.05
1	16.05	6.79	3.49	1.92	1.07	0.59	0.31	0.16	0.05
1	17.69	7.47	4.01	2.31	1.34	0.77	0.42	0.22	0.11
1	18.58	8.22	4.62	2.77	1.68	1.00	0.57	0.31	0.16
1	19.51	9.04	5.31	3.32	2.10	1.30	0.77	0.44	0.23
1	20.48	9.95	6.11	3.99	2.62	1.69	1.04	0.61	0.34
1	21.50	10.94	7.02	4.79	3.28	2.19	1.41	0.86	0.49
1	22.58	12.04	8.08	5.74	4.10	2.85	1.90	1.20	0.71
1	23.71	13.24	9.29	6.89	5.12	3.71	2.57	1.68	1.02
1	24.89	14.57	10.68	8.27	6.40	4.82	3.47	2.35	1.48
1	26.14	16.02	12.28	9.93	8.00	6.27	4.68	3.29	2.15
1	27.45	17.62	14.12	11.91	10.00	8.15	6.32	4.61	3.12
1	28.82	19.39	16.24	14.29	12.50	10.59	8.53	6.45	4.52
1	30.26	21.32	18.68	17.15	15.63	13.77	11.52	9.03	6.55
1	31.77	23.46	21.48	20.58	19.54	17.89	15.55	12.65	9.50
1	33.36	25.80	24.70	24.70	24.42	23.26	20.99	17.71	13.78
1	35.03	28.38	28.41	29.64	30.53	30.24	28.34	24.79	19.98
1	36.79	31.22	32.67	35.57	38.16	39.31	38.26	34.71	28.97
1	38.62	34.34	34.57	42.68	47.70	51.11	51.65	48.59	42.00
1	40.55	37.78	43.21	51.22	59.62	66.44	69.73	68.02	60.90
1	42.58	41.56	49.69	61.46	74.53	86.37	94.13	95.23	80.30
1	44.71	45.71	57.14	73.75	93.16	112.29	127.08	133.32	128.04
1	46.94	50.28	65.71	88.50	116.45	145.97	171.56	186.65	185.66
1	49.29	55.31	75.57	106.20	145.57	189.77	231.32	261.32	269.21
1	51.75	60.84	86.90	127.44	181.96	246.69	312.66	365.84	390.35
TWR →	2.59	6.08	13.04	25.49	45.49	74.01	109.43	146.34	175.66

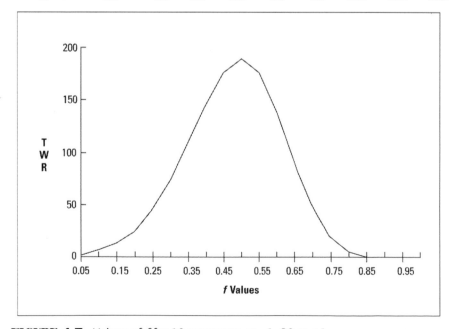

FIGURE 4.7   Values of f for 10 sequences at −1, 30 at +1

0.5	0.55	0.6	0.65	0.7	0.75	0.8	0.85	0.9	0.95	1
2.00	1.82	1.67	1.54	1.43	1.33	1.25	1.18	1.11	1.05	1.00
1.00	0.82	0.67	0.54	0.43	0.33	0.25	0.18	0.11	0.05	0.00
0.50	0.37	0.27	0.19	0.13	0.08	0.05	0.03	0.01	.00	0.00
0.25	0.17	0.11	0.07	0.04	0.02	0.01	.00	.00	.00	0.00
0.13	0.07	0.04	0.02	0.01	0.01	.00	.00	.00	.00	0.00
0.06	0.03	0.02	0.01	.00	.00	.00	.00	.00	.00	0.00
0.03	0.02	0.01	.00	.00	.00	.00	.00	.00	.00	0.00
0.02	0.01	.00	.00	.00	.00	.00	.00	.00	.00	0.00
0.01	.00	.00	.00	.00	.00	.00	.00	.00	.00	0.00
.00	.00	.00	.00	.00	.00	.00	.00	.00	.00	0.00
.00	.00	.00	.00	.00	.00	.00	.00	.00	.00	0.00
.00	.00	.00	.00	.00	.00	.00	.00	.00	.00	0.00
.00	.00	.00	.00	.00	.00	.00	.00	.00	.00	0.00
0.01	.00	.00	.00	.00	.00	.00	.00	.00	.00	0.00
0.01	.00	.00	.00	.00	.00	.00	.00	.00	.00	0.00
0.01	0.01	.00	.00	.00	.00	.00	.00	.00	.00	0.00
0.02	0.01	.00	.00	.00	.00	.00	.00	.00	.00	0.00
0.03	0.01	.00	.00	.00	.00	.00	.00	.00	.00	0.00
0.05	0.02	0.01	.00	.00	.00	.00	.00	.00	.00	0.00
0.08	0.03	0.01	.00	.00	.00	.00	.00	.00	.00	0.00
0.11	0.05	0.02	0.01	.00	.00	.00	.00	.00	.00	0.00
0.17	0.08	0.03	0.01	.00	.00	.00	.00	.00	.00	0.00
0.25	0.12	0.05	0.02	.00	.00	.00	.00	.00	.00	0.00
0.38	0.18	0.08	0.03	0.01	.00	.00	.00	.00	.00	0.00
0.57	0.29	0.13	0.05	0.01	.00	.00	.00	.00	.00	0.00
0.86	0.44	0.20	0.08	0.02	0.01	.00	.00	.00	.00	0.00
1.28	0.69	0.32	0.13	0.04	0.01	.00	.00	.00	.00	0.00
1.92	1.07	0.52	0.21	0.07	0.02	.00	.00	.00	.00	0.00
2.89	1.65	0.83	0.35	0.12	0.03	0.01	.00	.00	.00	0.00
4.33	2.56	1.32	0.58	0.20	0.05	0.01	.00	.00	.00	0.00
6.49	3.97	2.11	0.95	0.34	0.09	0.02	.00	.00	.00	0.00
9.74	6.15	3.38	1.57	0.58	0.16	0.03	.00	.00	.00	0.00
14.61	9.53	5.41	2.58	0.99	0.28	0.05	0.01	.00	.00	0.00
21.92	14.77	8.65	4.26	1.68	0.49	0.10	0.01	.00	.00	0.00
32.88	22.89	13.85	7.04	2.86	0.87	0.17	0.02	.00	.00	0.00
49.32	35.49	22.15	11.61	4.87	1.51	8.31	0.03	.00	.00	0.00
73.98	55.00	35.45	19.16	8.28	2.65	0.56	0.06	.00	.00	0.00
110.97	85.25	56.71	31.61	14.07	4.64	1.00	0.11	.00	.00	0.00
166.45	132.14	90.74	52.16	23.92	8.12	1.80	0.21	0.01	.00	0.00
249.68	204.82	145.19	86.06	40.66	14.21	3.24	0.38	0.01	.00	0.00
374.51	317.48	232.30	142.00	69.12	24.86	5.83	0.70	0.03	.00	0.00
187.26	174.61	139.38	92.30	48.38	18.64	4.66	0.60	0.02	.00	0.00

Just because we experienced one exact sequence of six coin flips wherein the drawdown was $2 doesn't mean we can use that as any kind of a meaningful benchmark, since the next exact sequence is equally likely to be any other possible sequence as it is to be the sequence we are basing this draw down figure on.

Return to the coin toss, whereby if we win, we win $1, and if we lose, we lose $1. Suppose 20 tosses have gone by and you have experienced a drawdown of $5 at one point. What does this mean? Does this mean that we can expect "about" a $5 drawdown on the next 20 tosses? Since coin tossing is an independent trials process (as trading is for the most part), the answer to all of these questions is no. The only estimating we can perform here is one based on the losing streaks involved. With a 20-coin toss we can figure probabilities of getting 20 tosses against us, 19 tosses, and so on. But what we are talking about with drawdown is absolute worst case—an extreme. What we are looking for is an answer to the question, "How far out on the tails of the distribution, to the adverse side, is the limit?" The answer is that there is no limit—all future coin tosses, the next 20 tosses and all sequences of 20 tosses, could go against us. It's highly unlikely, but it could happen. To assume that there is a maximum drawdown that we can expect is simply an illusion. The idea is propagated for a trader's peace of mind. Statistically, it has no significance. If we are trading on a fixed fractional basis (where the drawdown is also a function of when it happens—i.e., how big the account was when the drawdown started), then drawdown is absolutely meaningless.

Third, the drawdown under fixed fraction is not the drawdown we would encounter on a constant contract basis (i.e., nonreinvestment). This was demonstrated in the previous chapter. Fourth and finally, in this exercise we are trying to discern only how much to commit to the next trade, not the next sequence of trades. Drawdown is a sequence of trades—should the maximum drawdown occur on one trade, then that one trade would also be the biggest losing trade.

If you want to measure the downside of a system, then you should look at the biggest losing trade, since drawdown is arbitrary and, in effect, meaningless. This becomes even more so when you are considering fixed fractional (i.e., reinvestment of returns) trading. Many traders try to "limit their drawdown" either consciously (as when they are designing trading systems) or subconsciously. This is understandable, as drawdown is the trader's nemesis. Yet we see that, as a result of its arbitrary nature, drawdown is uncontrollable. What is controllable, at least to an extent, is the largest loss. As you have seen, optimal f is a function of the largest loss. It is possible to control your largest loss by many techniques, such as only day-trading, using options, and so on. The point here is that you can control your largest loss as well as your frequency of large losses (at least to some extent).

It is important to note at this point that the drawdown you can expect with fixed fractional trading, as a percentage retracement of your account equity, historically would have been at least as much as f percent. In other words, if f is .55, then your drawdown would have been at least 55% of your equity (leaving you with 45% at one point). This is so because if you are trading at the optimal f, as soon as your biggest loss is hit, you would experience the drawdown equivalent to f. Again, assuming f for a system is .55, and assuming that translates into trading one contract for every $10,000, your biggest loss would be $5,500. As should by now be obvious, when the biggest loss was encountered (again we're speaking historically, i.e., about what would have happened), you would have lost $5,500 for each contract you had on, and you would have had one contract on for every $10,000 in the account. Therefore, at that point your drawdown would have been 55% of equity. However, it is possible that the drawdown would continue, that the next trade or series of trades would draw your account down even more. Therefore, the better a system, the higher the f. The higher the f, generally the higher the drawdown, since the drawdown (as a percentage) can never be any less than the f. There is a paradox involved here, in that if a system is good enough to generate an optimal f that is a high percentage, then the drawdown for such a good system will also be quite high. While optimal f allows you to experience the greatest geometric growth, it also gives you enough rope to hang yourself.

CONSEQUENCES OF STRAYING TOO FAR FROM THE OPTIMAL f

The fact that the difference between being at the optimal value for f and being at any other value increases geometrically over time is particularly important to gamblers. Time in this sense is synonymous with action. For years, a simple system for blackjack has been to simply keep track of how many fives have fallen from the deck. The fewer the fives contained in the remaining deck, the greater is the player's advantage over the casino. Depending on the rules of the casino, this advantage could range to almost as high as 3.6% for a deck with no remaining fives. Roughly, then, the optimal f for this strategy would range from 0 to about .075 to .08 for each hand, depending on how many fives had fallen (i.e., you would use a different f value for each different number of remaining fives in a deck. This is a dependent trials process, and therefore your optimal betting strategy would be to trade variable fraction based on the optimal f for each scenario of the ratio of fives left in the deck). If you go into the casino and play through only one deck, you will not be penalized for deviating from the optimal f (as you would if you were to play 1,000 hands). It is incorrect to think that if you have an edge on a particular hand, you should simply increase the size of your wager. How much you increase it by is paramount.

To illustrate, if you have a stake of $500 and start playing at a table where $5 is the minimum bet, your minimum bet is therefore 1% of your stake. If you encounter, during the course of the deck, a situation where all fives are depleted from the deck, you then have an edge of anywhere from 3 to 3.6%, depending on the house rules. Say your optimal f now is .08, or one bet per every $62.50 in your stake ($5, the maximum possible loss on the next hand, divided by .08).

Suppose you had been breaking even to this point in the game and still had $500 in your stake. You would then bet $40 on the next hand ($500/$62.50 * $5). If you were to bet $45, you could expect a decrease in performance. There is no benefit to betting the extra $5 unit. This decrease in performance grows geometrically over time. If you calculate your optimal f on each hand, and slightly over- or underbet, you can expect a decrease in performance geometrically proportional to the length of the game (the action). If you were to bet, say, $100, on the situation described above, you would be at an f factor way out to the right of the optimal f. You wouldn't stand a chance over time—no matter how good a card counter you were! If you are too far to the right of the optimal f, even if you know exactly what cards remain in the deck, you are in a losing situation!

Next are four more charts, which, if you still do not see, drive home the importance of being near the optimal f. These are equity curve charts. An equity curve is simply the total equity of an account (plotted on the Y axis) over a period of time or series of trades (the X axis). On these four charts, we assume an account starts out with 10 units. Then the following sequence of 21 trades/bets is encountered:

$$1, 2, 1, -1, 3, 2, -1, -2, -3, 1, -2, 3, 1, 1, 2, 3, 3, -1, 2, -1, 3$$

If you have done the calculations yourself, you will find that the optimal f is .6, or bet one unit for every five in your stake (since the biggest losing trade is for three units).

The first equity curve (Figure 4.8) shows this sequence on a constant one-contract basis. Nice consistency. No roller-coaster drawdowns. No geometric growth, either.

Next comes an equity curve with f at .3, or bet one unit for every 10 units in your stake (Figure 4.9). Makes a little more than constant contract.

On the third equity curve graph you see the sequence at the optimal f value of .6, or one bet for every five in your stake (Figure 4.10). Notice how much more it has made than at $f = .3$.

The final equity curve shows the sequence of bets at $f = .9$, or one bet for every $3\frac{1}{3}$ units in your stake (Figure 4.11). Notice how quickly the equity took off until it hit the drawdown periods (7 through 12). When f is too high, the market systems get beaten down so low during a drawdown that it takes far longer to come out of them, if ever, than at the optimal values.

Even at the optimal values, the drawdowns can be quite severe for any market/system. It is not unusual for a market system trading one contract under optimal f to see 80 to 95% of its equity erased in the bad drawdowns. But notice how at the optimal values the equity curve is able to recover in short order and go on to higher ground. These four charts have all traded the same sequence of trades, yet look at how using the optimal f affects performance, particularly after drawdowns.

Obviously, the greater an account's capitalization, the more accurately its traders can stick to optimal f, as the dollars per single contract required are a smaller percentage of the total equity. For example, suppose optimal f for a given market system dictates we trade one contract for every $5,000 in an account. If an account starts out with $10,000 equity, then it can gain (or lose) 50% before a quantity adjustment is necessary. Contrast this to a $500,000 account, where there would be a contract adjustment for every 1% change in equity. Clearly, the larger account can take advantage of the benefits provided by optimal f better than a smaller account can. Theoretically, optimal f assumes you can trade in infinitely divisible quantities, which is not the case in real life, where the smallest quantity you can trade in is a single contract. In the asymptotic sense this does not matter. In the real-life integer-bet scenario, a good case could be presented for trading a market system that requires as small a percentage of the account equity as possible, especially for smaller accounts. But there is a trade-off here as well. Since we are striving to trade in markets that would require us to trade in greater multiples, we will be paying greater commissions, execution costs, and slippage. Bear in mind that the amount required per contract in real life is the greater of the initial margin requirement or the dollar amount per contract dictated by the optimal f.

As the charts bear out, you pay a substantial penalty for deviating from the optimal fixed dollar fraction. *Being at the right value for* f *is more important than how good your trading system is* (provided of course that the system is profitable on a single-contract basis)! Therefore, the finer you can cut it (i.e., the more frequently you adjust the size of the positions you are trading, so as to align yourself with what the optimal f dictates), the better off you are. Most accounts, therefore, would be better off trading the smaller markets. Corn may not seem like a very exciting market to you compared to the S&Ps. Yet, for most people, the corn market can get awfully exciting if they have a few hundred contracts on.

Throughout the text, we refigure the amount of contracts you should have on for the next trade based on the dictates of the optimal f for a given market system. However, the finer you can cut it, the better. If you refigure how many contracts you should have on every day as opposed to simply every week, you will be that much better off. If you refigure how many contracts you should have on every hour as opposed to every day, you will be even better off. However, there is the old trade-off of commissions, slippage, and fees, not to mention the cost of mistakes, which will be more likely the more frequently you realign yourself with the dictates of the optimal f. Bear in mind that realigning yourself on every trade is not the only way to do it, and the finer (more frequently) you can cut it—the more often you realign yourself with the dictates of the optimal f—the more the benefits of the optimal f will work for you. Ideally, you will realign yourself with optimal f on as close to a continuous basis as possible with respect to the trade-offs of commissions, fees, slippage, and the costs of human error.

It is doubtful whether anyone in the history of the markets has been able to religiously stick to trading on a constant contract basis. If someone quadrupled their money, would they still stick to trading in the same exact

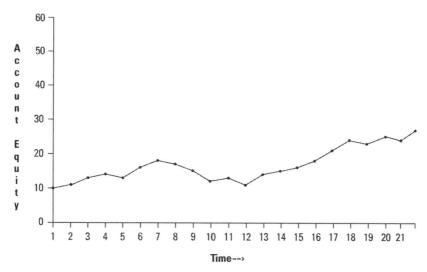

FIGURE 4.8 Equity curve for 21 trades on a constant contract basis

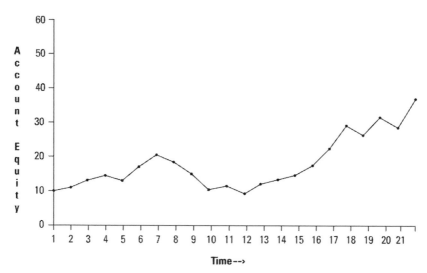

FIGURE 4.9 Equity curve for 21 trades with $f = .30$, or 1 contract for every 10 units in the stake

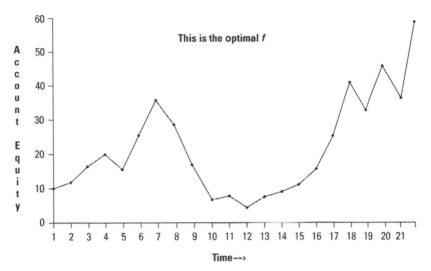

FIGURE 4.10 Equity curve for 21 trades with $f = .60$, or 1 contract for every 5 units in the stake

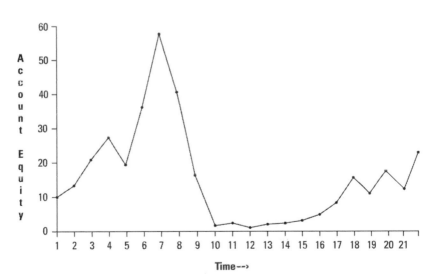

FIGURE 4.11 Equity curve for 21 trades with $f = .90$, or 1 contract for every 3.33 units in the stake

size? Conversely, would someone trading 10 contracts on every trade who was suddenly cut down to trading less than 10 contracts inject enough capital into the account to margin 10 contracts again on the next trade? It's quite unlikely. Any time a trader trading on a constant contract basis deviates from always trading the same constant contract size, the problem of what quantities to trade in arises. This is so whether the trader recognizes this problem or not. As you have seen demonstrated in this chapter, this is a problem for the trader. Constant contract trading is not the solution, because you can never experience geometric growth trading constant contract. So, like it or not, the question of what quantity to take on the next trade is inevitable for everyone. To simply select an arbitrary quantity is a costly mistake. Optimal f is factual; it is mathematically correct.

Are there traders out there who aren't planning on reinvesting their profits? Unless we're looking at optimal f via the highest TWR, we wouldn't know a good market system.

If a system is good enough, it is often possible to have a value for f that implies applying a dollar amount per contract that is less than the initial margin. Remember that f gives us the peak of the curve; to go off to the right of the peak (take on more contracts) provides no benefit. But the trader need not use that value for f that puts him at the peak; he may want to go to the left of the peak (i.e., apply more dollars in equity to each contract he puts on). You could, for instance, divide your account into two equal parts and resolve to keep one part cash and one part as dollars to apply to trading positions and use f on that half. This in effect would amount to a half f or fractional f strategy.

By now it should be obvious that we have a working range for usable values of f, that range being from zero to the optimal value. The higher you go within this range, the greater the return (up to but not beyond the optimal f) and the greater the risk (the greater the expected drawdowns in size—not, however, in frequency). The lower you go in this range, the less the risk (again in terms of extent but not frequency of drawdowns), and the less the potential returns. However, as you move down this range toward zero, the greater the probability is that an account will be profitable (remember that a constant-contract-based account has a greater probability of being profitable than a fractional f one). Ziemba's *Gambling Times* articles on Kelly demonstrated that *at smaller profit targets the half Kelly was more apt to reach these levels before halving than was a full Kelly bet. In other words, the fractional Kelly (fractional* f*) bet is safer—it has less variance in the final outcome after X bets. This ability to choose a fraction of the optimal* f *(choosing a value between 0 and the optimal* f*) allows you to have any desired risk/return trade-off that you like.*

Referring back to our four equity curve charts where the optimal $f = .60$, notice how nice and smooth the half f chart of $f = .30$ is. Half f makes for a much smoother equity curve than does full *f*. Of course, the trade-off is less return—again, a difference that grows as time passes.

Here, a word of caution is in order. Just as there is a price to be paid (in reduced return and greater drawdowns) for being too far to the right of the peak of the *f* curve (i.e., too many contracts on), there is also a price to be paid for being to the left of the peak of the *f* curve (i.e., too few contracts on). This price is not as steep as being too far to the right, so if you must err, err to the left.

As you move to the left of the peak of the curve (i.e., allocate more dollars per contract) you reduce your drawdowns arithmetically. However, you also reduce your returns geometrically. Reducing your returns geometrically is the price you pay for being to the left of the optimal *f* on the *f* curve. However, using the fractional *f* still makes good sense in many cases. When viewed from the perspective of time required to reach a specific goal (as opposed to absolute gain), the fractional *f* strategy makes a great deal of sense. Very often, a fractional *f* strategy will not take much longer to reach a specific goal than will the full *f* (the height of the goal and what specific fraction of *f* you choose will determine how much longer). If minimizing the time required to reach a specific goal times the potential drawdown as a percentage of equity retracement is your priority, then the fractional *f* strategy is most likely for you.

Aside from, or in addition to, diluting the optimal *f* by using a percentage or fraction of the optimal *f*, you could diversify into other markets and systems (as was just done by putting 50% of the account into cash, as if cash were another market or system).

EQUALIZING OPTIMAL *f*

Optimal *f* will yield the greatest geometric growth on a stream of outcomes. This is a mathematical fact. Consider the hypothetical stream of outcomes:

$$+2, -3, +10, -5$$

This is a stream from which we can determine our optimal *f* as .17, or to bet one unit for every $29.41 in equity. Doing so on such a stream will yield the greatest growth on our equity.

Consider for a moment that this stream represents the trade profits and losses (P&Ls) on one share of stock. Optimally, we should buy one share of stock for every $29.41 that we have in account equity, regardless of what the current stock price is. But suppose the current stock price is $100 per share. Further, suppose the stock was $20 per share when the first two trades occurred and was $50 per share when the last two trades occurred.

Recall that with optimal f we are using the stream of past trade P&Ls as a proxy for the distribution of expected trade P&Ls currently. Therefore, we can preprocess the trade P&L data to reflect this by converting the past trade P&L data to reflect a commensurate percentage gain or loss based upon the current price.

For our first two trades, which occurred at a stock price of $20 per share, the $2 gain corresponds to a 10% gain and the $3 loss corresponds to a 15% loss. For the last two trades, taken at a stock price of $50 per share, the $10 gain corresponds to a 20% gain and the $5 loss corresponds to a 10% loss.

The formulas to convert raw trade P&Ls to percentage gains and losses for longs and shorts are as follows:

$$P\&L\% = \text{ Exit Price/Entry Price} -1 \quad \text{(for longs)} \quad (4.10a)$$

$$P\&L\% = \text{ Entry Price/Exit Price} -1 \quad \text{(for shorts)} \quad (4.10b)$$

or we can use the following formula to convert both longs and shorts:

$$P\&L\% = \text{ P\&L in Points/Entry Price} \quad (4.11)$$

Thus, for our four hypothetical trades, we now have the following stream of *percentage* gains and losses (assuming all trades are long trades):

$$+.1, -.15, +.2, -.1$$

We call this new stream of translated P&Ls the *equalized data*, because it is equalized to the price of the underlying instrument when the trade occurred.

To account for commissions and slippage, you must adjust the exit price downward in Equation (4.10a) for an amount commensurate with the amount of the commissions and slippage. Likewise, you should adjust the exit price upward in (4.10b). If you are using (4.11), you must deduct the amount of the commissions and slippage (in points again) from the numerator P&L in Points.

Next, we determine our optimal f on these percentage gains and losses. The f that is optimal is .09. We must now convert this optimal f of .09 into a dollar amount based upon the current stock price. This is accomplished by the following formula:

$$f\$ = \text{Biggest \% Loss} * \text{Current Price} * \$\text{per Point}/-f \quad (4.12)$$

Thus, since our biggest percentage loss was $-.15$, the current price is $100 per share, and the number of dollars per full point is 1 (since we are dealing with buying only one share), we can determine our $f\$$ as:

$$f\$ = -.15 * 100 * 1/-.09$$
$$= -15/-.09$$
$$= 166.67$$

Thus, we would optimally buy one share for every $166.67 in account equity. If we used 100 shares as our unit size, the only variable affected would have been the number of dollars per full point, which would have been 100. The resulting $f\$$ would have been $16,666.67 in equity for every 100 shares.

Suppose now that the stock went down to $3 per share. Our $f\$$ equation would be exactly the same except for the current price variable, which would now be 3. Thus, the amount to finance one share by becomes:

$$f\$ = -.15 * 3 * 1/-.09$$
$$= -.45/-.09$$
$$= 5$$

We optimally would buy one share for every $5 we had in account equity.

Notice that the optimal f does not change with the current price of the stock. It remains at .09. However, the $f\$$ changes continuously as the price of the stock changes. This doesn't mean that you must alter a position you are already in on a daily basis, but it does make it more likely to be beneficial that you do so. As an example, if you are long a given stock and it declines, the dollars that you should allocate to one unit (100 shares in this case) of this stock will decline as well, with the optimal f determined off of equalized data. If your optimal f is determined off of the raw trade P&L data, it will not decline. In both cases, your daily equity is declining. Using the equalized optimal f makes it more likely that adjusting your position size daily will be beneficial.

Equalizing the data for your optimal f necessitates changes in the by-products. We have already seen that both the optimal f and the geometric mean (and hence the TWR) change. The arithmetic average trade changes because now it, too, must be based on the idea that all trades in the past must be adjusted as if they had occurred from the current price. Thus, in our hypothetical example of outcomes on one share of +2, −3, +10, and −5, we have an average trade of $1. When we take our percentage gains and losses of +.1, −15, +.2, and −.1, we have an average trade (in percent) of +.5. At $100 per share, this translates into an average trade of 100 * .05 or $5 per trade. At $3 per share, the average trade becomes $.15(3 * .05).

The geometric average trade changes as well.

$$\text{GAT} = G * (\text{Biggest Loss}/-f)$$

where: G = Geometric mean − 1.
          f = Optimal fixed fraction.

(and, of course, our biggest loss is always a negative number). This equation is the equivalent of:

$$\text{GAT} = (\text{geometric mean} - 1) * f\$$$

We have already obtained a new geometric mean by equalizing the past data. The $f\$$ variable, which is constant when we do not equalize the past data, now changes continuously, as it is a function of the current underlying price. Hence, our geometric average trade changes continuously as the price of the underlying instrument changes.

Our threshold to the geometric also must be changed to reflect the equalized data.

$$T = AAT/GAT * \text{Biggest Loss}/-f \qquad (4.13)$$

where: T = The threshold to the geometric.
AAT = The arithmetic average trade.
GAT = The geometric average trade.
f = The optimal f (0 to 1).

This equation can also be rewritten as:

$$T = AAT/GAT * f\$ \qquad (4.13a)$$

Now, not only do the AAT and GAT variables change continuously as the price of the underlying changes, so too does the $f\$$ variable.

Finally, when putting together a portfolio of market systems we must figure daily HPRs. These too are a function of $f\$$:

$$\text{Daily HPR} = D\$/f\$ + 1 \qquad (4.14)$$

where: D\$ = The dollar gain or loss on 1 unit from the previous day. This is equal to (Tonight's Close – Last Night's Close) * Dollars per Point.
$f\$$ = The current optimal f in dollars, calculated from Equation (4.12). Here, however, the current price variable is last night's close.

For example, suppose a stock tonight closed at $99 per share. Last night it was $102 per share. Our biggest percentage loss is -15. If our f is .09, then our $f\$$ is:

$$f\$ = -.15 * 102 * 1/-.09$$
$$= -15.3/-.09$$
$$= 170$$

Since we are dealing with only 1 share, our dollars per point value is $1. We can now determine our daily HPR for today as:

$$\text{Daily HPR} = (99 - 102) * 1/170 + 1$$
$$= -3/170 + 1$$
$$= -.01764705882 + 1$$
$$= .9823529412$$

Return now to what was said at the outset of this discussion. Given a stream of trade P&Ls, the optimal f will make the greatest geometric growth on that stream (provided it has a positive arithmetic mathematical expectation). We use the stream of trade P&Ls as a proxy for the distribution of possible outcomes on the next trade. Along this line of reasoning, it may be advantageous for us to equalize the stream of past trade profits and losses to be what they would be if they were performed at the current market price. In so doing, we may obtain a more realistic proxy of the distribution of potential trade profits and losses on the next trade. Therefore, we should figure our optimal f from this adjusted distribution of trade profits and losses.

This does not mean that we would have made more by using the optimal f off of the equalized data. We would *not* have, as the following demonstration shows:

P&L	Percentage	Underlying Price	f$	Number of Shares	Cumulative
At $f = .09$, trading the equalized method:					$10,000
+2	.1	20	$33.33	300	$10,600
−3	−.15	20	$33.33	318	$9,646
+10	.2	50	$83.33	115.752	$10,803.52
−5	−.1	50	$83.33	129.642	$10,155.31

P&L	Percentage	Underlying Price	f$	Number of Shares	Cumulative
At $f = .17$, trading the nonequalized method:					$10,000
+2	.1	20	$29.41	340.02	$10,680.04
−3	−.15	20	$29.41	363.14	$9,590.61
+10	.2	50	$29.41	326.1	$12,851.61
−5	−.1	50	$29.41	436.98	$10,666.71

However, if all of the trades were figured off of the current price (say $100 per share), the equalized optimal f would have made more than the raw optimal f.

Which, then, is the better to use? Should we equalize our data and determine our optimal f (and its by-products), or should we just run everything as it is? This is more a matter of your beliefs than it is mathematical fact. It is a matter of what is more pertinent in the item you are trading, percentage changes or absolute changes. Is a $2 move in a $20 stock the same as a $10 move in a $100 stock? What if we are discussing dollars and euros? Is a .30-point move at .4500 the same as a .40-point move at .6000?

My personal opinion is that you are probably better off with the equalized data. Often, the matter is moot, in that if a stock has moved from $20 per share to $100 per share and we want to determine the optimal f, we want to use current data. The trades that occurred at $20 per share may not be representative of the way the stock is presently trading, regardless of whether they are equalized or not.

Generally, then, you are better off not using data where the underlying was at a dramatically different price than it presently is, as the characteristics of the way the item trades may have changed as well. In that sense, the optimal f off of the raw data and the optimal f off of the equalized data will be identical if all trades occurred at the same underlying price.

So we can state that if it does matter a great deal whether you equalize your data or not, then you're probably using too much data anyway. You've gone so far into the past that the trades generated back then probably are not very representative of the next trade. In short, we can say that it doesn't much matter whether you use equalized data or not, and if it does, there's probably a problem. If there isn't a problem, and there is a difference between using the equalized data and the raw data, you should opt for the equalized data. This does not mean that the optimal f figured off of the equalized data would have been optimal in the past. It would not have been. The optimal f figured off of the raw data would have been the optimal in the past. However, in terms of determining the as-yet-unknown answer to the question of what will be the optimal f (or closer to it tomorrow), the optimal f figured off of the equalized data makes better sense, as the equalized data is a fairer representation of the distribution of possible outcomes on the next trade.

Equations (4.10a) through (4.11) will give different answers depending upon whether the trade was initiated as a long or a short. For example, if a stock is bought at 80 and sold at 100, the percentage gain is 25. However, if a stock is sold short at 100 and covered at 80, the gain is only 20%. In both cases, the stock was bought at 80 and sold at 100, but the sequence—the chronology of these transactions—must be accounted for. As the chronology of transactions affects the distribution of percentage gains and losses, we assume that the chronology of transactions in the future will be more like the chronology in the past than not. Thus, Equations (4.10a) through (4.11) will give different answers for longs and shorts.

Of course, we could ignore the chronology of the trades (using 4.01c for longs and using the exit price in the denominator of 4.01c for shorts), but to do so would be to reduce the information content of the trade's history.

Further, the risk involved with a trade is a function of the chronology of the trade, a fact we would be forced to ignore.

FINDING OPTIMAL *f* VIA PARABOLIC INTERPOLATION

Originally, I had hoped to find a method of finding the optimal f by way of a single equation like the Kelly formula. In finding the optimal f we are looking for that value for f which generates the highest TWR in the domain 0 to 1.0 for f. Since f is the only variable we have to maximize the TWR for, we say that we are *maximizing in one dimension*.

We can use another technique to iterate to the optimal f with a little more style than the brute methods already described. Recall that in the iterative technique we bracket an intermediate point (A, B); test a point within the bracket (X); and obtain a new, smaller bracketing interval (either A, X or X, B). This process continues until the answer is converged upon. This is still brutish, but not so brutish as the simple 0 to 1 by .01 loop method.

The best (i.e., fastest and most elegant) way to find a maximum in one dimension, when you are certain that only one maximum exists, that each successive point to the left of the maximum lessens, and that each successive point to the right of the maximum lessens (as is the case with the shape of the f curve), is to use *parabolic interpolation*. When there is only one local extreme (be it a maximum or a minimum) in the range you are searching, parabolic interpolation will work. If there is more than one local extreme, parabolic interpolation will not work (see Figure 4.12).

With this technique we simply input three coordinate points. The axes of these points are the TWRs (Y axis) and the f values (X axis). We can find the abscissa (the X axis, or f value corresponding to the peak of a parabola) by the following formula, given the three coordinates:

$$\text{ABSCISSA} = X2 - .5 * \frac{(X2 - X1)^2 * (Y2 - Y3) - (X2 - X3)^2 * (Y2 - Y1)}{(X2 - X1) * (Y2 - Y3) - (X2 - X3) * (Y2 - Y1)} \quad (4.15)$$

The result returned by this equation is the value for f (or X if you will) that corresponds to the abscissa of a parabola where the three coordinates (X1, Y1), (X2, Y2), (X3, Y3) lie on the parabola.

The object now is to superimpose a parabola over the f curve, change one of the input coordinates to draw an amended parabola, and keep on doing this until the abscissa of the most recent parabola converges with the

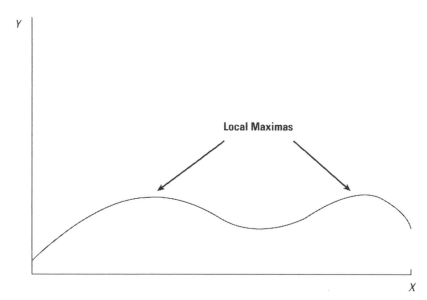

FIGURE 4.12 A function with two local extremes

previous parabola's abscissa. Convergence is determined when the absolute value of the difference between two abscissas is less than a prescribed amount called the tolerance, or TOL for short. This amount should be chosen with respect to how accurate you want your f to be. Generally, I use a value of .005 for TOL. This gives the same accuracy in searching for the optimal f as the brute force techniques described earlier.

We can start with two of the three coordinate points as $(0, 0)$, $(1.0, 0)$. The third coordinate point must be a point that lies on the actual f curve itself. Let us choose the X value here to be $1 - \text{TOL}$, or .995. To make sure that the coordinate lies on the f curve, we determine our Y value by finding what the TWR is at $f = .995$. Assume we are looking for the optimal f for the four-trade example $-1, -3, 3, 5$. For these four trades the TWR at $f = .995$ is .017722. Now we have the three coordinates: $(0, 0)$, $(.995, .017722)$, $(1.0, 0)$. We plug them into the above described equation to find the abscissa of a parabola that contains these three points, and our result is .5.

Now we compute the TWR corresponding to this abscissa; this equals 1.145833. Since the X value here now $(.5)$ is to the left of the value for X2 previously $(.995)$, we move our three points over to the left, and compute a new abscissa to the parabola that contains the three points $(0, 0)$, $(.5, 1.145833)$, $(.995, .017722)$.

This abscissa is at .499439. The TWR corresponding to this f value is 1.146363. When we encounter a difference in abscissas that is less than or equal to TOL, we will have converged to the optimal f.

Shown here are the full seven passes and the values used in each pass so that you may better understand this technique.

PARABOLIC INTERPOLATION

Pass#	x1	y1	x2	y2	x3	y3	abscissa
1	0	0	0.995	0.017722	1	0	0.5
2	0	0	0.5	1.145833	0.995	0.017722	0.499439
3	0	0	0.499439	1.146363	0.5	1.145833	0.426923
4	0	0	0.426923	1.200415	0.499439	1.146363	0.410853
5	0	0	0.410853	1.208586	0.426923	1.200415	0.387431
6	0	0	0.387431	1.218059	0.410853	1.208586	0.375727
7	0	0	0.375727	1.22172	0.387431	1.218059	0.364581
8	0	0	0.364581	1.224547	0.375727	1.22172	0.356964
9	0	0	0.356964	1.226111	0.364581	1.224547	0.350489

Convergence is extremely rapid. Typically, the more peaked the curve for the TWR (i.e., the more plays which comprise the TWR) the faster convergence is attained.

Refer now to Figure 4.13. This graphically shows the parabolic interpolation process for the coin-toss example with a 2:1 payoff, where the optimal f is .25. On the graph, notice the familiar f curve, which peaks out at .25. The first step here is to draw a parabola through three points: A, B, and C. The coordinates for A are (0, 0). For C the coordinates are (1, 0). For point B we now pick a point whose coordinates lie on the f curve itself. Once parabola ABC is drawn, we obtain its abscissa (the f value corresponding to the peak of the parabola ABC). We find what the TWR is for this f value. This gives us coordinates for point D. We repeat the process, this time drawing a parabola through points A, B, and D. Once the abscissa to parabola ABD is found, we can find the TWR that corresponds to an f value of the abscissa of parabola ABD. These coordinates (f value, TWR) give us point E.

Notice how quickly we are converging to the peak of the f curve at $f = .25$. If we were to continue with the exercise in Figure 4.12, we would next draw a parabola through points E, B, and D, and continue until we converged upon the peak of the f curve.

One potential problem with this technique from a computer standpoint is that the denominator in the equation that solves for the abscissa might

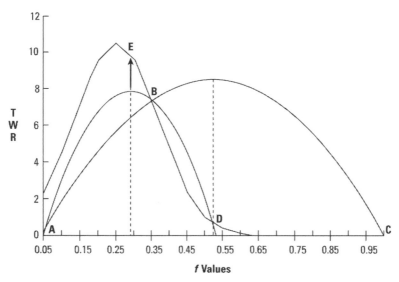

FIGURE 4.13 Parabolic interpolation performed on TWRs of 20 sequences of +2, −1

equal zero while running. One possible solution is the following fast and dirty patch in Java:

$$\text{dm} \ = \ (x2 - x1) * (y2 - y3) - (x2 - x3) * (y2 - y1);$$
$$\text{If (dm} == 0.0)$$
$$\text{dm} = .00001;$$
$$\text{abscissa} \ = \ x2 - 0.5 * (((x2 - x1) * (x2 - x1) * (y2 - y3) - (x2 - x3)$$
$$(x2 - x3) * (y2 - y1)/\text{dm};$$

This patch will not detract from the integrity of the results.

Note that this method can be used to find a local maximum for a given function, provided only one maximum exists within the range. The same technique could be used to find a local minimum for a function that opened upward (for example, the function Y equals X squared is such a function). Again, the technique will work provided there is only one local minimum (as is the case with our example). The only change from looking for a local maximum is in the equation for finding the abscissa:

$$\text{ABSCISSA} =$$
$$X2 + .5 * \frac{(X2 - X1)^2 * (Y2 - Y3) - (X2 - X3)^2 * (Y2 - Y1)}{(X2 - X1) * (Y2 - Y3) - (X2 - X3) * (Y2 - Y1)}$$

Note that here, for a local minimum, the first operator is a plus (+) sign, not a minus (−) sign as when we were looking for a local maximum.

THE NEXT STEP

The real problem with the formula to this point is that it makes the assumption that all HPRs have an equal probability of occurrence. What is needed is a new formula that allows for different probabilities associated with different HPRs. Such a formula would allow you to find an optimal f given a description of a probability distribution of HPRs. To accommodate this, we need to rework (4.06) to:

$$\text{HPR} = \left(1 + \left(\frac{A}{\left(\frac{W}{f} \right)} \right) \right)^P \tag{4.16}$$

where A = outcome of the scenario
          P = probability of the scenario
          W = worst outcome of all n scenarios
          f = value for f which we are testing

Now, we obtain the terminal wealth relative, or TWR[4], originally given by (3.03) and (4.07) to:

$$\text{TWR} = \prod_{i=1}^{T} \text{HPR}_i$$

or

$$\text{TWR} = \prod_{i=1}^{T} \left(1 + \left(\frac{A_i}{\left(\frac{W}{f} \right)} \right) \right)^{P_i} \tag{4.17}$$

Finally, if we take Equation (4.18) to the Σ_{p_i} root, we can find our average compound growth per play, also called the geometric mean HPR, and replace that given in (4.08) which will become more important later on:

$$G = \text{TWR}^{1/\Sigma_{p_i}} \tag{4.18}$$

---

[4]In this formulation, unlike the 1990 formulations, the TWR has no special meaning. In this instance, it is simply an interim value used to find G, and it does *not* represent the multiple made on our starting stake. The variable named "TWR" is maintained solely for consistency's sake.

or

$$G = \left(\prod_{i=1}^{T} \left(\left(1 + \left(\frac{A_i}{\frac{W}{f}} \right) \right)^{P_i} \right) \right)^{1/\Sigma P_i} \tag{4.18a}$$

where: T = Number of different scenarios.
          TWR = Terminal wealth relative.
          HPR_i = Holding period return of the ith scenario.
          A_i = Outcome of the ith scenario.
          P_i = Probability of the ith scenario.
          W = Worst outcome of all n scenarios.
          f = Value for f which we are testing.

Just as you could use Equation (4.04) to solve Equation (4.03), likewise you can use Equation (4.18a) to solve *any* optimal f problem. It will yield the same answers as the Kelly formulas when the data correctly has a Bernoulli distribution. It will yield the same answers as previously mentioned formulas if you pump a distribution of trades through it (where the probability of each trade is $1/T$). This formula can be used to maximize the expected value of the logarithm of any starting quantity of anything when there is exponential growth involved. We will now see how to employ this formula in the context of *scenario planning*.

SCENARIO PLANNING

People who forecast for a living, be they economists, stock market forecasters, meteorologists, government agencies, or the like, have a notorious history for incorrect forecasts. Most decisions anyone must make in life usually require that the individual make a forecast about the future.

There are a couple of pitfalls that immediately crop up. To begin with, people generally make more optimistic assumptions about the future than the actual probabilities. Most people feel that they are far more likely to win the lottery this month than they are to die in an auto accident, even though the probabilities of the latter are greater. This is true not only on the level of the individual; it is even more pronounced at the group level. When people work together, they tend to see a favorable outcome as the most likely result.

The second pitfall—and the more harmful—is that people make straight-line forecasts into the future. People predict what the price of a gallon of gas will be two years from now; they predict what will happen with their jobs, who the next president will be, what the next styles will be, and on and on. Whenever we think of the future, we tend to think in terms of a single most likely outcome. As a result, whenever we must make decisions, whether as an individual or a group, we tend to make these decisions based on what we think will be the single most likely outcome in the future. As a consequence, we are extremely vulnerable to unpleasant surprises.

Scenario planning is a partial solution to this problem. A scenario is simply a possible forecast, a story about one way that the future might unfold. Scenario planning is a collection of scenarios, to cover the spectrum of possibilities. Of course, the complete spectrum can never be covered, but the scenario planner wants to cover as many possibilities as he or she can. By acting in this manner, as opposed to using a straight-line forecast of the most likely outcome, the scenario planner can prepare for the future as it unfolds. Furthermore, scenario planning allows the planner to be prepared for what might otherwise be an unexpected event. Scenario planning is tuned to reality in that it recognizes that *certainty is an illusion.*

Suppose you are in a position where you are involved in the long-run planning for your company. Let's say you make a particular product. Rather than making a single most likely straight-line forecast, you decide to exercise scenario planning. You will need to sit down with the other planners and brainstorm for possible scenarios. What if you cannot get enough of the raw materials to make your product? What if one of your competitors fails? What if a new competitor emerges? What if you have severely underestimated demand for this product? What if a war breaks out on such and such a continent? What if it is a nuclear war?

Because each scenario is only one of several possible, each scenario can be considered seriously. But what do you do once you have defined these scenarios?

To begin with, you must determine what goal you would like to achieve for each given scenario. Depending upon the scenario, the goal need not be a positive one. For instance, under a bleak scenario, your goal may simply be damage control. Once you have defined a goal for a given scenario, you then need to draw up the contingency plans pertaining to that scenario to achieve the desired goal. For instance, in the rather unlikely bleak scenario where your goal is damage control, you need to have plans to go to should this scenario manifest itself so that you can minimize the damage. Scenario planning, above all else, provides the planner with a course of action to take should a certain scenario develop. It forces you to make plans before the fact; it forces you to be prepared for the unexpected.

Scenario planning, however, can do a lot more. There is a hand-in-glove fit between scenario planning and optimal *f*. Optimal *f* allows us to determine the optimal quantity to allocate to a given set of possible scenarios. Our existence limits us to existing in only one scenario at a time, even though we are planning for multiple futures, multiple scenarios.

Therefore, oftentimes, scenario planning puts us in a position where we must make a decision regarding how much of a resource to allocate today, given the possible scenarios of tomorrow. This is the true heart of scenario planning: quantifying it.

First, we must define each unique scenario. Second, we must assign a probability of that scenario's occurrence. Being a probability means that this number is between 0 and 1. We need not consider any further scenarios with a probability of 0. Note that these probabilities are not cumulative. In other words, the probability assigned to a given scenario is unique to that scenario. Suppose we are decision makers for XYZ Manufacturing Corporation. Two of the many scenarios we have are as follows. In one scenario, we have the probability of XYZ Manufacturing filing for bankruptcy with a probability of .15, and, in another scenario, we have XYZ being put out of business by intense foreign competition with a probability of .07. Now, we must ask if the first scenario, filing for bankruptcy, includes filing for bankruptcy due to the second scenario, intense foreign competition. If it does, then the probabilities in the first scenario must not take the probabilities of the second scenario into account, and we must amend the probabilities of the first scenario to be .08 (.15 − .07).

Just as important as the uniqueness of each probability to each scenario is that the sum of the probabilities of all of the scenarios we are considering must equal 1 exactly. They must equal not 1.01 nor .99, but 1.

For each scenario, we now have a probability of just that scenario assigned. We must now also assign an outcome result. This is a numerical value. It can be dollars made or lost as a result of a scenario's manifesting itself; it can be units of utility or medication or anything. However, our output is going to be in the same units that we put in.

You must have at least one scenario with a negative outcome in order to use this technique. This is mandatory.

A last prerequisite to using this technique is that the arithmetic mathematical expectation, the sum of all of the outcome results times their respective probabilities [Equation (1.01a)], must be greater than zero. If the arithmetic mathematical expectation equals zero or is negative, the following technique cannot be used.[5] That is not to say that scenario planning itself cannot be used. It can and should. However, optimal f can be incorporated with scenario planning only when there is a positive, mathematical expectation.

Lastly, you must try to cover as much of the spectrum of outcomes as possible. In other words, you really want to account for 99% of the possible outcomes. This may sound nearly impossible, but many scenarios can be

---

[5]However, later in the text we will be using scenario planning for portfolios, and, therein, a negative arithmetic mathematical expectation will be allowed and can possibly benefit the portfolio as a whole.

made broader so that you don't need 10,000 scenarios to cover 99% of the spectrum.

In making your scenarios broader, you must avoid the common pitfall of three scenarios: an optimistic one, a pessimistic one, and a third in which things remain the same. This is too simple, and the answers derived therefrom are often too crude to be of any value. Would you want to find your optimal *f* for a trading system based on only three trades?

So, even though there may be an unknowably large number of scenarios to cover the entire spectrum, we can cover what we believe to be about 99% of the spectrum of outcomes. If this makes for an unmanageably large number of scenarios, we can make the scenarios broader to trim down their number. However, by trimming down their number, we lose a certain amount of information. When we trim down the number of scenarios (by broadening them) to only three (a common pitfall), we have effectively eliminated so much information that the effectiveness of this technique is severely hampered.

What, then, is a good number of scenarios to have? As many as you can and still manage them.

Think of the two-to-one coin toss as a spectrum of two scenarios. Each has a probability, and that probability is .5 for each scenario, labeled heads and tails. Each has an outcome, +2 and −1, respectively:

Scenario	Probability	Outcome
Heads	.5	2
Tails	.5	−1

Assume again that we are decision making for XYZ. We are looking at marketing a new product of ours in a primitive, remote little country. Assume we have five possible scenarios we are looking at (in reality, you would have many more than this, but we'll use five for the sake of simplicity). These five scenarios portray what we perceive as possible futures for this primitive remote country, their probabilities of occurrence, and the gain or loss of investing there.

Scenario	Probability	Result
War	.1	−$500,000
Trouble	.2	−$200,000
Stagnation	.2	0
Peace	.45	$500,000
Prosperity	.05	$1,000,000
Sum	1.00	

The sum of our probabilities equals 1. We have at least one scenario with a negative result, and our mathematical expectation is positive:

$$(.1 * -500,000) + (.2 * -200,000) + \dots \text{etc.} = 185,000$$

We can, therefore, use the technique on this set of scenarios.

Notice first, however, that if we used the single most likely outcome method, we would conclude that peace will be the future of this country, and we would then act as though peace were to occur, as though it were a certainty, only vaguely remaining aware of the other possibilities.

Returning to the technique, we must determine the optimal f. The optimal f is that value for f (between zero and one) which maximizes the geometric mean, using Equations (4.16 to 4.18). Now, we obtain the terminal wealth relative, or TWR using Equation (4.17). Finally, if we take Equation (4.17) to the Σ_{p_i} root, we can find our average compound growth per play, also called the geometric mean HPR, which will become more important later on. We use Equation (4.18) for this.

Here is how to perform these equations. To begin with, we must decide on an optimization scheme, a way of searching through the f values to find that f which maximizes our equation. Again, we can do this with a straight loop with f from .01 to 1, through iteration, or through parabolic interpolation.

Next, we must determine the worst possible result for a scenario among all of the scenarios we are looking at, regardless of how small the probability of that scenario's occurrence are. In the example of XYZ Corporation, this is −$500,000.

Now, for each possible scenario, we must first divide the worst possible outcome by negative f. In our XYZ Corporation example, we will assume that we are going to loop through f values from .01 to 1. Therefore, we start out with an f value of .01. Now, if we divide the worst possible outcome of the scenarios under consideration by the negative value for f, we get the following:

$$\frac{-\$500,000}{-.01} = 50,000,000$$

Notice how negative values divided by negative values yield positive results, and vice versa. Therefore, our result in this case is positive. Now, as we go through each scenario, we will divide the outcome of the scenario by the result just obtained. Since the outcome to the first scenario is also the worst scenario—a loss of $500,000—we now have:

$$\frac{-\$500,000}{50,000,000} = -.01$$

The next step is to add this value to 1. This gives us:

$$1 + (-.01) = .99$$

Last, we take this answer to the power of the probability of its occurrence, which in our example is .1:

$$.99^{.1} = .9989954713$$

Next, we go to the next scenario labeled *Trouble*, where there is a .2 loss of $200,000. Our worst-case result is still −$500,000. The f value we are working on is still .01, so the value we want to divide this scenario's result by is still 50 million:

$$\frac{-200,000}{50,000,000} = -.004$$

Working through the rest of the steps to obtain our HPR:

$$1 + (-.004) = .996$$
$$.996^{.2} = .9991987169$$

If we continue through the scenarios for this test value of .01 for f, we will find the three HPRs corresponding to the last three scenarios:

Stagnation	1.0
Peace	1.004487689
Prosperity	1.000990622

Once we have turned each scenario into an HPR for the given f value, we must multiply these HPRs together:

$$
\begin{array}{r}
.9989954713 \\
*\ \ .9991987169 \\
*\ 1.0 \\
*\ 1.004487689 \\
*\ 1.000990622 \\
\hline
1.003667853
\end{array}
$$

This gives us the interim TWR, which in this case is 1.003667853. Our next step is to take this to the power of 1 divided by the sum of the probabilities. Since the sum of the probabilities will always equal 1 the way we are calculating this, we can state that we must raise the TWR to the power of 1 to give us the geometric mean. Since anything raised to the power of 1 equals itself, we can say that, in this case, our geometric mean equals the TWR. We therefore have a geometric mean of 1.003667853.

The answer we have just obtained in our example is our geometric mean corresponding to an f value of .01. Now we move on to an f value of .02, and repeat the whole process until we have found the geometric mean corresponding to an f value of .02. We will proceed as such until we arrive at that value for f which yields the highest geometric mean.

In the case of our example, we find that the highest geometric mean is obtained at an f value of .57, which yields a geometric mean of 1.1106. Dividing our worst possible outcome to a scenario ($-500,000$) by the negative optimal f yields a result of $877,192.35. In other words, if XYZ Corporation wants to commit to marketing this new product in this remote country, they will optimally commit this amount to this venture at this time. As time goes by and things develop, the scenarios, their resultant outcomes, and probabilities will likewise change. This f amount will then change as well. The more XYZ Corporation keeps abreast of these changing scenarios, as well as the more accurate the scenarios they develop as input are, the more accurate their decisions will be. Note that if XYZ Corporation cannot commit this $877,192.35 to this undertaking at this time, then they are too far beyond the peak of the f curve. It is the equivalent of the guy who has too many commodity contracts with respect to what the optimal f says he should have. If XYZ Corporation commits more than this amount to this project at this time, the situation would be analogous to a commodity trader with too few contracts.

There is an important point to note about scenarios and trading. What you use for a scenario can be any of a number of things:

1. It can be, as in the previous example, the outcomes that a given trade may take. This is useful if you are trading only one item. However, when you trade a portfolio of items, you violate the rule that all holding period lengths must be uniform.

2. If you know what the distribution of price outcomes will be, you can use that for scenarios. For example, suppose you have reason to believe that prices for the next day for a given item are normally distributed. Therefore, you can discern your scenarios based on the normal distribution. For example, in the normal distribution, 97.72% of the time, prices will not exceed 2 standard deviations to the upside, and 99.86% of the time they will not exceed 3 standard deviations to the upside. Therefore, as one scenario, you can have as the result something

between 2 and 3 standard deviations in price to the upside (whatever dollar amount that would be to you trading one unit over the next day, holding period), whose probability would be .9986 − .9772 = .0214, or 2.14% probability.

3. You can use the distributions of possible monetary outcomes for trading one unit with the given market approach over the next holding period. This is my preferred method, and it lends itself well to portfolio construction under the new framework.

Although I strongly recommend using the third item from the preceding list, whichever method you use, remember that *you want to be constantly updating your scenarios, their outcomes, and the probability of occurrences as conditions change. Then, you always want to go into the next holding period with what the formulas presently tell you is optimal.* The situation is analogous to that of a blackjack player. As the composition of the deck changes with each card drawn, so, too, do the player's probabilities. However, he must always adjust to what the probabilities currently dictate.

Although the quantity discussed here is a quantity of money, it can be a quantity of anything and the technique is just as valid.

If you create different scenarios for the stock market, the optimal *f* derived from this methodology will give you the correct percentage to be invested in the stock market at any given time. For instance, if the *f* returned is .65, then that means that 65% of your equity should be in the stock market, with the remaining 35% in, say, cash. This approach will provide you with the greatest geometric growth of your capital in a long-run sense. Of course, again, the output is only as accurate as the input you have provided the system with in terms of scenarios, their probabilities of occurrence, and resultant payoffs and costs.

This same process can be used as an alternative parametric technique for determining the optimal *f* for a given trade. Suppose you are making your trading decisions based on fundamentals. You could, if you wanted, outline the different scenarios that the trade may take. The more scenarios, and the more accurate the scenarios, the more accurate your results would be. Let's say you are looking to buy a municipal bond for income, but you're not planning on holding the bond to maturity. You could outline numerous different scenarios of how the future might unfold. Now, you can use these scenarios to determine how much to invest in this particular bond issue.

Suppose a trader is presented with a decision to buy soybeans. He may be using Elliot Wave, he may be using weather forecasts, but whatever he is using, let's say he can discern the following scenarios for this potential trade:

Scenario	Probability	Result
Best-case outcome	.05	150/cent bushel (profit)
Quite likely	.4	10/cent bushel (profit)
Typical	.45	−5/cent bushel (loss)
Not good	.05	−30/cent bushel (loss)
Disastrous	.05	−150/cent bushel (loss)

Now, when our Elliot Wave soybean trader (or weather forecaster soybean trader) paints this set of scenarios, this set of possible outcomes to this trade, and, in order to maximize his long-run growth (and survival), assumes that he must make this same trading decision an infinite number of times into the future, he will find, using this scenario planning approach, that optimally he should bet .02 (2%) of his stake on this trade. This translates into putting on one soybean contract for every $375,000 in equity, since the scenario with the largest loss, −150/cent bushel, divided by the optimal f for this scenario set, .02, results in $7,500/.02 = $375,000. Thus, at one contract for every $375,000 in equity, the trader can be said to be risking 2% of his stake on the next trade.

For each trade, regardless of the basis the trader uses for making the trade (i.e., Elliot Wave, weather, etc.), the scenario parameters may change. Yet the trader must maximize the long-run geometric growth of his account by assuming that the same scenario parameters will be infinitely repeated. Otherwise, the trader pays a severe price. Notice in our soybean trader example, if the trader were to go *to the right of the peak of the f curve* (that is, have slightly too many contracts), he gains no benefit. In other words, if our soybean trader were to put on one contract for every $300,000 in account equity, he would actually make less money in the long run than putting on one contract for every $375,000.

When we are presented with a decision in which there is a different set of scenarios for each facet of the decision, selecting the scenario whose geometric mean corresponding to its optimal f is greatest will maximize our decision in an asymptotic sense.

For example, suppose we are presented with a decision that involves two possible choices. It could have many possible choices, but for the sake of simplicity we will say it has two possible choices, which we will call "white" and "black." If we choose the decision labeled white, we determine that it will present the possible future scenarios to us:

Scenario	Probability	Result
A	.3	−20
B	.4	0
C	.3	30

Mathematical expectation = $3.00
Optimal f= .17
Geometric mean = 1.0123

It doesn't matter what these scenarios are, they can be anything. To further illustrate this, they will simply be assigned letters, A, B, C in this discussion. Further, it doesn't matter what the result is; it can be just about anything.

Our analysis determines that the black decision will present the following scenarios:

Scenario	Probability	Result
A	.3	−10
B	.4	5
C	.15	6
D	.15	20

Mathematical expectation = $2.90
Optimal $f = .31$
Geometric mean = 1.0453

Many people would opt for the white decision, since it is the decision with the higher mathematical expectation. With the white decision, you can expect, *on average*, a $3.00 gain versus black's $2.90 gain. Yet the black decision is actually the correct decision because it results in a greater geometric mean. With the black decision, you would expect to make 4.53% (1.0453 − 1) *on average* as opposed to white's 1.23% gain. When you consider the effects of reinvestment, the black decision makes more than three times as much, on average, as does the white decision!

The reader may protest at this point that, "We're not doing this thing over again; we're only doing it once. We're not reinvesting back into the same future scenarios here. Won't we come out ahead if we always select the highest arithmetic mathematical expectation for each set of decisions that present themselves to us?"

The only time we want to be making decisions based on greatest arithmetic mathematical expectation is if we are planning on not reinvesting the money risked on the decision at hand. Since, in almost every case, the money risked on an event today will be risked again on a different event in the future, and money made or lost in the past affects what we have available to risk today, we should decide, based on geometric mean, to maximize the long-run growth of our money. Even though the scenarios that present themselves tomorrow won't be the same as those today, by always deciding based on greatest geometric mean, we are maximizing our decisions. It is analogous to a dependent trials process, like a game of blackjack. In each hand, the probabilities change and, therefore, the optimal fraction to bet changes as well. By always betting what is optimal for that hand, however, we maximize our long-run growth. Remember that, to maximize long-run growth, we must look at the current contest as one that expands infinitely into the future. In other words, we must look at each individual event as though we were to play it an infinite number of times if we wanted to maximize growth over many plays of different contests.

As a generalization, whenever the outcome of an event has an effect on the outcome(s) of subsequent event(s), we are better off to maximize for greatest geometric expectation. In the rare cases where the outcome of an event has no effect on subsequent events, we are then better off to maximize for greatest arithmetic expectation.

Mathematical expectation (arithmetic) does not take the dispersion between the outcomes of the different scenarios into account and, therefore, can lead to incorrect decisions when reinvestment is considered.

Using this method of scenario planning gets you quantitatively positioned with respect to the possible scenarios, their outcomes, and the likelihood of their occurrence. The method is inherently more conservative than positioning yourself per the greatest arithmetic mathematical expectation. The geometric mean of a data set is never greater than the arithmetic mean. Likewise, this method can never have you position yourself (have a greater commitment) otherwise than selecting by the greatest arithmetic mathematical expectation would. In the asymptotic sense (the long-run sense), this is not only the superior method of positioning yourself as it achieves greatest geometric growth; it is also a more conservative one than positioning yourself per the greatest arithmetic mathematical expectation.

Since reinvestment is almost always a fact of life (except on the day before you retire)—that is, you reuse the money that you are using today—we must make today's decision under the assumption that the same decision will present itself a thousand times over, in order to maximize the results of our decision. We must make our decisions and position ourselves in order to maximize geometric expectation. Further, since the outcomes of most events do, in fact, have an effect on the outcomes of subsequent events, we should make our decisions and position ourselves based on maximum geometric expectation. This tends to lead to decisions and positions that are not always obvious.

Note that we have created our own binned distribution in creating our scenarios here. Similarly, if we know the distributional form of the data, we can use that and the probabilities associated, with that distribution with this technique for finding the optimal f. Such techniques we call "parametric techniques," as opposed to the "Empirical Techniques" described prior to this section in the text. The Scenario Planning Approach, as described here, where we create the data bins from empirical data, being therefore a hybrid approach between an empirical means of determining optimal f and a parametric one.

SCENARIO SPECTRUMS

We now must become familiar with the notion of a *scenario spectrum*. A scenario spectrum is a set of scenarios, aligned in succession, left to right, from worst outcome to best, which range in probability from 0% to 100%. For example, consider the scenario spectrum for a simple coin toss whereby we lose on heads and win on tails, and both have a .5 probability of occurrence (Figure 4.14).

A scenario spectrum can have more than two scenarios—you can have as many scenarios as you like (see Figure 4.15).

This scenario spectrum corresponds to the following scenarios, taken from the previous section pertaining to XYZ Manufacturing Corporation's assessment of marketing a new product in a remote little country:

Scenario	Probability	Result	Prob × Result
War	.1	−$500,000	−$50,000
Trouble	.2	−$200,000	−$40,000
Stagnation	.2	$0	$0
Peace	.45	$500,000	$225,000
Prosperity	.05	$1,000,000	$50,000
	Sum 1.00		Expectation $185,000

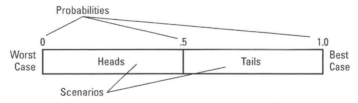

FIGURE 4.14 Scenario spectrum for a simple coin toss in which tails wins

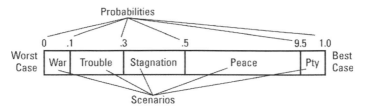

FIGURE 4.15 Scenario spectrum with multiple scenarios

Notice that this is a *valid scenario spectrum* since:

A. There is at least one scenario with a negative result.
B. The sum of the probabilities equals 1.00.
C. The scenarios within the spectrum do not overlap.

For example the stagnation scenario implies peace. However, the stagnation scenario implies peace with zero economic growth. The peace scenario is separate and apart from this, and implies peace with at least some economic growth. In other words, the stagnation scenario is *not* encapsulated in the peace scenario, nor is any scenario encapsulated in another.

One last point about scenario spectrums, and this is very important: All scenarios within a given spectrum must pertain to outcomes of a given holding period. Again, the length of the holding period can be any length you choose—it can be one day, one week, quarter, month, year, whatever, but the holding period must be decided upon. Once decided upon, all scenarios in a given spectrum must pertain to possible outcomes over the *next* holding period, and all scenario spectrums must be for the same length holding period. This is critical. Thus, if you decide upon one day for the holding period length, then all of your scenarios in all of your scenario spectrums must pertain to possible outcomes for the next day.

CHAPTER 5

Characteristics of Optimal f

How does a very small account, an account that is going to start out trading one contract, use the optimal f approach? One suggestion is that such an account start out by trading one contract not for every optimal f amount in dollars (biggest loss/$-f$), but rather that the drawdown and margin must be considered in the initial phase. The amount of funds allocated toward the first contract should be the greater of the optimal f amount in dollars or the margin plus the maximum historic drawdown (on a one-unit basis):

$$A = MAX \{(Biggest\ Loss/-f), (Margin + ABS(Drawdown))\} \qquad (5.01)$$

where: $A =$ The dollar amount to allocate to the first contract.
       $f =$ The optimal f (0 to 1).
   Margin $=$ The initial speculative margin for the given contract.
Drawdown $=$ The historic maximum drawdown.
  MAX$\{\ \} =$ The maximum value of the bracketed values.
  ABS$(\ ) =$ The absolute value function.

With this procedure an account can experience the maximum drawdown again and still have enough funds to cover the initial margin on another

175

trade. Although we cannot expect the worst-case drawdown in the future not to exceed the worst-case drawdown historically, it is rather unlikely that we will start trading right at the beginning of a new historic drawdown.

A trader utilizing this idea will then subtract the amount in Equation (5.01) from his or her equity each day. With the remainder, he or she will then divide by (Biggest Loss/−f). The answer obtained will be rounded down to the integer, and 1 will be added. The result is how many contracts to trade.

An example may help clarify. Suppose we have a system where the optimal f is .4, the biggest historical loss is −\$3,000, the maximum drawdown was −\$6,000, and the margin is \$2,500. Employing Equation (5.01) then:

$$A = \text{MAX}\{(-\$3,000/-.4), (\$2,500 + \text{ABS}(-\$6,000))\}$$
$$= \text{MAX}\{(\$7,500), (\$2,500 + \$6,000)\}$$
$$= \text{MAX}\{\$7,500, \$8,500)\}$$
$$= \$8,500$$

We would thus allocate \$8,500 for the first contract. Now suppose we are dealing with \$22,500 in account equity. We therefore subtract this first contract allocation from the equity:

$$\$22,500 - \$8,500 = \$14,000$$

We then divide this amount by the optimal f in dollars:

$$\$14,000/\$7,500 = 1.867$$

Then we take this result down to the integer:

$$\text{INT}(1.867) = 1$$

and add 1 to the result (the one contract represented by the \$8,500 we have subtracted from our equity):

$$1 + 1 = 2$$

We therefore would trade two contracts. If we were just trading at the optimal f level of one contract for every \$7,500 in account equity, we would have traded three contracts (\$22,500/\$7,500). As you can see, this technique can be utilized no matter how large an account's equity is (yet the larger the equity, the closer the two answers will be). Further, the larger the equity, the less likely it is that we will eventually experience a drawdown that will have us eventually trading only one contract. For smaller accounts, or for accounts just starting out, this is a good idea to employ.

THRESHOLD TO GEOMETRIC

Here is another good idea for accounts just starting out, one that may not be possible if you are employing the technique just mentioned. This technique makes use of another by-product calculation of optimal f called the *threshold to geometric.* The by-products of the optimal f calculation include calculations, such as the TWR, the geometric mean, and so on, that were derived in obtaining the optimal f, and that tell us something about the system. The threshold to the geometric is another of these by-product calculations. Essentially, *the threshold to geometric tells us at what point we should switch over to fixed fractional trading, assuming we are starting out constant-contract trading.*

Refer back to the example of a coin toss where we win $2 if the toss comes up heads and we lose $1 if the toss comes up tails. We know that our optimal f is .25, or to make one bet for every $4 we have in account equity. If we are starting out trading on a constant-contract basis, we know we will average $.50 per unit per play. However, if we start trading on a fixed fractional basis, we can expect to make the geometric average trade of $.2428 per unit per play.

Assume we start out with an initial stake of $4, and therefore we are making one bet per play. Eventually, when we get to $8, the optimal f would have us step up to making two bets per play. However, two bets times the geometric average trade of $.2428 is $.4856. Wouldn't we be better off sticking with one bet at the equity level of $8, whereby our expectation per play would still be $.50? The answer is "Yes." The reason is that the optimal f is figured on the basis of contracts that are infinitely divisible, which may not be the case in real life.

We can find that point where we should move up to trading two contracts by the formula for the threshold to the geometric, T:

$$T = AAT/GAT * \text{Biggest Loss}/-f \qquad (5.02)$$

where: T = The threshold to the geometric.
AAT = The arithmetic average trade.
GAT = The geometric average trade.
f = The optimal f (0 to 1).

In our example of the 2-to-1 coin toss:

$$T = .50/.2428 * -1/-.25$$
$$= 8.24$$

Therefore, we are better off switching up to trading two contracts when our equity gets to $8.24 rather than $8. Figure 5.1 shows the threshold to the

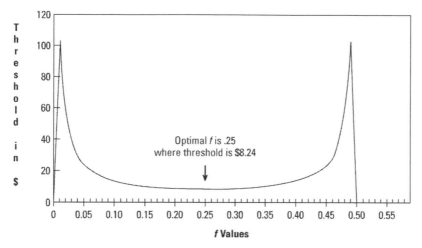

FIGURE 5.1   Threshold to the geometric for 2:1 coin toss

geometric for a game with a 50% chance of winning $2 and a 50% chance of losing $1.

Notice that the trough of the threshold to the geometric curve occurs at the optimal f. This means that since the threshold to the geometric is the optimal level of equity to go to trading two units, you go to two units at the lowest level of equity, optimally, when incorporating the threshold to the geometric at the optimal f.

Now the question is, "Can we use a similar approach to know when to go from two cars to three cars?" Also, "Why can't the unit size be 100 cars starting out, assuming you are starting out with a large account, rather than simply a small account starting out with one car?" To answer the second question first, it is valid to use this technique when starting out with a unit size greater than one. However, it is valid only if you do *not* trim back units on the downside before switching into the geometric mode. The reason is that before you switch into the geometric mode you are assumed to be trading in a constant-unit size.

Assume you start out with a stake of 400 units in our 2-to-1 coin-toss game. Your optimal f in dollars is to trade one contract (make one bet) for every $4 in equity. Therefore, you will start out trading 100 contracts (making 100 bets) on the first trade. Your threshold to the geometric is at $8.24, and therefore you would start trading 101 contracts at an equity level of $404.24. You can convert your threshold to the geometric, which is computed on the basis of advancing from one contract to two, as:

$$\text{Converted T} = \text{EQ} + \text{T} - (\text{Biggest Loss}/-f) \qquad (5.02a)$$

where: EQ = The starting account equity level.
          T = The threshold to the geometric for going from one car
             to two.
          f = The optimal f (0 to 1).

Therefore, since your starting account equity is \$400, your T is \$8.24, your biggest loss −\$1, and your f is .25:

$$\text{Converted T} = 400 + 8.24 - (-1/-.25)$$
$$= 400 + 8.24 - 4$$
$$= 404.24$$

Thus, you would progress to trading 101 contracts (making 101 bets) if and when your account equity reached \$404.24. We will assume you are trading in a constant-contract mode until your account equity reaches \$404.24, at which point you will begin the geometric mode. Therefore, until your account equity reaches \$404.24, you will trade 100 contracts on the next trade regardless of the remaining equity in your account. If, after you cross the geometric threshold (that is, after your account equity hits \$404.24), you suffer a loss and your equity drops below \$404.24, you will go back to trading on a constant 100-contract basis if and when you cross the geometric threshold again.

This inability to trim back contracts on the downside when you are below the geometric threshold is the drawback to using this procedure when you are at an equity level of trading more than two contacts. If you are only trading one contract, the geometric threshold is a very valid technique for determining at what equity level to start trading two contracts (since you cannot trim back any further than one contract should you experience an equity decline). However, it is not a valid technique for advancing from two contracts to three, because the technique is predicated upon the fact that you are currently trading on a constant-contract basis. That is, if you are trading two contracts, unless you are willing not to trim back to one contract if you suffer an equity decline, the technique is not valid, and likewise if you start out trading 100 contracts. You could do just that (not trim back the number of contracts you are presently trading if you experience an equity decline), in which case the threshold to the geometric, or its converted version in Equation (5.02a), would be the valid equity point to add the next contract. The problem with doing this (not trimming back on the downside) is that you will make less (your TWR will be less) in an asymptotic sense. You will not make as much as if you simply traded the full optimal f. Further, your drawdowns will be greater and your risk of ruin higher. Therefore, the threshold to the geometric is beneficial only if you are starting out in the lowest denomination of bet size (one contract) and advancing to two, and it is a benefit only if the arithmetic average trade is more than twice the size of the geometric average trade. Furthermore, it is beneficial to use only when you cannot trade fractional units. In Chapter 10 we will see that the concept of a "Threshold" to the geometric is a precursor to the larger notion of Continuous Dominance.

ONE COMBINED BANKROLL VERSUS SEPARATE BANKROLLS

Some very important points regarding fixed fractional trading must be covered before we discuss the parametric techniques. First, when trading more than one market system simultaneously, you will generally do better in an asymptotic sense using only one combined bankroll from which to figure your contract sizes, rather than separate bankrolls for each.

It is for this reason that we "recapitalize" the subaccounts on a daily basis as the equity in an account fluctuates. What follows is a run of two similar systems, System A and System B. Both have a 50% chance of winning, and both have a payoff ratio of 2:1. Therefore, the optimal f dictates that we bet $1 for every $4 units in equity. The first run we see shows these two systems with positive correlation to each other. We start out with $100, splitting it into two subaccount units of $50 each. After a trade is registered, it affects only the cumulative column for that system, as each system has its own separate bankroll. The size of each system's separate bankroll is used to determine bet size on the subsequent play:

	System A			System B	
Trade	**P&L**	**Cumulative**	**Trade**	**P&L**	**Cumulative**
		50.00			50.00
2	25.00	75.00	2	25.00	75.00
−1	−18.75	56.25	−1	−18.75	56.25
2	28.13	84.38	2	28.13	84.38
−1	−21.09	63.28	−1	−21.09	63.28
2	31.64	94.92	2	31.64	94.92
−1	−23.73	71.19	−1	−23.73	71.19
		−50.00			−50.00
Net Profit		21.19140			21.19140
	Total net profit of the two banks =				$42.38

Now we will see the same thing, only this time we will operate from a combined bank starting at 100 units. Rather than betting $1 for every $4 in the combined stake for each system, we will bet $1 for every $8 in the combined bank. Each trade for either system affects the combined bank, and it is the combined bank that is used to determine bet size on the subsequent play:

System A		System B		
Trade	**P&L**	**Trade**	**P&L**	**Combined Bank**
				100.00
2	25.00	2	25.00	150.00
−1	−18.75	−1	−18.75	112.50
2	28.13	2	28.13	168.75
−1	−21.09	−1	−21.09	126.56
2	31.64	2	31.64	189.84
−1	−23.73	−1	−23.73	142.38
				−100.00
	Total net profit of the combined bank =			$42.38

Notice that using either a combined bank or a separate bank in the preceding example shows a profit on the $100 of $42.38. Yet what was shown is the case where there is positive correlation between the two systems. Now we will look at negative correlation between the same two systems, first with both systems operating from their own separate bankrolls:

System A			System B		
Trade	**P&L**	**Cumulative**	**Trade**	**P&L**	**Cumulative**
		50.00			50.00
2	25.00	75.00	−1	−12.50	37.50
−1	−18.75	56.25	2	18.75	56.25
2	28.13	84.38	−1	−14.06	42.19
−1	−21.09	63.28	2	21.09	63.28
2	31.64	94.92	−1	−15.82	47.46
−1	−23.73	71.19	2	23.73	71.19
		−50.00			−50.00
Net Profit		21.19140			21.19140
	Total net profit of the two banks =				$42.38

As you can see, when operating from separate bankrolls, both systems net out making the same amount regardless of correlation. However, with the combined bank:

System A		System B		
Trade	P&L	Trade	P&L	Combined Bank
				100.00
2	25.00	−1	−12.50	112.50
−1	−14.06	2	28.12	126.56
2	31.64	−1	−15.82	142.38
−1	−17.80	2	35.59	160.18
2	40.05	−1	−20.02	180.20
−1	−22.53	2	45.00	202.73
				−100.00
	Total net profit of the combined bank =			$102.73

With the combined bank, the results are dramatically improved. *When using fixed fractional trading you are best off operating from a single combined bank.*

TREAT EACH PLAY AS IF INFINITELY REPEATED

The next axiom of fixed fractional trading regards maximizing the current event as though it were to be performed an infinite number of times in the future. We have determined that for an independent trials process, you should always bet that f which is optimal (and constant) and likewise when there is dependency involved, only with dependency f is not constant.

Suppose we have a system where there is dependency in like begetting like, and suppose that this is one of those rare gems where the confidence limit is at an acceptable level for us, that we feel we can safely assume that there really is dependency here. For the sake of simplicity we will use a payoff ratio of 2:1. Our system has shown that, historically, if the last play was a win, then the next play has a 55% chance of being a win. If the last play was a loss, our system has a 45% chance of the next play's being a loss. Thus, if the last play was a win, then from the Kelly formula,

Equation (4.03), for finding the optimal f (since the payoff ratio is Bernoulli distributed):

$$f = ((2+1) * .55 - 1)/2$$
$$= (3 * .55 - 1)/2$$
$$= .65/2$$
$$= .325$$

After a losing play, our optimal f is:

$$f = ((2+1) * .45 - 1)/2$$
$$= (3 * .45 - 1)/2$$
$$= .35/2$$
$$= .175$$

Now dividing our biggest losses (-1) by these negative optimal fs dictates that we make one bet for every 3.076923077 units in our stake after a win, and make one bet for every 5.714285714 units in our stake after a loss. In so doing we will maximize the growth over the long run. Notice that we treat each individual play as though it were to be performed an infinite number of times.

Notice in this example that betting after both the wins and the losses still has a positive mathematical expectation individually. What if, after a loss, the probability of a win was .3? In such a case, the mathematical expectation is negative, hence there is no optimal f and as a result you shouldn't take this play:

$$ME = (.3 * 2) + (.7 * -1)$$
$$= .6 - .7$$
$$= -.1$$

In such circumstances, you would bet the optimal amount only after a win, and you would not bet after a loss. If there is dependency present, you must segregate the trades of the market system based upon the dependency and treat the segregated trades as separate market systems.

The same principle, namely that *asymptotic growth is maximized if each play is considered to be performed an infinite number of times into the future*, also applies to simultaneous wagering (or trading a portfolio). Consider two betting systems, A and B. Both have a 2:1 payoff ratio, and both win 50% of the time. We will assume that the correlation coefficient between the two systems is zero, but that is not relevant to the point being illuminated here. The optimal fs for both systems (if they were being traded alone, rather than simultaneously) are .25, or to make one bet for every four units in equity. The optimal fs for trading both systems simultaneously are .23, or one bet for every 4.347826087 units in account equity.[1] System B trades only two-thirds of the time, so some trades will be done when the two systems are not trading simultaneously. This first sequence is demonstrated with a starting combined bank of 1,000 units, and each bet for each system is performed with an optimal f of one bet per every 4.347826087 units:

	A		B	Combined Bank
				1,000.00
−1	−230.00			770.00
2	354.20	−1	−177.10	947.10
−1	−217.83	2	435.67	1,164.93
2	535.87			1,700.80
−1	−391.18	−1	−391.18	918.43
2	422.48	2	422.48	1,763.39

Next, we see the same exact thing, the only difference being that when A is betting alone (i.e., when B does not have a bet at the same time as A), we make one bet for every four units in the combined bank for System A, since that is the optimal f on the single, individual play. On the plays where the bets are simultaneous, we are still betting one unit for every 4.347826087 units in account equity for both A and B. Notice that in so doing we are taking each bet, whether it is individual or simultaneous, and applying that optimal f which would maximize the play as though it were to be performed an infinite number of times in the future.

	A		B	Combined Bank
				1,000.00
−1	−250.00			750.00
2	345.00	−1	−172.50	922.50
−1	−212.17	2	424.35	1,134.67
2	567.34			1,702.01
−1	−391.46	−1	−391.46	919.09
2	422.78	2	422.78	1,764.65

As can be seen, there is a slight gain to be obtained by doing this and the more trades that elapse, the greater the gain. Although we are not yet

---

[1]The method we are using here to arrive at these optimal bet sizes is described later in the text in Chapter 9.

discussing multiple simultaneous plays (i.e., "portfolios"), we invoke them here to illuminate the point. The same principle applies to trading a portfolio where not all components of the portfolio are in the market all the time. You should trade at the optimal levels for the combination of components (or single component) that results in the optimal growth as though that combination of components (or single component) were to be traded an infinite number of times in the future.

EFFICIENCY LOSS IN SIMULTANEOUS WAGERING OR PORTFOLIO TRADING

Let's again return to our 2:1 coin-toss game. Let's again assume that we are going to play two of these games, which we'll call System A and System B, simultaneously and that there is zero correlation between the outcomes of the two games. We can determine our optimal *f*s for such a case as betting one unit for every 4.347826 in account equity when the games are played simultaneously. When starting with a bank of 100 units, notice that we finish with a bank of 156.86 units:

	System A		System B		
	Trade	P&L	Trade	P&L	Bank
Optimal *f* is 1 unit for every 4.347826 in equity:					
					100.00
	−1	−23.00	−1	−23.00	54.00
	2	24.84	−1	−12.42	66.42
	−1	−15.28	2	30.55	81.70
	2	37.58	2	37.58	156.86

Now let's consider System C. This would be the same as Systems A and B, only we're going to play this game alone, without another game going simultaneously. We're also going to play it for eight plays—as opposed to the previous endeavor, where we played two games for four simultaneous plays. Now our optimal *f* is to bet one unit for every four units in equity. What we have is the same eight outcomes as before, but a different, better end result:

	System C		
	Trade	**P&L**	**Bank**
Optimal f is 1 unit for every 4.00 in equity:			
			100.00
	−1	−25.00	75.00
	2	37.50	112.50
	−1	−28.13	84.38
	2	42.19	126.56
	2	63.28	189.84
	2	94.92	284.77
	−1	−71.19	213.57
	−1	−53.39	160.18

The end result here is better not because the optimal fs differ slightly (both are at their respective optimal levels), but because there is a small efficiency loss involved with simultaneous wagering. *This inefficiency is the result of not being able to recapitalize your account after every single wager as you could betting only one market system.* In the simultaneous two-bet case, you can recapitalize only three times, whereas in the single eight-bet case you recapitalize seven times. Hence, the efficiency loss in simultaneous wagering (or in trading a portfolio of market systems).

We just witnessed the case where the simultaneous bets were not correlated. Let's look at what happens when we deal with positive (+1.00) correlation:

	System A		System B		
	Trade	**P&L**	**Trade**	**P&L**	**Bank**
Optimal f is 1 unit for every 8.00 in equity:					
					100.00
	−1	−12.50	−1	−12.50	75.00
	2	18.75	2	18.75	112.50
	−1	−14.06	−1	−14.06	84.38
	2	21.09	2	21.09	126.56

Notice that after four simultaneous plays where the correlation between the market systems employed is +1.00, the result is a gain of 126.56 on a starting stake of 100 units. This equates to a TWR of 1.2656, or a geometric

mean, a growth factor per play (even though these are combined plays) of
$1.2656 \wedge (1/4) = 1.06066$.

Now refer back to the single-bet case. Notice here that after four plays, the outcome is 126.56, again on a starting stake of 100 units. Thus, the geometric mean of 1.06066. This demonstrates that the rate of growth is the same when trading at the optimal fractions for perfectly correlated markets. As soon as the correlation coefficient comes down below +1.00, the rate of growth increases. Thus, we can state that *when combining market systems, your rate of growth will never be any less than with the single-bet case, no matter how high the correlations are, provided that the market system being added has a positive arithmetic mathematical expectation.*

Recall the first example in this section, where there were two market systems that had a zero correlation coefficient between them. This market system made 156.86 on 100 units after four plays, for a geometric mean of $(156.86/100) \wedge (1/4) = 1.119$. Let's now look at a case where the correlation coefficients are −1.00. Since there is never a losing play under the following scenario, the optimal amount to bet is an infinitely high amount (in other words, bet one unit for every infinitely small amount of account equity). But, rather than getting that greedy, we'll just make one bet for every four units in our stake so that we can make the illustration here:

	System A		System B		
	Trade	P&L	Trade	P&L	Bank
Optimal *f* is 1 unit for every 0.00 in equity (shown is 1 for every 4):					
					100.00
	−1	−12.50	2	25.00	112.50
	2	28.13	−1	−14.06	126.56
	−1	−15.82	2	31.64	142.38
	2	35.60	−1	−17.80	160.18

There are two main points to glean from this section. The first is that there is a small efficiency loss with simultaneous betting or portfolio trading, a loss caused by the inability to recapitalize after every individual play. The second point is that combining market systems, provided they have a positive mathematical expectation, and even if they have perfect positive correlation, never decreases your total growth per time period. However, as you continue to add more and more market systems, the efficiency loss becomes considerably greater. If you have, say, 10 market systems and they all suffer a loss simultaneously, that loss could be terminal to the account, since you have not been able to trim back size for each loss as you would have had the trades occurred sequentially.

Therefore, we can say that there is a gain from adding each new market system to the portfolio provided that the market system has a correlation coefficient less than one and a positive mathematical expectation, or a negative expectation but a low enough correlation to the other components in the portfolio to more than compensate for the negative expectation. There is a marginally decreasing benefit to the geometric mean for each market system added. That is, each new market system benefits the geometric mean to a lesser and lesser degree. Further, as you add each new market system, there is a greater and greater efficiency loss caused as a result of simultaneous rather than sequential outcomes. At some point, to add another market system may do more harm then good.

TIME REQUIRED TO REACH A SPECIFIED GOAL AND THE TROUBLE WITH FRACTIONAL f

Suppose we are given the arithmetic average HPR and the geometric average HPR for a given system. We can determine the standard deviation (SD) in HPRs from the formula for estimated geometric mean:

$$EGM = \sqrt{AHPR^2 - SD^2}$$

where: AHPR = The arithmetic mean HPR.
          SD = The population standard deviation in HPRs.

Therefore, we can estimate the SD as:

$$SD^2 = AHPR^2 - EGM^2$$

Returning to our 2:1 coin-toss game, we have a mathematical expectation of $.50, and an optimal f of betting $1 for every $4 in equity, which yields a geometric mean of 1.06066. We can use Equation (5.03) to determine our arithmetic average HPR:

$$AHPR = 1 + (ME/f\$) \qquad (5.03)$$

where: AHPR = The arithmetic average HPR.
          ME = The arithmetic mathematical expectation in units.
          $f\$$ = The biggest loss/$-f$.
          f = The optimal f (0 to 1).

Thus, we would have an arithmetic average HPR of:

$$AHPR = 1 + (.5/(-1/-.25))$$
$$= 1 + (.5/4)$$
$$= 1 + .125$$
$$= 1.125$$

Now, since we have our AHPR and our EGM, we can employ Equation (5.04) to determine the estimated SD in the HPRs:

$$SD^2 = AHPR^2 - EGM^2$$
$$= 1.125^2 - 1.06066^2$$
$$- 1.265625 - 1.124999636$$
$$= .140625364$$

Thus, SD ^ 2, which is the variance in HPRs, is .140625364. Taking the square root of this yields an SD in these HPRs of .140625364 ^ (1/2) = .3750004853. You should note that this is the estimated SD because it uses the estimated geometric mean as input. It is probably not completely exact, but it is close enough for our purposes.

However, suppose we want to convert these values for the SD (or variance), arithmetic, and geometric mean HPRs to reflect trading at the fractional f. These conversions are now given:

$$FAHPR = (AHPR - 1) * FRAC + 1 \qquad (5.04)$$
$$FSD = SD * FRAC \qquad (5.05)$$
$$FGHPR - \sqrt{FAHPR^2 - FSD^2} \qquad (5.06)$$

where: FRAC = The fraction of optimal f we are solving for.
         AHPR = The arithmetic average HPR at the optimal f.
         SD = The standard deviation in HPRs at the optimal f.
         FAHPR = The arithmetic average HPR at the fractional f.
         FSD = The standard deviation in HPRs at the fractional f.
         FGHPR = The geometric average HPR at the fractional f.

For example, suppose we want to see what values we would have for FAHPR, FGHPR, and FSD at half the optimal f (FRAC = .5) in our 2:1 coin-toss game. Here, we know our AHPR is 1.125 and our SD is .3750004853. Thus:

$$FAHPR = (AHPR - 1) * FRAC + 1$$
$$= (1.125 - 1) * .5 + 1$$
$$= .125 * .5 + 1$$
$$= .0625 + 1$$
$$= 1.0625$$

$$FSD = SD * FRAC$$
$$= .3750004853 * .5$$
$$= .1875002427$$

$$FGHPR = \sqrt{FAHPR^2 - FSD^2}$$
$$= \sqrt{1.065^2 - .1875002427^2}$$
$$= \sqrt{1.12890625 - .03515634101}$$
$$= \sqrt{1.093749909}$$
$$= 1.04582499$$

Thus, for an optimal f of .25, or making one bet for every \$4 in equity, we have values of 1.125, 1.06066, and .3750004853 for the arithmetic average, geometric average, and SD of HPRs, respectively. Now we have solved for a fractional (.5) f of .125 or making one bet for every \$8 in our stake, yielding values of 1.0625, 1.04582499, and .1875002427 for the arithmetic average, geometric average, and SD of HPRs, respectively.

We can now take a look at what happens when we practice a fractional f strategy. We have already determined that under fractional f we will make geometrically less money than under optimal f. Further, we have determined that the drawdowns and variance in returns will be less with fractional f. What about time required to reach a specific goal?

We can quantify the expected number of trades required to reach a specific goal. This is not the same thing as the expected time required to reach a specific goal, but since our measurement is in trades we will use the two notions of time and trades elapsed interchangeably here:

$$T = \ln(\text{Goal}) / \ln(\text{Geometric Mean}) \tag{5.07}$$

where: T = The expected number of trades to reach a
           specific goal.
    Goal = The goal in terms of a multiple on our starting stake,
           a TWR.
    $\ln()$ = The natural logarithm function.

or:

$$T = \text{Log}_{\text{Geometric Mean}} \text{Goal}$$

(i.e. The 'Log base Geoemetric Mean' of the Goal) $\tag{5.07a}$

Returning to our 2:1 coin-toss example, at optimal f we have a geometric mean of 1.06066, and at half f this is 1.04582499. Now let's calculate the expected number of trades required to double our stake (goal $= 2$). At full f:

$$T = \ln(2)/\ln(1.06066)$$
$$= .6931471/.05889134$$
$$= 11.76993$$

Thus, at the full f amount in this 2:1 coin-toss game, we anticipate it will take us 11.76993 plays (trades) to double our stake.

Now, at the half f amount:

$$T = \ln(2)/\ln(1.04582499)$$
$$= .6931471/.04480602$$
$$= 15.46996$$

Thus, at the half f amount, we anticipate it will take us 15.46996 trades to double our stake. In other words, trading half f in this case will take us 31.44% longer to reach our goal.

Well, that doesn't sound too bad. By being more patient, allowing 31.44% longer to reach our goal, we eliminate our drawdown by half and our variance in the trades by half. Half f is a seemingly attractive way to go. The smaller the fraction of optimal f that you use, the smoother the equity curve, and hence the less time you can expect to be in the worst-case drawdown.

Now, let's look at it in another light. Suppose you open two accounts, one to trade the full f and one to trade the half f. After 12 plays, your full f account will have more than doubled to 2.02728259 (1.06066^{12}) times your starting stake. After 12 trades your half f account will have grown to 1.712017427 (1.04582499^{12}) times your starting stake. This half f account will double at 16 trades to a multiple of 2.048067384 (1.04582499^{16}) times your starting stake. So, by waiting about one-third longer, you have achieved the same goal as with full optimal f, only with half the commotion. However, by trade 16 the full f account is now at a multiple of 2.565777865 (1.06066^{16}) times your starting stake. Full f will continue to pull out and away. By trade 100, your half f account should be at a multiple of 88.28796546 times your starting stake, but the full f will be at a multiple of 361.093016!

So anyone who claims that the only thing you sacrifice with trading at a fractional versus full f is time required to reach a specific goal is completely correct. Yet time is what it's all about. We can put our money in Treasury bills and they will reach a specific goal in a certain time with an absolute minimum of drawdown and variance! Time truly is of the essence.

COMPARING TRADING SYSTEMS

We have seen that two trading systems can be compared on the basis of their geometric means at their respective optimal *f*s. Further, we can compare systems based on how high their optimal *f*s themselves are, with the higher optimal *f* being the riskier system. This is because the least the drawdown may have been is at least an *f* percent equity retracement. So, there are two basic measures for comparing systems, the geometric means at the optimal *f*s, with the higher geometric mean being the superior system, and the optimal *f*s themselves, with the lower optimal *f* being the superior system. Thus, rather than having a single, one-dimensional measure of system performance, we see that performance must be measured on a two-dimensional plane, one axis being the geometric mean, the other being the value for *f* itself. *The higher the geometric mean at the optimal* f, *the better the system. Also, the lower the optimal* f, *the better the system.*

Geometric mean does not imply anything regarding drawdown. That is, a higher geometric mean does not mean a higher (or lower) drawdown. The geometric mean pertains only to return. The optimal *f* is the measure of minimum expected historical drawdown as a percentage of equity retracement. A higher optimal *f* does not mean a higher (or lower) return. We can also use these benchmarks to compare a given system at a fractional *f* value and another given system at its full optimal *f* value.

Therefore, when looking at systems, you should look at them in terms of how high their geometric means are and what their optimal *f*s are. For example, suppose we have System A, which has a 1.05 geometric mean and an optimal *f* of .8. Also, we have System B, which has a geometric mean of 1.025 and an optimal *f* of .4. System A at the half *f* level will have the same minimum historical worst-case equity retracement (drawdown) of 40%, just as System B's at full *f*, but System A's geometric mean at half *f* will still be higher than System B's at the full *f* amount. Therefore, System A is superior to System B.

"Wait a minute," you say. "I thought the only thing that mattered was that we had a geometric mean greater than one, that the system need be only marginally profitable, that we can make all the money we want through money management!" That's still true. However, the rate at which you will make the money is still a function of the geometric mean at the *f* level you are employing. The expected variability will be a function of how high the *f* you are using is. So, although it's true that you *must* have a system with a geometric mean at the optimal *f* that is greater than one (i.e., a positive mathematical expectation) and that you can still make virtually an unlimited amount with such a system after enough trades, the rate of growth (the number of trades required to reach a specific goal) is dependent upon the geometric mean at the *f* value employed. The variability en route to that goal is also a function of the *f* value employed.

Yet these considerations, the degree of the geometric mean and the *f* employed, are *secondary* to the fact that you must have a positive mathematical expectation, although they are useful in comparing two systems or techniques that have positive mathematical expectations and an equal confidence of their working in the future.

TOO MUCH SENSITIVITY TO THE BIGGEST LOSS

A recurring criticism with the entire approach of optimal *f* is that it is too dependent on the biggest losing trade. This seems to be rather disturbing to many traders. They argue that the amount of contracts you put on today should not be so much a function of a single bad trade in the past.

Numerous different algorithms have been worked up by people to alleviate this apparent oversensitivity to the largest loss. Many of these algorithms work by adjusting the largest loss upward or downward to make the largest loss be a function of the current volatility in the market. The relationship seems to be a quadratic one. That is, the absolute value of the largest loss seems to get bigger at a faster rate than the volatility. (Volatility is usually defined by these practitioners as the average daily range of the last few weeks, or average absolute value of the daily net change of the last few weeks, or any of the other conventional measures of volatility.) However, this is not a deterministic relationship. That is, just because the volatility is X today does not mean that our largest loss *will* be X^Y. It simply means that it usually is *somewhere near* X^Y.

If we could determine in advance what the largest possible loss would be going into today, we could then have a much better handle on our money management.[2] Here again is a case where we must consider the worst-case

---

[2]This is where using options in a trading strategy is so useful. Either buying a put or call outright in opposition to the underlying position to limit the loss to the strike price of the options, or simply buying options outright in lieu of the underlying, gives you a floor, an absolute maximum loss. Knowing this is extremely handy from a money-management, particularly an optimal *f*, standpoint. Futher, if you know what your maximum possible loss is in advance (e.g., a day trade), then you can always determine what the *f* is in dollars perfectly for any trade by the relation dollars at risk per unit/optimal *f*. For example, suppose a day trader knew his optimal *f* was .4. His stop today, on a one-unit basis, is going to be $900. He will therefore optimally trade one unit for every $2,250 ($900/.4) in account equity.

scenario and build from there. The problem is that we do not know exactly what our largest loss can be going into today. An algorithm that can predict this is really not very useful to us because of the one time that it fails.

Consider, for instance, the possibility of an exogenous shock occurring in a market overnight. Suppose the volatility were quite low prior to this overnight shock, and the market then went locked-limit against you for the next few days. Or suppose that there were no price limits, and the market just opened an enormous amount against you the next day. These types of events are as old as commodity and stock trading itself. They can and do happen, *and they are not always telegraphed in advance* by increased volatility.

Generally, then, you are better off not to "shrink" your largest historical loss to reflect a current low-volatility marketplace. Furthermore, *there is the concrete possibility of experiencing a loss larger in the future than what was the historically largest loss.* There is no mandate that the largest loss seen in the past is the largest loss you can experience today. This is true regardless of the current volatility coming into today.

The problem is that, empirically, the f that has been optimal in the past is a function of the largest loss of the past. There's no getting around this. However, as you shall see when we get into the parametric techniques, you can budget for a greater loss in the future. In so doing, you will be prepared if the almost inevitable larger loss comes along. Rather than trying to adjust the largest loss to the current climate of a given market so that your empirical optimal f reflects the current climate, you will be much better off learning the parametric techniques.

The scenario planning techniques, which are a parametric technique, are a possible solution to this problem, and it can be applied whether we are deriving our optimal f empirically or, as we shall learn later, parametrically.

THE ARC SINE LAWS AND RANDOM WALKS

Now we turn the discussion toward drawdowns. First, however, we need to study a little bit of theory in the way of the first and second arc sine laws. These are principles that pertain to random walks. The stream of trade profits and losses (P&Ls) that you are dealing with may not be truly random. The degree to which the stream of P&Ls you are using differs from being purely random is the degree to which this discussion will not pertain to your stream of P&Ls. Generally, though, most streams of trade P&Ls are nearly random as determined by the runs test and the linear correlation coefficient (serial correlation).

Furthermore, not only do the arc sine laws assume that you know in advance the amount you can win or lose; they also assume that the amount you can win is equal to the amount you can lose, and that this is always a constant amount. In our discussion, we will assume that the amount you can win or lose is $1 on each play. The arc sine laws also assume that you have a 50% chance of winning and a 50% chance of losing. Thus, the arc sine laws assume a game where the mathematical expectation is zero.

These caveats make for a game that is considerably different, and considerably simpler, than trading is. However, the first and second arc sine laws are exact for the game just described. To the degree that trading differs from the game just described, the arc sine laws do not apply. For the sake of learning the theory, however, we will not let these differences concern us for the moment.

Imagine a truly random sequence such as coin tossing[3] where we win one unit when we win and we lose one unit when we lose. If we were to plot out our equity curve over X tosses, we could refer to a specific point (X,Y), where X represented the Xth toss and Y our cumulative gain or loss as of that toss.

We define *positive territory* as anytime the equity curve is above the X axis or on the X axis when the previous point was above the X axis. Likewise, we define *negative territory* as anytime the equity curve is below the X axis or on the X axis when the previous point was below the X axis. We would expect the total number of points in positive territory to be close to the total number of points in negative territory. But this is not the case.

If you were to toss the coin N times, your probability (Prob) of spending K of the events in positive territory is:

$$\text{Prob} \sim 1/\pi * \sqrt{K} * \sqrt{(N - K)} \tag{5.08}$$

The symbol $\sim$ means that both sides tend to equality in the limit. In this case, as either K or $(N - K)$ approaches infinity, the two sides of the equation will tend toward equality.

---

[3] Although empirical tests show that coin tossing is not a truly random sequence due to slight imperfections in the coin used, we will assume here, and elsewhere in the text when referring to coin tossing, that we are tossing an ideal coin with exactly a .5 chance of landing heads or tails.

Thus, if we were to toss a coin 10 times (N = 10) we would have the following probabilities of being in positive territory for K of the tosses:

K	Probability[4]
0	.14795
1	.1061
2	.0796
3	.0695
4	.065
5	.0637
6	.065
7	.0695
8	.0796
9	.1061
10	.14795

You would expect to be in positive territory for 5 of the 10 tosses, yet that is the least likely outcome! In fact, the most likely outcome is that you will be in positive territory for all of the tosses or for none of them!

This principle is formally detailed in the *first arc sine law*, which states:

For a Fixed A (0 < A < 1) and as N approaches infinity, the probability that K/N spent on the positive side is < A tends to:

$$\text{Prob}\{(K/N) \; < \; A\} = 2/\pi * \sin^{-1} \sqrt{A} \qquad (5.09)$$

Even with N as small as 20, you obtain a very close approximation for the probability.

Equation (5.09), the first arc sine law, tells us that with probability .1, we can expect to see 99.4% of the time spent on one side of the origin, and with probability .2, the equity curve will spend 97.6% of the time on the same side of the origin! With a probability of .5, we can expect the equity curve to spend in excess of 85.35% of the time on the same side of the origin. That is just how perverse the equity curve of a fair coin is!

---

[4]Note that since neither K nor N may equal 0 in Equation (5.08) (as you would then be dividing by 0), we can discern the probabilities corresponding to K = 0 and K = N by summing the probabilities from K = 1 to K = N − 1 and subtracting this sum from 1. Dividing this difference by 2 will give us the probabilities associated with K = 0 and K = N.

Now here is the *second arc sine law,* which also uses Equation (5.09) and hence has the same probabilities as the first arc sine law, but applies to an altogether different incident, the maximum or minimum of the equity curve. The second arc sine law states that the maximum (or minimum) point of an equity curve will most likely occur at the endpoints, and least likely at the center. The distribution is exactly the same as the amount of time spent on one side of the origin!

If you were to toss the coin N times, your probability of achieving the maximum (or minimum) at point K in the equity curve is also given by Equation (5.08):

$$\text{Prob} \sim 1/\pi * \sqrt{K} * \sqrt{(N-K)}$$

Thus, if you were to toss a coin 10 times (N = 10), you would have the following probabilities of the maximum (or minimum) occurring on the Kth toss:

K	Probability
0	.14795
1	.1061
2	.0796
3	.0695
4	.065
5	.0637
6	.065
7	.0695
8	.0796
9	.1061
10	.14795

In a nutshell, the second arc sine law states that the maximum or minimum is most likely to occur near the endpoints of the equity curve and least likely to occur in the center.

TIME SPENT IN A DRAWDOWN

Recall the caveats involved with the arc sine laws. That is, the arc sine laws assume a 50% chance of winning and a 50% chance of losing. Further, they assume that you win or lose the exact same amounts and that the generating stream is purely random. Trading is considerably more complicated than this. Thus, the arc sine laws don't apply in a pure sense, but they do apply in spirit.

Consider that the arc sine laws worked on an arithmetic mathematical expectation of zero. Thus, with the first law, we can interpret the percentage of time on either side of the zero line as the percentage of time on either side of the arithmetic mathematical expectation. Likewise with the second law, where, rather than looking for an absolute maximum and minimum, we were looking for a maximum above the mathematical expectation and a minimum below it. The minimum below the mathematical expectation could be greater than the maximum above it if the minimum happened later and the arithmetic mathematical expectation was a rising line (as in trading) rather than a horizontal line at zero.

However we can interpret the spirit of the arc sine laws as applying to trading in the following ways. First, each trade, regardless of the amount won or lost, must be considered as winning one unit or losing one unit respectively. Thus, we now therefore have a line whose slope is the ratio of the difference between the number of wins and losses, and the sum of the number of wins and number of losses, rather than the horizontal line whose slope is zero in the arc sine laws.

For example, suppose I had four trades, three of which were winning trades. The slope of my line therefore equals $(3 - 1)/(3 + 1) = 2/4 = .5$. This is our slope and our mathematical expectation (given that all wins are figured as $+1$, all losses as -1).

We can interpret the first arc sine law as stating that we should expect to be on one side of the mathematical expectation line for far more trades than we spend on the other side of the mathematical expectation line. Regarding the second arc sine law, we should expect the maximum deviations from the mathematical expectation line, either above or below it, as being most likely to occur near the beginning or the end of the equity curve graph and least likely near the center of it.

THE ESTIMATED GEOMETRIC MEAN (OR HOW THE DISPERSION OF OUTCOMES AFFECTS GEOMETRIC GROWTH)

This discussion will use a gambling illustration for the sake of simplicity. Let's consider two systems: System A, which wins 10% of the time and has a twenty-eight-to-one win/loss ratio, and System B, which wins 70% of the time and has a one-to-one ratio. Our mathematical expectation, per unit bet, for A is 1.9 and for B is .4. Therefore, we can say that for every unit bet, System A will return, on average, 4.75 times as much as System B. But let's examine this under fixed fractional trading. We can find our optimal fs by dividing the mathematical expectations by the win/loss ratios [per

Equation (4.05)]. This gives us an optimal f of .0678 for A and .4 for B. The geometric means for each system at their optimal f levels are then:

$$A = 1.044176755$$
$$B = 1.0857629$$

System	% Wins	Win:Loss	ME	f	Geomean
A	.1	28:1	1.9	.0678	1.0441768
B	.7	1:1	.4	.4	1.0857629

As you can see, System B, although less than one-fourth the mathematical expectation of A, makes almost twice as much per bet (returning 8.57629% of your entire stake per bet, on average, when reinvesting at the optimal f levels) as does A (returning 4.4176755% of your entire stake per bet, on average, when reinvesting at the optimal f levels).

Now, assuming a 50% drawdown on equity will require a 100% gain to recoup, then:

1.044177 to the power of x is equal to 2.0 at approximately x equals 16.5, or more than 16 trades to recoup from a 50% drawdown for System A. Contrast this to System B, where 1.0857629 to the power of x is equal to 2.0 at approximately x equals 9, or nine trades for System B to recoup from a 50% drawdown.

What's going on here? Is this because System B has a higher percentage of winning trades? The reason B is outperforming A has to do with the dispersion of outcomes and its effect on the growth function. Most people have the mistaken impression that the growth function, the TWR, is:

$$\text{TWR} = (1 + R)^T$$

where: R = Interest rate per period, e.g., 7% = .07.
T = Number of periods.

Since $1 + R$ is the same thing as an HPR, we can say that most people have the mistaken impression that the growth function,[5] the TWR, is:

$$\text{TWR} = \text{HPR}^T$$

---

[5]Many people mistakenly use the arithmetic average HPR in the equation for HPR^T. As is demonstrated here, this will not give the true TWR after T plays. What you must use is the geometric average HPR, rather than the arithmetic in HPR^T. This will give you the true TWR. If the standard deviation in HPRs is 0, then the arithmetic average HPR and the geometric average HPR are equivalent, and it matters not which you use, arithmetic or geometric average HPR, in such a case.

This function is true only when the return (i.e., the HPR) is constant, which is not the case in trading.

The real growth function in trading (or any event where the HPR is not constant) is the multiplicative product of the HPRs. Assume we are trading coffee, and our optimal f is one contract for every \$21,000 in equity, and we have two trades, a loss of \$210 and a gain of \$210, for HPRs of .99 and 1.01, respectively. In this example, our TWR would be:

$$\text{TWR} = 1.01 * .99$$
$$= .9999$$

An insight can be gained by using the estimated geometric mean (EGM), which very closely approximates the geometric mean:

$$G = \sqrt{A^2 - S^2}$$

or:

$$G = \sqrt{A^2 - V}$$

where: G = geometric mean HPR
   A = arithmetic mean HPR
   S = standard deviation in HPRs
   V = variance in HPRs

Now we take Equations (4.18) and (3.04) to the power of n to estimate the TWR. This will very closely approximate the *multiplicative* growth function, the actual TWR, of Equation (4.17):

$$\text{TWR} = (\sqrt{A^2 - S^2})^T \qquad\qquad (5.10)$$

where: T = Number of periods.
   A = Arithmetic mean HPR.
   S = Population standard deviation in HPRs.

The insight gained is that we can see, mathematically, the trade-off between an increase in the arithmetic average trade (the HPR) versus the dispersion in the HPRs (the standard deviations or the variance), hence the reason that the 70% one-to-one system did better than the 10% twenty-eight-to-one system.

Our goal should be to maximize the coefficient of this function, to maximize Equation (3.04): Expressed literally, to maximize *the square root of the quantity HPR squared minus the variance in HPRs*.

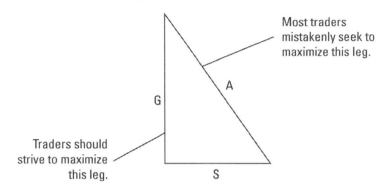

FIGURE 5.2   Pythagorean Theorem in money management

The exponent of the estimated TWR, T, will take care of itself. That is to say that increasing T is not a problem, as we can increase the number of markets we are following, trading more short-term types of systems, and so on.

We can rewrite Equation [3.04] to appear as:

$$A^2 = G^2 + S^2$$

This brings us to the point where we can envision exactly what the relationships are. Notice that this equation is the familiar Pythagorean Theorem: The hypotenuse of a right-angle triangle squared equals the sum of the squares of its sides (Figure 5.2). But here, the hypotenuse is A, and we want to maximize one of the legs, G.

In maximizing G, any increase in S will require an increase in A to offset. When S equals zero, then A equals G, thus conforming to the misconstrued growth function TWR $= (1 + R)^T$.

So, in terms of their relative effect on G, we can state that an increase in A is equal to a decrease of the same amount in S, and vice versa. Thus, any amount by which the dispersion in trades is reduced (in terms of reducing the standard deviation) is equivalent to an increase in the arithmetic average HPR. This is true regardless of whether or not you are trading at optimal f!

If a trader is trading on a fixed fractional basis, then he wants to maximize G, not necessarily A. In maximizing G, the trader should realize that the standard deviation, S, affects G in directly the same proportion as does A, per the Pythagorean Theorem! Thus, when the trader reduces the standard deviation (S) of his trades, it is equivalent to an equal increase in the arithmetic average HPR (A), and vice versa!

THE FUNDAMENTAL EQUATION OF TRADING

We can glean a lot more than just how trimming the size of our losses, or reducing our dispersion in trades, improves our bottom line. Return now to Equation (5.10), the estimated TWR. Since $(X^Y)^Z = X^{(Y*Z)}$, we can further simplify the exponents in the equation, thus simplifying Equation (5.10) to:

$$\text{TWR} = (A^2 - S^2)^{T/2} \qquad (5.10a)$$

This last equation, the simplification for the estimated TWR, we will call the fundamental equation for trading, since it describes how the different factors, A, S, and T, affect our bottom line in trading.

There are a few things that are readily apparent. The first of these is that if A is less than or equal to one, then regardless of the other two variables, S and T, our result can be no greater than one. If A is less than one, then as T approaches infinity, A approaches zero. This means that if A is less than or equal to one (mathematical expectation less than or equal to zero since mathematical expectation $= A - 1$), we do not stand a chance at making profits. In fact, if A is less than one, it is simply a matter of time until we go broke.

Provided that A is greater than one, we can see that increasing T increases our total profits. For each increase of one trade, the coefficient is further multiplied by its square root.

Each time we can increase T by one, we increase our TWR by a factor equivalent to the square root of the coefficient (which is the geometric mean). Thus, each time a trade occurs or an HPR elapses, each time T is increased by one, the coefficient is multiplied by the geometric mean.

An important point to note about the fundamental trading equation is that it shows that if you reduce your standard deviation to a greater extent than you reduce your arithmetic average HPR, you are better off. It stands to reason, therefore, that cutting your losses short, if possible, benefits you. But the equation demonstrates that at some point you no longer benefit by cutting your losses short. That is the point where you would be getting out of too many trades with a small loss that later would have turned profitable, thus reducing your A to a greater extent than your S.

Along these same lines, reducing big winning trades can help your program if it reduces your S greater than it reduces your A. This can be accomplished, in many cases, by incorporating options into your trading program. Having an option position that goes against your position in the underlying (either by buying long an option or writing an option) can possibly help.

As you can see, the fundamental trading equation can be utilized to dictate many changes in our trading. These changes may be in the way of tightening (or loosening) our stops, setting targets, and the like. These changes are the result of inefficiencies in the way we are carrying out our trading, as well as inefficiencies in our trading program or methodology.

WHY IS f OPTIMAL?

To see that f is optimal in the sense of maximizing wealth:

$$\text{since } G = \left(\prod_{i=1}^{T} \text{HPR}_i \right)^{1/T}$$

$$\text{and} \quad \left(\prod_{i=1}^{T} \text{HPR}_i \right)^{1/T} = \exp\left(\frac{\sum_{i=1}^{T} \ln(\text{HPR}_i)}{T} \right)$$

Then, if one acts to maximize the geometric mean at every holding period, if the trial is sufficiently long, by applying either the weaker law of large numbers or the Central Limit Theorem to the sum of *independent* variables (i.e., the numerator on the right side of this equation), almost certainly higher terminal wealth will result than from using any other decision rule.

Furthermore, we can also apply Rolle's Theorem to the problem of the proof of f's optimality. Recall that we are defining *optimal* here as meaning that which will result in the greatest geometric growth as the number of trials increases. The TWR is the measure of average geometric growth; thus, we wish to prove that there is a value for f that results in the greatest TWR.

Rolle's Theorem states that if a *continuous* function crosses a line parallel to the X-axis at two points, a and b, and the function is continuous throughout the interval a,b, then there exists at least one point in the interval where the first derivative equals zero (i.e., at least one relative extremum).

Given that all functions with a positive arithmetic mathematical expectation cross the X-axis twice[6] (the X being the f axis), at $f = 0$ and at that point to the right where f results in computed HPRs where the variance in those HPRs exceeds the difference of the arithmetic mean of those HPRs minus one, we have our a,b interval on X, respectively. Furthermore, the

---

[6]Actually, at $f = 0$, the TWR = 0, and thus we cannot say that it crosses 0 to the upside here. Instead, we can say that at an f value which is an infinitesimally small amount beyond 0, the TWR crosses a line an infinitesimally small amount above 0. Likewise to the right but in reverse, the line, the f curve, the TWR, crosses this line which is an infinitesimally small amount above the X-axis as it comes back down to the X-axis.

first derivative of the fundamental equation of trading (i.e., the estimated TWR) is continuous for all f within the interval, since f results in AHPRs and variances in those HPRs, within the interval, which are differentiable in the function in that interval; thus, the function, the estimated TWR, is continuous within the interval. Per Rolle's Theorem, it must, therefore, have at least one relative extremum in the interval, and since the interval is positive, that is, above the X-axis, the interval must contain at least one maximum.

In fact, there can be only one maximum in the interval given that the change in the geometric mean HPR (a transformation of the TWR, given that the geometric mean HPR is the Tth root of the TWR) is a direct function of the change in the AHPR and the variance, both of which vary in *opposite directions to each other as* f *varies*, per the Pythagorean theorem. This guarantees that there can be only one peak. Thus, there must be a peak in the interval, and there can be only one peak. There is an f that is optimal at only one value for f, where the first derivative of the TWR with respect to f equals zero.

Let us go back to Equation (4.07). Now, we again consider our two-to-one coin toss. There are two trades, two possible scenarios. If we take the first derivative of (4.07) with respect to f, we obtain:

$$\frac{d\text{TWR}}{df} = \left(\left(1 + f^* \left(\frac{-\text{trade}_1}{\text{biggest loss}}\right)\right) * \left(\frac{-\text{trade}_2}{\text{biggest loss}}\right)\right)$$
$$+ \left(\left(\frac{-\text{trade}_1}{\text{biggest loss}}\right) * \left(1 + f * \left(\frac{-\text{trade}_2}{\text{biggest loss}}\right)\right)\right) \quad (5.11)$$

If there were more than two trades, the same basic form could be used, only it would grow monstrously large in short order, so we'll use only two trades for the sake of simplicity. Thus, for the sequence $+2, -1$ at $f = .25$:

$$\frac{d\text{TWR}}{df} = \left(\left(1 + .25 * \left(\frac{-2}{-1}\right)\right) * \left(\frac{-1}{-1}\right)\right) + \left(\left(\frac{-2}{-1}\right) * \left(1 + .25 * \left(\frac{-1}{-1}\right)\right)\right)$$

$$\frac{d\text{TWR}}{df} = ((1 + .25 * 2) * - 1) + (2 * (1 + .25 * - 1))$$

$$\frac{d\text{TWR}}{df} = ((1 + .5) * - 1) + (2 * (1 - .25))$$

$$\frac{d\text{TWR}}{df} = (1.5 * - 1) + (2 * .75)$$

$$\frac{d\text{TWR}}{df} = -1.5 + 1.5 = 0$$

And we see that the function peaks at .25, where the slope of the tangent is zero, exactly at the optimal f, and no other local extremum can exist because of the restriction caused by the Pythagorean Theorem.

Lastly, we will see that optimal f is indifferent to T. We can take the first derivative of the estimated TWR, Equation (5.10a) with respect to T as:

$$\frac{d\text{TWR}}{dT} = \left(A^2 - S^2\right)^{T/2} * \ln\left(A^2 - S^2\right) \tag{5.12}$$

Since $\ln(1) = 0$, then if $A^2 - S^2 = 1$, that is, $A^2 - 1 = S^2$ (or variance), the function peaks out and the single optimal maximum TWR is found with respect to f. Notice, though, that both A, the arithmetic average HPR, and S, the standard deviation in those HPRs, are not functions of T. Instead, they are indifferent to T; thus, (5.10a) is indifferent to T at the optimal f. The f that is optimal in the sense of maximizing the estimated TWR will always be the same value regardless of T.

CHAPTER 6

Laws of Growth, Utility, and Finite Streams

Since this book deals with the mathematics involving growth, we must discuss the laws of growth. When dealing with growth in mathematical terms, we can discuss it in terms of growth functions or of the corresponding growth rates.

We can speak of growth functions as falling into three distinct categories, where each category is associated with a growth *rate*. Figure 6.1 portrays these three categories as lines B, C, and D, and their growth rates as A, B, and C, respectively. Each growth function has its growth rate immediately to its left.

Thus, for growth function B, the linear growth function, its growth rate is line A. Further, although B is a growth function itself, it also represents the growth rate for function C, the exponential growth rate.

Notice that there are three growth functions, *linear, exponential,* and *hyperbolic.* Thus, the hyperbolic growth function has an exponential growth rate, the exponential growth function has a linear growth rate, and the linear growth function has a flat-line growth rate.

The X and Y-axes are important here. If we are discussing growth functions (B, C, or D), the Y-axis represents quantity and the X-axis represents time. If we are discussing growth rates, the Y-axis represents quantity change with respect to time, and the X-axis represents quantity.

When we speak of growth rates and functions in general, we often speak of the growth of a population of something. The first of the three major growth functions is the linear growth function, line B, and its rate, line A. Members of a population characterized by linear growth tend to easily find a level of coexistence.

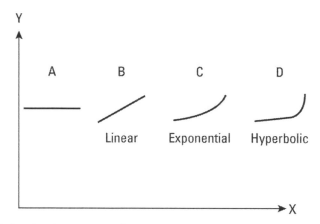

FIGURE 6.1 The three growth functions

Next, we have the exponential growth function, line C, and its growth rate, which is linear, line B. Here, we find competition among the members of the population and a survival-of-the-fittest principle setting in. In the exponential growth function, however, it is possible for a mutation to appear, which has a selective advantage, and establish itself.

Finally, in the hyperbolic growth function, line D, and its (exponential) growth rate, line C, we find a different story. Unlike the exponential growth function, which has a linear growth rate, this one's growth rate is itself exponential. That is, the greater the quantity, the faster the growth rate! Thus, the hyperbolic function, unlike the exponential function, reaches a point that we call a *singularity*. That is, it reaches a point where it becomes infinitely large, a vertical asymptote. This is not true of the exponential growth function, which simply becomes larger and larger. In the hyperbolic function, we also find competition among the members of the population, and a survival-of-the-fittest characteristic. However, at a certain point in the evolution of a hyperbolic function, it becomes nearly impossible for a mutation with a selective advantage to establish itself, since the rest of the population is growing at such a rapid rate.

In either the exponential or hyperbolic growth functions, if there are functional links between the competing species within the population, it can cause any of the following:

1. Increased competition among the partners; or
2. Mutual stabilization among the partners; or
3. Extinction of all members of the population.

The notion of populations is also a recurring theme throughout this book, and it is nearly impossible to discuss the mathematics of growth without discussing populations. The mathematics of growth is the corpus callosum between population growth and the new framework presented in this book.

Trading is exponential, *not* hyperbolic. However, if you had someone who would give you money to trade if your performance came in as promised, and that person had virtually unlimited funds, then your trading would be hyperbolic. This sounds like managed money. The problem faced by money managers is the caveat laid on the money manager by the individual of unlimited wealth: *if your performance comes in as promised.* In the later chapters in this book, we will discuss techniques to address this caveat.

MAXIMIZING EXPECTED AVERAGE COMPOUND GROWTH

Thus far, in this book, we have looked at finding a value for f that was asymptotically dominant. That is, we have sought a single value for f for a given market system, which, if there truly was independence between the trades, would maximize geometric growth with certainty as the number of trades (or holding periods) approached infinity. That is, we would end up with greater wealth in the very long run, with a probability that approached certainty, than we would using any other money management strategy.

Recall that if we have only one play, we maximize growth by maximizing the arithmetic average holding period return (i.e., $f = 1$). If we have an infinite number of plays, we maximize growth by maximizing the geometric average holding period return (i.e., $f = $ optimal f). However, *the f that is truly optimal is a function of the length of time—the number of finite holding period returns—that we are going to play.*

For one holding period return, the optimal f will always be 1.0 for a positive arithmetic mathematical expectation game. If we bet at any value for f other than 1.0, and quit after only one holding period, we will not have maximized our expected average geometric growth. What we regard as the optimal f would be optimal only if you were to play for an infinite number of holding periods. The f that is truly optimal starts at one for a positive arithmetic mathematical expectation game, and converges toward what we call the optimal f as the number of holding periods approaches infinity.

To see this, consider again our two-to-one coin-toss game, where we have determined the optimal f to be .25. That is, if the coin tosses are independent of previous tosses, by betting 25% of our stake on each and every play, we will maximize our geometric growth with certainty as the length of this game, the number of tosses (i.e., the number of holding periods) approaches infinity. That is, our expected average geometric growth—what we would expect to end up with, as an expected value, given every possible combination of outcomes—would be greatest if we bet 25% per play.

Consider the first toss. There is a 50% probability of winning $2 and a 50% probability of losing $2. At the second toss, there is a 25% chance of winning $2 on the first toss and winning $2 on the second, a 25% chance of winning $2 on the first and losing $1 on the second, a 25% chance of losing $1 on the first and winning $2 on the second, and a 25% chance of losing $1 on the first and losing $1 on the second (we know these probabilities to be true because we have already stated the prerequisite that these events are independent). The combinations bloom out in time in a tree-like fashion. Since we had only two scenarios (heads and tails) in this scenario spectrum, there are only two branches off of each node in the tree. If we had more scenarios in this spectrum, there would be that many more branches off of each node in this tree:

Toss#		
1	2	3
		Heads
	Heads	
		Tails
Heads		
		Heads
	Tails	
		Tails
		Heads
	Heads	
		Tails
Tails		
		Heads
	Tails	
		Tails

If we bet 25% of our stake on the first toss and quit, we will not have maximized our expected average compound growth (EACG).

What if we quit after the second toss? What, then, should we optimally bet, knowing that we maximize our expected average compound gain by betting at $f = 1$ when we are going to quit after one play, and betting at the optimal f if we are going to play for an infinite length of time?

If we go back and optimize f, allowing there to be a different f value used for the first play as well as the second play, with the intent of maximizing what our average geometric mean HPR would be at the end of the second play, we would find the following: First, the optimal f for quitting after two plays in this game approaches the asymptotic optimal, going from 1.0 if we quit after one play to .5 for both the first play and the second. That is, if we were to quit after the second play, we should optimally bet .5 on both the first and second plays to maximize growth. (Remember, we allowed for the first play to be an f value different from the second, yet they both came out the same: .5 in this case. It is a fact that if you are looking to maximize growth, the f that is optimal—for finite as well as infinite streams—is uniform.)

We can see this if we take the first two possible combinations of tosses:

Toss#	
1	**2**
	Heads
Heads	
	Tails
	Heads
Tails	
	Tails

Which can be represented by the following outcomes:

Toss#	
1	**2**
	2
2	
	−1
	2
−1	
	−1

These outcomes can be expressed as holding period returns for various f values. In the following, it is shown for an f of .5 for the first toss, as well as for an f of .5 for the second:

Toss#	
1	**2**
	2
2	
	.5
	2
.5	
	.5

Now, we can express all tosses subsequent to the first toss as TWRs by multiplying by the subsequent tosses on the tree. The numbers following the last toss on the tree (the numbers in parentheses) are the last TWRs taken to the root of $1/n$, where n equals the number of HPRs, or tosses—in this case two—and represents the geometric mean HPR for that terminal node on the tree:

	Toss#
1	**2**
	4 (2.0)
2	
	1 (1.0)
	1 (1.0)
.5	
	.25 (.5)

Now, if we total up the geometric mean HPRs and take their arithmetic average, we obtain the *expected average compound growth*, in this case:

$$
\begin{array}{r}
2.0 \\
1.0 \\
1.0 \\
\underline{.5} \\
\dfrac{4.5}{4} = 1.125
\end{array}
$$

Thus, if we were to quit after two plays, and yet do this same thing over an infinite number of times (i.e., quit after two plays), we would optimally bet .5 of our stake on each and every play, thus maximizing our EACG.

Notice that we did not bet with an f of 1.0 on the first play, even though that is what would have maximized our EACG if we had quit at one play. Instead, if we are planning on quitting after two plays, we maximize our EACG growth by betting at .5 on both the first play and the second play.

Notice that the f that is optimal in order to maximize growth is uniform for all plays, yet it is a function of how long you will play. If you are to quit after only one play, the f that is optimal is the f that maximizes the arithmetic mean HPR (which is always an f of 1.0 for a positive expectation game, 0.0 for a negative expectation game). If you are playing a positive expectation game, the f that is optimal continues to decrease as the length of time after which you quit grows, and, asymptotically, if you play for an infinitely long time, the *f* that is optimal is that which maximizes the geometric mean HPR. In a negative expectation game, the *f* that is optimal simply stays at zero.

However, the *f* that you use to maximize growth is always uniform, and that uniform amount is a function of where you intend to quit the game. If you are playing the two-to-one coin-toss game, and you intend to quit after one play, you have an *f* value that provides for optimal growth of 1.0. If you intend to quit after two plays, you have an *f* that is optimal for maximizing growth of .5 on the first toss and .5 on the second. Notice that you do not bet 1.0 on the first toss if you are planning on maximizing the EACG by quitting at the end of the second play. Likewise, if you are planning on playing for an infinitely long period of time, you would optimally bet .25 on the first toss and .25 on each subsequent toss.

Note the key word there is *infinitely*, not *indefinitely*. All streams are finite—we are all going to die eventually. Therefore, when we speak of the optimal *f* as the *f* that maximizes expected average compound return, we are speaking of that value which maximizes it if played for an infinitely long period of time. Actually, it is slightly suboptimal because none of us will be able to play for an infinitely long time. And, the *f* that will maximize EACG will be slightly above—will have us take slightly heavier positions—than what we are calling the optimal *f*.

What if we were to quit after three tosses? Shouldn't the *f* which then maximizes expected average compound growth be lower still than the .5 it is when quitting after two plays, yet still be greater than the .25 optimal for an infinitely long game?

Let's examine the tree of combinations here:

Toss#		
1	**2**	**3**
		Heads
	Heads	
		Tails
Heads		
		Heads
	Tails	
		Tails
		Heads
	Heads	
		Tails
Tails		
		Heads
	Tails	
		Tails

Converting these to outcomes yields:

	Toss#	
1	**2**	**3**
		2
	2	
		−1
2		
		2
	−1	
		−1
		2
	2	
		−1
−1		
		2
	−1	
		−1

If we go back with a computer and iterate to that value for f which maximizes expected average compound growth when quitting after three tosses, we find it to be .37868. Therefore, converting the outcomes to HPRs based upon a .37868 value for f at each toss yields:

	Toss#	
1	**2**	**3**
		1.757369
	1.757369	
		.621316
1.757369		
		1.757369
	.621316	
		.621316
		1.757369
	1.757369	
		.621316
.621316		
		1.757369
	.621316	
		.621316

Now we can express all tosses subsequent to the first toss as TWRs by multiplying by the subsequent tosses on the tree. The numbers following the last toss on the tree (the numbers in parentheses) are the last TWRs taken to the root of $1/n$, where n equals the number of HPRs, or tosses, in this case three, and represent the geometric mean HPR for that terminal node on the tree:

Toss#		
1	**2**	**3**
		5.427324 (1.757365)
	3.088329	
		1.918831 (1.242641)
1.757369		
		1.918848 (1.242644)
	1.09188	
		.678409 (.87868)
		1.918824 (1.242639)
	1.091875	
		.678401 (.878676)
.621316		
		.678406 (.878678)
	.386036	
		.239851 (.621318)

$$\frac{8.742641}{8} = 1.09283 \text{ is the expected average compound growth (EACG)}$$

If you are the slightest bit skeptical of this, I suggest you go back over the last few examples, either with pen and pencil or computer, and find a value for f that results in a greater EACG than the values presented. Allow yourself the liberty of a nonuniform f—that is, an f that is allowed to change at each play. You'll find that you get the same answers as we have, and that f is uniform, although a function of the length of the game.

From this, we can summarize the following conclusions:

1. To maximize the EACG, we always end up with a uniform f. That is, the value for f is uniform from one play to the next.

2. The f that is optimal in terms of maximizing the EACG is a function of the length of the game. For positive expectation games, it starts at 1.0, the value that maximizes the arithmetic mean HPR, diminishes slightly each play, and asymptotically approaches that value which maximizes the geometric mean HPR (which we have been calling—and will call throughout the sequel—the optimal f).

3. Since all streams are finite in length, regardless of how long, we will always be ever-so-slightly suboptimal by trading at what we call the optimal f, regardless of how long we trade. Yet, the difference diminishes with each holding period. Ultimately, we are to the left of the peak of what was truly optimal. This is not to say that everything mentioned about the $n+1$ dimensional landscape of leverage space, to be discussed later in the text—the penalties and payoffs of where you are with respect to the optimal f for each market system—aren't true. It is true, however, that the landscape is a function of the number of holding periods at which you quit. The landscape we project with the techniques in this book is the asymptotic altitudes—what the landscape approaches as we continue to play.

To see this, let's continue with our two-to-one coin toss. In the graph (Figure 6.2), we can see the value for f, which optimally maximizes our EACG for quitting at one play through eight plays. Notice how it approaches the optimal f of .25, the value that maximizes growth asymptotically, as the number of holding periods approaches infinity.

Two-to-One Coin-Toss Game	
Quitting after HPR #	**f that Maximizes EACG**
1	1.0
2	.5
3	.37868
4	.33626
5	.3148
6	.3019
7	.2932
8	.2871
.	.
.	.
.	.
Infinity	.25 (this is the value we refer to as the optimal f)

In reality, if we trade with what we are calling in this text the optimal f, we will always be slightly suboptimal, the degree of which diminishes as more and more holding periods elapse. If we knew exactly how many holding periods we were going to trade for, we could then use that value for f which maximizes EACG (which would be slightly greater than the optimal f) and be truly optimal. Unfortunately, we rarely know exactly how many holding periods we are going to play for, and there is consolation in

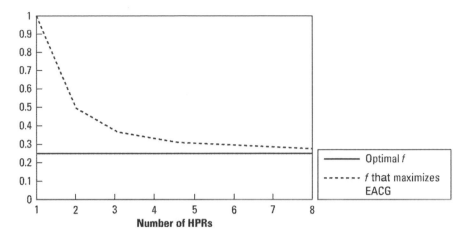

FIGURE 6.2 Optimal *f* as an asymptote

the fact that what we are calling the optimal *f* approaches what would be optimal to maximize EACG as more holding periods elapse. Later we will see the *continuous dominance* techniques, which allow us to approximate the notion of maximizing EACG when there is an active/inactive equity split (i.e., anytime someone is trading less aggressively than optimal *f*).

Note that none of these notions is addressed or even alluded to in the older mean-variance, risk-return models, which are next to be discussed in the beginning of the following chapter. The older models disregard leverage and its workings almost entirely. This is one more reason to opt for the new model of portfolio construction to be mentioned later in the text.

UTILITY THEORY

The discussion of utility theory is brought up in this book since, oftentimes, geometric mean maximizers are criticized for being able to maximize only the ln case of utility; that is, they seek to maximize only wealth, not investor satisfaction. This book attempts to show that geometric mean maximization can be applicable, regardless of one's utility preference function. Therefore, we must, at this point, discuss utility theory, in general, as a foundation.

Utility theory is often attacked as being an ivory-tower, academic construct to explain investor behavior. Unfortunately, most of these attacks come from people who have made the a priori assumption that all investor utility functions are ln; that is, they seek to maximize wealth. While this author is not a great proponent of utility theory, I accept it for lack of a better explanation for investor preferences. However, I strongly feel that if an investor's utility function is other than ln, the markets, and investing in general, are poor places to deal with this or to try to maximize one's utility—you're on the n + 1 dimensional landscape to be discussed in Chapter 9 regardless of your utility preference curve, and you will pay the consequences in real currency for being suboptimal. In short, the markets are a bad place to find out you are not a wealth maximizer. The psychiatrist's couch may be a more gentle environment in which to deal with that.

THE EXPECTED UTILITY THEOREM

A guy in an airport has $500, but needs $600 for a ticket he *must* have. He is offered a bet with a 50% probability of winning $100 and a 50% probability of losing $500. Is this a good bet? In this instance, where we assume it to be a life-and-death situation where he must have the ticket, it *is* a good bet.

The mathematical expectation of utility is vastly different in this instance than the mathematical expectation of wealth. Since, if we subscribe to utility theory, we determine *good bets* based on their mathematical expectation of *utility* rather than *wealth*, we assume that the mathematical expectation of utility in this instance is positive, even though wealth is not. Think of the words *utility* and *satisfaction* as meaning the same thing in this discussion.

Thus, we have what is called the *Expected Utility Theorem*, which states that *investors possess a utility-of-wealth function,* U(x), *where* x *is wealth, that they will seek to maximize. Thus, investors will opt for those investment decisions that maximize their utility-of-wealth function.* Only when the utility preference function $U(x) = \ln x$, that is, when the utility, or satisfaction, of wealth equals the wealth, will the expected utility theorem yield the same selection as wealth maximization.

CHARACTERISTICS OF UTILITY PREFERENCE FUNCTIONS

There are five main characteristics of utility preference functions:

1. Utility functions are unique up to a positive linear transformation. Thus, a utility preference function, such as the preceding one, $\ln x$, will lead to the same investments being selected as a utility function of $25 + \ln x$, as it would be a utility function of $71^* \ln x$ or one of the form $(\ln x)/1.453456$. That is, a utility function that is affected by a positive

constant (added, subtracted, multiplied, or divided) will result in the same investments being selected. Thus, it will lead to the same set of investments maximizing utility as before the positive constant affects the function.

2. More is preferred to less. In economic literature, this is often referred to as *nonsatiation*. In other words, a utility function must never result in a choice for less wealth over more wealth when the outcomes are certain or their probabilities equal. Since utility must, therefore, increase as wealth increases, the first derivative of utility, with respect to wealth, must be positive. That is:

$$U'(x) > -0 \tag{6.01}$$

Given utility as the vertical axis and wealth as the horizontal axis, then the utility preference curve must never have a negative slope.

The ln x case of utility preference functions shows a first derivative of x^{-1}.

3. There are three possible assumptions regarding an investor's feelings toward risk, also called his *risk aversion*. He is either averse to, neutral to, or seeks risk. These can all be defined in terms of a fair gamble. If we assume a fair game, such as coin tossing, winning $1 on heads and losing $1 on tails, we can see that the arithmetic expectation of wealth is zero. A risk-averse individual would not accept this bet, whereas a risk seeker would accept it. The investor who is risk-neutral would be indifferent to accepting this bet.

Risk aversion pertains to the second derivative of the utility preference function, or $U''(x)$. A risk-averse individual will show a negative second derivative, a risk seeker a positive second derivative, and one who is risk-neutral will show a zero second derivative of the utility preference function.

Figure 6.3 depicts the three basic types of utility preference functions, based on $U''(x)$, the investor's level of risk aversion. The ln x case of utility preference functions shows neutral risk aversion. The investor is indifferent to a fair gamble.[1] The ln x case of utility preference functions shows a second derivative of $-x^{-2}$.

---

[1]Actually, investors should reject a fair gamble. Since the amount of money an investor has to work with is finite, there is a lower absorbing barrier. It can be shown that if an investor accepts fair gambles repeatedly, it is simply a matter of time before the lower absorbing barrier is met. That is, if you keep on accepting fair gambles, eventually you will go broke with a probability approaching certainty.

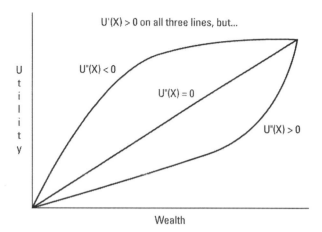

FIGURE 6.3   Three basic types of utility functions

4. The fourth characteristic of utility preference functions pertains to how
    an investor's levels of risk aversion change with changes in wealth. This
    is referred to as *absolute risk aversion*. Again, there are three possi-
    ble categories. First is the individual who exhibits increasing absolute
    risk aversion. As wealth increases, he holds fewer dollars in risky as-
    sets. Next is the individual with constant absolute risk aversion. As his
    wealth increases, he holds the same dollar amount in risk assets. Last
    is the individual who displays decreasing absolute risk aversion. As this
    individual's wealth increases, he is willing to hold more dollars in risky
    assets.

    The mathematical formulation for defining absolute risk aversion,
    $A(x)$, is as follows:

$$A(x) = \frac{-U''(x)}{U'(x)} \tag{6.02}$$

Now, if we want to see how absolute risk aversion changes with
a change in wealth, we would take the first derivative of $A(x)$ with respect to x (wealth), obtaining $A'(x)$. Thus, an individual with increasing absolute risk aversion would have $A'(x) > 0$, constant absolute risk
aversion would see $A'(x) = 0$, and decreasing absolute risk aversion has
$A'(x) < 0$.

The ln x case of utility preference functions shows *decreasing* absolute risk aversion. For the ln x case:

$$A(x) = \frac{-(-x^{-2})}{x^{-1}} = x^{-1} \quad \text{and} \quad A'(x) = -x^{-2} < 0$$

5. The fifth characteristic of utility preference functions pertains to how the percentage of wealth invested in risky assets changes with changes in wealth. This is referred to as *relative risk aversion*. That is, this pertains to how your percentages in risky assets change, rather than how your dollar amounts change, with respect to changes in wealth. Again, there are three possible categories: increasing, constant, and decreasing relative risk aversion, where the percentages invested in risky assets increase, stay the same, or decline, respectively.

The mathematical formulation for defining relative risk aversion, $R(x)$, is as follows:

$$R(x) = \frac{(-x * U''(x))}{U'(x)} = x * A(x) \qquad (6.03)$$

Therefore, $R'(x)$, the first derivative of relative risk aversion, indicates how relative risk aversion changes with respect to changes in wealth. So, individuals who show increasing, constant, or decreasing relative risk aversion will then show positive, zero, and negative $R'(x)$, respectively.

The $\ln x$ case of utility preference functions shows *constant* relative risk aversion. For the $\ln x$ case:

$$R(x) = \frac{(-x^*(-x^{-2}))}{x^{-1}} = 1 \qquad \text{and} \qquad R'(x) = 0$$

ALTERNATE ARGUMENTS TO CLASSICAL UTILITY THEORY

Readers should be aware that utility theory, although broadly accepted, is not universally accepted as an explanation of investor behavior. For example, R. C. Wentworth contends, with reference to the Expected Utility Theorem, that the use of the mean is an ad hoc, unjustified assumption. His theory is that players assume that the mode, rather than the mean, will prevail, and will act to maximize this.

I personally find Wentworth's work in this area particularly interesting.[2] There are some rather interesting aspects to these papers. First, classical utility theory is directly attacked, which automatically alienates every professor in every management science department in the world. The theoretical foundation paradigm of the nonlinear *utility-of-wealth* function is sacred to these people. Wentworth draws parallels between *mode*

---

[2]See "Utility, Survival, and Time: Decision Strategies under Favorable Uncertainty," and "A Theory of Risk Management under Favorable Uncertainty," both by R. C. Wentworth, unpublished. 8072 Broadway Terrace, Oakland, CA 94611.

maximizers and evolution; hence, Wentworth calls his the *survival hypothesis*. A thumbnail sketch of the comparison with classical utility theory would appear as:

	Utility Theory		
"One-shot," risky decision making	+ Nonlinear utility-of-wealth function	=>	Observed behavior

	Survival Hypothesis		
"One-shot," risky decision making	+ Expansion into equivalent time series	=>	Identical observed behavior

Furthermore, there are some interesting experiments in biology that tend to support Wentworth's ideas, which ask the question why, for instance, should bumblebees search for nectar, in a controlled experiment, according to the dictates of classical utility theory?

So, why mention classical utility theory at all? It is not the purpose of this book to presuppose anything regarding utility theory. However, there is an interrelationship between utility and this new framework in asset allocation, and if one does subscribe to a utility framework notion, then they will be shown how this applies. This portion of the book is directed toward those readers unfamiliar with the notion of utility preference curves. However, it does not take a position on the validity of utility functions, and the reader should be made aware that there are other non-utility-based criteria that may explain investor behavior.

FINDING YOUR UTILITY PREFERENCE CURVE

Whether one subscribes to classical utility theory, considering that it is better to know yourself than not know yourself, we will now detail a technique for determining your own utility preference function. What follows is an adaptation from *The Commodity Futures Game, Who Wins? Who Loses? Why?* by Tewles, Harlow, and Stone.[3]

To begin with, you should determine two extreme values, one positive and the other negative, which should represent extreme trade outcomes. Typically, you should make this value be three to five times greater than the largest amounts you would typically expect to win or lose on the next trade.

---

[3] Richard J. Tewles, Charles V. Harlow, and Herbert L. Stone, *The Commodity Futures Game, Who Wins? Who Loses? Why?* New York: McGraw-Hill Book Company, 1977.

Let's suppose you expect, in the best case, to win $5,000 on a trade, and lose $3,000. Thus, we can make our extremes $20,000 on the upside and −$10,000 on the downside.

Next, set up a table as follows, with a leftmost column called *Probabilities of Best Outcome*, and give it 10 rows with values progressing from 1.0 to 0 by increments of .1. Your next column should be called *Probabilities of Worst Outcome*, and those probabilities are simply 1 minus the probabilities of the best outcome on that row. The third column will be labeled *Certainty Equivalent*. In the first row, you will put the value of the best outcome, and in the last row, the value of the worst outcome. Thus, your table should look like this:

P (Best Outcome)	P (Worst Outcome)	Certainty Equivalent	Computed Utility
1.0	0	$20,000	
.9	.1		
.8	.2		
.7	.3		
.6	.4		
.5	.5		
.4	.6		
.3	.7		
.2	.8		
.1	.9		
0	1.0	−$10,000	

Now, we introduce the notion of *certainty equivalents*. A certainty equivalent is an amount you would accept in lieu of a trading opportunity or an amount you might pay to sidestep a trade opportunity.

You should now fill in column three, the certainty equivalents. For the first row, the one where we entered $20,000, this simply means you would accept $20,000 in cash right now, rather than take a trade with a 100% probability of winning $20,000. Likewise, with the last row where we have filled in $10,000, this simply means you would be willing to pay $10,000 not to have to take a trade with a 100% chance of losing $10,000.

Now, on the second row, you must determine a certainty equivalent for a trade with a 90% chance of winning $20,000 and a 10% chance of losing $10,000. What would you be willing to accept in cash instead of taking this trade? Remember, this is real money with real buying power, and the rewards or consequences of this transaction will be immediate and in cash. Let's suppose it's worth $15,000 to you. That is, for $15,000 in cash, handed to you right now, you will forego this opportunity of a 90% chance of winning $20,000 and 10% chance of losing $10,000.

You should complete the table for the certainty equivalent columns. For instance, when you are on the second to last row, you are, in effect, asking yourself how much you would be willing to pay not to have to accept a 10% chance of winning $20,000 with a 90% chance of losing $10,000. Since you are willing to pay, you should enter this certainty equivalent as a negative amount.

When you have completed the third column, you must now calculate the fourth column, the *Computed Utility* column. The formula for computed utility is simply:

$$\text{Computed Utility} = U * P(\text{best outcome}) + V * P(\text{worst outcome})$$

(6.04)

where: U = Given constant, equal to 1.0 in this instance.
V = Given constant, equal to -1.0 in this instance.

Thus, for the second row in the table:

$$\text{Computed utility} = 1 * .9 - 1 * .1$$
$$= .9 - .1$$
$$= .8$$

When you are finished calculating the computed utility columns, your table might look like this:

P (Best Outcome)	P (Worst Outcome)	Certainty Equivalent	Computed Utility
1.0	0	20,000	1.0
.9	.1	15,000	.8
.8	.2	10,000	.6
.7	.3	7,500	.4
.6	.4	5,000	.2
.5	.5	2,500	0
.4	.6	800	−.2
.3	.7	−1,500	−.4
.2	.8	−3,000	−.6
.1	.9	−4,000	−.8
0	1.0	−10,000	−1.0

You then graphically plot the certainty equivalents as the X-axis and the computed utilities as the Y-axis. Our completed utility function looks as is shown in Figure 6.4.

Now you should repeat the test, only with different best and worst outcomes. Select a certainty equivalent from the preceding table to act as *best*

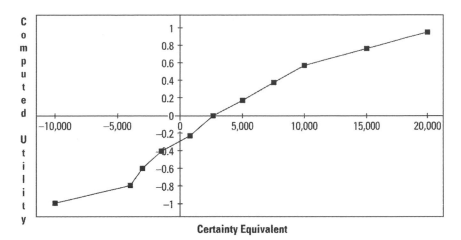

FIGURE 6.4 Example utility function

outcome, and one for *worst outcome* as well. Suppose we choose $10,000 and −$4,000. Notice that the computed utilities associated with certainty equivalents are .6 with $10,000 and −.8 with −$4,000. Thus, U and V, in determining computed utilities in this next table, will be .6 and −.8, respectively. Again, assign certainty equivalents and calculate the corresponding computed utilities:

P (Best Outcome)	P (Worst Outcome)	Certainty Equivalent	Computed Utility
1.0	0	10,000	.6
.9	.1	8,000	.46
.8	.2	6,000	.32
.7	.3	5,000	.18
.6	.4	4,000	.04
.5	.5	2,500	−.10
.4	.6	500	−.24
.3	.7	−1,000	−.38
.2	.8	−2,000	−.52
.1	.9	−3,000	−.66
0	1.0	−4,000	−.80

And, again, you should plot these values. You should repeat the process a few times, and keep plotting all the values on the same chart. What you will probably begin to see is some scattering to the values; that is, they will not all neatly fit on the same line. The scattering of these values reveals information about yourself, in that the scattering represents inconsistencies in your decisions. Usually, scattering is more pronounced near the extremes (left and right) of the chart. This is normal and simply indicates areas where you have probably not had a lot of experience winning and losing money.

The shape of the curve is also important, and should be looked at with respect to the earlier section entitled "Characteristics of Utility Preference Functions." It is not at all uncommon for the curve to be imperfect, not simply the textbook concave-up, concave-down, or straight-line shape. Again, this reveals information about yourself, and warrants careful analysis.

Ultimately, the most conducive form of utility preference function for maximizing wealth is a straight line pointing upwards, decreasing absolute risk aversion, constant relative risk aversion, and near indifference to a fair gamble; i.e., we are indifferent to a gamble with anything less than the very slightest positive arithmetic mathematical expectation. If your line is anything less than this, then this may be the time for you to reflect upon what you want as well as *why*, and perhaps make necessary personal changes.

UTILITY AND THE NEW FRAMEWORK

This book does not take a stand regarding utility theory, other than this: Regardless *of your utility preference curve, you are somewhere in the leverage space, described later in the text, of Figure 9.2 for individual games, and somewhere in the n + 1 dimensional leverage space for multiple simultaneous games, and you reap the benefits of this as well as pay the consequences* no matter what your utility preference.

Oftentimes, the geometric mean criterion is criticized as it only strives to maximize wealth, and it maximizes utility only for the ln function.

Actually, if someone does not subscribe to an ln utility preference function, they can still maximize utility much as we are maximizing wealth with optimal f, except they will have a different value for optimal f at each holding period. That is, if someone's utility preference function is other than ln (wealth maximization), then their optimal f to (asymptotically) maximize utility is uniform, while at the same time, their optimal f to maximize wealth is nonuniform. In other words, if, as you make more money, your utility is such that you are willing to risk less, then your optimal f will decrease as each holding period elapses.

Do not get this confused with the notion, presented earlier, that the f that is optimal for maximizing expected average compound growth is a function of the number of holding periods at which you quit. It still is, but the idea presented here is that the f that is optimal to maximize utility is not uniform throughout the time period. For example, we have seen in our two-to-one coin toss game that if we were planning on quitting after three plays, three holding periods, we would maximize growth by betting .37868 on each and every play. That is, we uniformly bet .37868 for all three plays.

Now, if we're looking to maximize utility, and our utility function were other than that of maximizing wealth, we would not have a uniform f value to bet on each and every play. Rather, we would have a different f value to bet on each and every play.

Thus, it is possible to maximize utility with the given approach (for utility preference functions other than ln), provided you use a *nonuniform* value for f from one holding period to the next. When utility preference is ln—that is, when one prefers wealth maximization—the f that is optimal is always uniform. Thus, the optimal f is the same from one play to the next. When utility preference is other than ln, wealth maximization, a nonuniform optimal f value from one holding period to the next is called for.

Like maximizing wealth, utility can also be maximized in the very same fashion that we are maximizing wealth. This can be accomplished by assigning *utils*, rather than a dollar value for the outcomes to each scenario. A util is simply a unit of satisfaction. The scenario set must also contain negative util scenarios, just as in wealth maximization, you must have a scenario that encompasses losing money. Also, the (arithmetic) mathematical expectation of the scenario set must be positive in terms of utils, or negative if it improves the overall mix of components.

But, how do you determine the nonuniform value for f as you go through holding periods when your utility preference curve is other than ln? As each new holding period is encountered, and you update the outcome values (specified in utils) as your account equity itself changes, you will get a new optimal f value, which, divided by the largest losing scenario (specified in utils), yields an optimal $f\$$ value (also specified in utils), and you will know how many contracts to trade. The process is simple; you simply substitute utils in lieu of dollars. The only other thing you need to do is keep track of your account's cumulative utils (i.e., the surrogate for equity). Notice that, if you do this and your utility preference function is other than ln, you will actually end up with a nonuniform optimal f, in terms of *dollars*, from one holding period to the next.

For example, if we are again faced with a coin-toss game that offers us \$2 on heads being tossed, and a loss of \$1 if tails is tossed, how much should we bet? We know that if we want to maximize wealth, and we are going to play this game repeatedly, and we have to play each subsequent play with money that we started the game with, we should optimally bet 25% of our stake on each and every play. Not only would this maximize wealth; it would also maximize utility if we determined that a win of \$2 were twice as valuable to us as a loss of \$1.

But what if a win of $2 were only one-and-a-half times as valuable to us as a loss of $1? To determine how to maximize utility then, we assign a util value of -1 to the losing scenario, the tails scenario, and a utils value of 1.5 to the winning scenario, the heads scenario. Now, we determine the optimal f based upon these utils rather than dollars, and we find it to be .166666, or to bet $16\frac{2}{3}\%$ on each and every play to maximize our geometric average utility. That means we divide our total cumulative utils to this point by .166666 to determine the number of contracts.

We can then translate this into how many contracts we have per dollars in our account, and, from there, figure what the f value (between zero and one) is that we are really using (based on dollars, not utils).

If we do this, then the original two-to-one coin-toss curve of wealth maximization, which peaks at .25 (Figure 9.2), still applies, and we are at the .166666 f abscissa. Thus, we pay the consequences of being suboptimal in terms of f on our wealth. However, there is a second f curve—one based on our utility—which peaks at .166666, and we are at the optimal f on this curve. Notice that, if we were to accept the .25 optimal f on this curve, we would be way right of the peak and would pay the concomitant consequences of being right of the peak with respect to our utility.

Now, suppose we were profitable in this holding period, and we go in and update the outcomes of the scenarios based on utils, only this time, since we have more wealth, the utility of a winning scenario in the next holding period is only 1.4 utils. Again, we would find our optimal f based on utils. Again, once we determined how many units to trade in the next holding period based on our cumulative utils, we could translate it into what the f (between zero and one) is for dollars, and we would find it to be nonuniform from the previous holding period.

The example shown is one in which we assume a sequence of more than one play, where we are reusing the same money we started with. If there was only one play, one holding period, or we received new money to play at each holding period, maximizing the arithmetic expected utility would be the optimal strategy. However, in most cases we must reuse the money on the next play, the next holding period, that we have used on this last play, and, therefore, we must strive to maximize geometric expected growth. To some, this might mean maximizing the geometric expected growth of wealth; to others, the geometric expected growth of utility. The mathematics is the same for both. Both have two curves in $n + 1$ space: a wealth maximization curve and a utility maximization curve. For those maximizing the expected growth of wealth, the two are the same.

If the reader has a different utility preference curve other than ln (wealth maximization), he may apply the techniques herein, provided he substitutes a *utils* quantity for the outcome of each scenario rather than a monetary value, which will then yield a nonuniform optimal f value (one whose value changes from one holding period to the next).

Such readers are forewarned, however, that they will still pay the consequences, in terms of their wealth, for being suboptimal in the $n + 1$ dimensional leverage space of wealth maximization. Again, this is so because, regardless of your utility preference curve, you are somewhere in the leverage space of Figure 9.2 for individual games, and somewhere in the $n + 1$ dimensional leverage space for multiple simultaneous games. You reap the benefits of this, as well as pay the consequences, no matter what your utility preference function. Ideally, you will have a utility preference function and it will be ln.

CHAPTER 7

Classical Portfolio Construction

MODERN PORTFOLIO THEORY

MODERN PORTFOLIO THEORY

Recall from Chapter 4 the paradox of the optimal f and a market system's drawdown. The better a market system is, the higher the value for f. Yet the drawdown (historically), if you are trading the optimal f, can never be lower than f. Generally speaking, then, the better the market system, the greater the drawdown will be as a percent of account equity (if you are trading optimal f). That is, if you want to have the greatest geometric growth in an account, then you can count on severe drawdowns along the way.

Diversification among other market systems is an effective way to buffer this drawdown while still staying close to the peak of the f curve (i.e., without having to trim back to half f, and so on). When one market system goes into a drawdown, another one that is being traded in the account will come on strong, thus canceling the drawdown of the other. This also provides for a catalytic effect on the entire account. The market system that just experienced the drawdown (and now is getting back to performing well) will have no less funds to start with than it did when the drawdown began (thanks to the other market system canceling out the drawdown).

Given a group of market systems and their respective optimal f's, a quantifiable, optimal portfolio mix does exist. Although we cannot be certain that what was the optimal portfolio mix in the past will be optimal in the future, it is more likely to be optimal or near optimal than is the case for the optimal system parameters of the past. Whereas optimal system parameters change quite quickly from one time period to another, optimal portfolio mixes change very slowly (as do optimal f values). Generally, the correlations between market systems tend to remain constant. This is good news to a trader who has found the optimal portfolio mix, the optimal diversification among market systems.

THE MARKOWITZ MODEL

The basic concepts of modern portfolio theory emanate from a monograph written by Dr. Harry Markowitz.[1] Essentially, Markowitz proposed that sound portfolio management has to do with composition, not individual stock selection.

Markowitz argued that diversification is effective only to the extent that the correlation coefficient between the markets involved is negative. Recall the linear correlation coefficient from Chapter 1. If we have a portfolio composed of one stock, our best diversification is obtained if we choose another stock such that the correlation between the two stock prices is as low as possible. The net result would be that the portfolio, as a whole (composed of these two stocks with negative correlation), would have less variation in price than either one of the stocks alone (see Figure 7.1).

The portfolio shown in Figure 7.1 (the combination of Market Systems A and B) will have variance at least as high as the individual market systems since the market systems have a correlation of $+1.00$ to each other.

The portfolio shown in Figure 7.2 (the combination of Market Systems A and C) will have less variance than either Market System A or Market System C since there is a negative correlation between Market Systems A and C.

Markowitz proposed that investors act in a rational manner and, given the choice, would opt for a portfolio with the same return as the one they have, but with less risk, or opt for a portfolio with a higher return than the one they have but with the same risk. Further, for a given level of risk there is an optimal portfolio with the highest yield; likewise, for a given yield there is an optimal portfolio with the lowest risk. Investors with portfolios where the yield could be increased with no resultant increase in risk or investors with portfolios where the risk could be lowered with no resultant decrease in yield are said to have *inefficient* portfolios.

---

[1]Markowitz, Harry. *Portfolio Selection—Efficient Diversification of Investments.* New Haven, CT: Yale University Press, 1959.

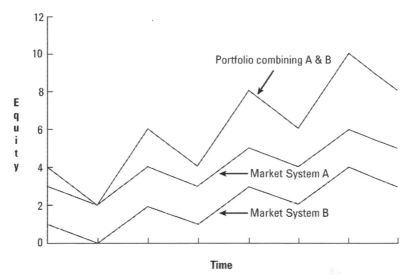

FIGURE 7.1 A portfolio of two positively correlated market systems—a poor choice

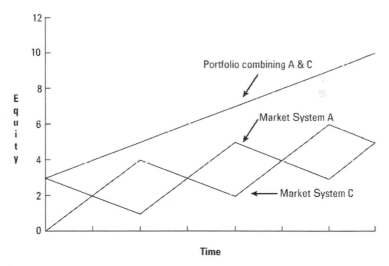

FIGURE 7.2 A portfolio of two negatively correlated market systems—a good choice

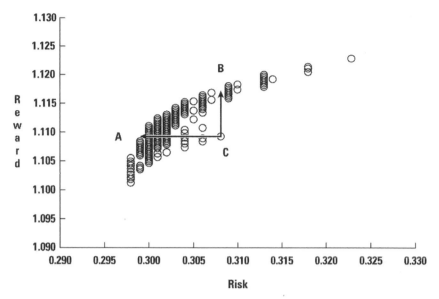

FIGURE 7.3 Risk/reward relationships for various portfolios according to modern portfolio theory

Figure 7.3 shows all of the available portfolios under a given study. If you hold Portfolio C, you would be better off with Portfolio A, as you would have the same return with less risk, or Portfolio B, where you would have more return with the same risk.

In describing this, Markowitz described what is called the *efficient frontier*. This is the set of portfolios that lie on the upper and left sides of the graph. These are portfolios where the yield can no longer be increased without increasing the risk and the risk cannot be lowered without lowering the yield. Portfolios lying on the efficient frontier are said to be *efficient* portfolios.

Portfolios lying high up and off to the right and low down and to the left are generally not very well diversified among very many issues. Portfolios lying in the middle of the efficient frontier are usually very well diversified. Which portfolio a particular investor chooses is a function of the investor's risk aversion, his willingness to assume risk. In the Markowitz model, any portfolio that lies upon the efficient frontier is said to be a good portfolio choice; where on the efficient frontier is a matter of personal preference (later on, we'll see that there is an exact optimal spot on the efficient frontier).

In Markowitz's work, risk was quantified for the first time. He described risk as the variation in a portfolio's returns, a definition many people have challenged.

DEFINITION OF THE PROBLEM

For the moment we are dropping the entire idea of optimal f; it will catch up with us later. It is easier to understand the derivation of the efficient frontier if we begin from the assumption that we are discussing a portfolio of stocks. These stocks are in a cash account and are paid for completely. That is, they are not on margin.

Under such a circumstance, we derive the efficient frontier of portfolios.[2] That is, for given stocks we want to find those with the lowest level of expected risk for a given level of expected gain, the given levels being determined by the particular investor's aversion to risk. Hence, this basic theory of Markowitz (aside from the general reference to it as Modern Portfolio Theory) is often referred to as $E–V$ theory (Expected return – Variance of return). Note that the inputs are based on returns. That is, the inputs to the derivation of the efficient frontier are the returns we would expect on a given stock and the variance we would expect of those returns. Generally, returns on stocks can be defined as the dividends expected over a given period of time plus the capital appreciation (or minus depreciation) over that period of time, expressed as a percentage gain (or loss).

Consider four potential investments, three of which are stocks and one a savings account paying $8\frac{1}{2}\%$ per year. Notice that we are defining the length

---

[2]In this chapter, an important assumption is made regarding these techniques. The assumption is that the generating distributions (the distribution of returns) have finite variance. These techniques are effective only to the extent that the input data used has finite variance. For more on this, see Fama, Eugene F., "Portfolio Analysis in a Stable Paretian Market," *Management Science* 11, pp. 404–419, 1965. Fama has demonstrated techniques for finding the efficient frontier parametrically for stably distributed securities possessing the same characteristic exponent, A, when the returns of the components all depend upon a single underlying market index. Readers should be aware that other work has been done on determining the efficient frontier when there is infinite variance in the returns of the components in the portfolio. These techniques are not covered here other than to refer interested readers to pertinent articles. For more on the stable Paretian distribution, see Chapter 2. For a discussion of infinite variance, see "The Student's Distribution" in Chapter 2.

of a holding period, the period we measure returns and their variances, as one year in this example:

Investment	Expected Return	Expected Variance of Return
Toxico	9.5%	10%
Incubeast Corp.	13%	25%
LA Garb	21%	40%
Savings Account	8.5%	0%

We can express expected returns as HPRs by adding 1 to them. Also, we can express expected variance of return as expected standard deviation of return by taking the square root of the variance. In so doing, we transform our table to:

Investment	Expected Return as an HPR	Expected Standard Deviation of Return
Toxico	1.095	.316227766
Incubeast Corp.	1.13	.5
LA Garb	1.21	.632455532
Savings Account	1.085	0

The time horizon involved is irrelevant so long as it is consistent for all components under consideration. That is, when we discuss expected return, it doesn't matter if we mean over the next year, quarter, five years, or day, as long as the expected returns and standard deviations for all of the components under consideration all have the same time frame. (That is, they must all be for the next year, or they must all be for the next day, and so on.)

Expected return is synonymous with *potential gains*, while variance (or standard deviation) in those expected returns is synonymous with *potential risk*. Note that the model is two-dimensional. In other words, we can say that the model can be represented on the upper right quadrant of the Cartesian plane (see Figure 7.4) by placing expected return along one axis (generally the vertical or Y axis) and expected variance or standard deviation of returns along the other axis (generally the horizontal or X axis).

There are other aspects to potential risk, such as potential risk of (probability of) a catastrophic loss, which E–V theory does not differentiate from

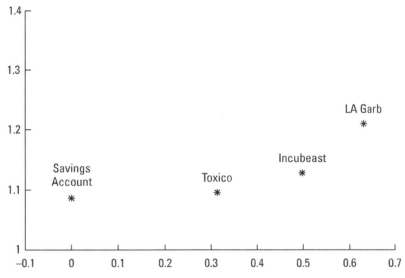

FIGURE 7.4 The upper-right quadrant of the Cartesian plane

variance of returns in regards to defining potential risk. While this may very well be true, we will not address this concept any further in this chapter so as to discuss E–V theory in its *classic* sense. However, Markowitz himself clearly stated that a portfolio derived from E–V theory is optimal only if the utility, the "satisfaction," of the investor is a function of expected return and variance in expected return only. Markowitz indicated that investor utility may very well encompass moments of the distribution higher than the first two (which are what E–V theory addresses), such as skewness and kurtosis of expected returns.

Potential risk is still a far broader and more nebulous thing than what we have tried to define it as. Whether potential risk is simply variance on a contrived sample, or is represented on a multidimensional hypercube, or incorporates further moments of the distribution, we try to define potential risk to account for our inability to really put our finger on it. That said, we will go forward defining potential risk as the variance in expected returns. However, we must not delude ourselves into thinking that risk is simply defined as such. Risk is far broader, and its definition far more elusive. There will be more on this in Chapter 12.

So the first step that an investor wishing to employ E–V theory must make is to quantify his or her beliefs regarding the expected returns and variance in returns of the securities under consideration for a certain time horizon (holding period) specified by the investor. These *parameters* can be arrived at empirically. That is, the investor can examine the past history of the securities under consideration and calculate the returns and their variances over the specified holding periods. Again the term *returns* means not only the dividends in the underlying security, but any gains in the value of the security as well. This is then specified as a percentage. *Variance* is the statistical variance of the percentage returns. A user of this approach would often perform a linear regression on the past returns to determine the return (the expected return) in the next holding period. The variance portion of the input would then be determined by calculating the variance of each past data point from what would have been predicted for that past data point (and not from the regression line calculated to predict the next expected return). Rather than gathering these figures empirically, the investor can also simply estimate what he or she believes will be the future returns and variances in those returns. Perhaps the best way to arrive at these parameters is to use a combination of the two. The investor should gather the information empirically, then, if need be, interject his or her beliefs about the future of those expected returns and their variances.

The next parameters the investor must gather in order to use this technique are the linear correlation coefficients of the returns. Again, these figures can be arrived at empirically, by estimation, or by a combination of the two.

In determining the correlation coefficients, it is important to use data points of the same time frame as was used to determine the expected returns and variance in returns. In other words, if you are using yearly data to determine the expected returns and variance in returns (on a yearly basis), then you should use yearly data in determining the correlation coefficients. If you are using daily data to determine the expected returns and variance in returns (on a daily basis), then you should use daily data in determining the correlation coefficients.

It is also very important to realize that we are determining the correlation coefficients of *returns* (gains in the stock price plus dividends), not of the underlying price of the stocks in question.

Consider our example of four alternative investments—Toxico, Incubeast Corp., LA Carb, and a savings account. We designate these with the symbols T, I, L, and S, respectively. Next, we construct a grid of the linear correlation coefficients as follows:

	I	L	S
T	−.15	.05	0
I		.25	0
L			0

From the parameters the investor has input, we can calculate the *covariance* between any two securities as:

$$COV_{a,b} = R_{a,b} * S_a * S_b \qquad (7.01)$$

where: $COV_{a,b}$ = The covariance between the ath security and the bth one.

          $R_{a,b}$ = The linear correlation coefficient between a and b.

          S_a = The standard deviation of the ath security.

          S_b = The standard deviation of the bth security.

The standard deviations, S_a and S_b, are obtained by taking the square root of the variances in expected returns for securities a and b.

Returning to our example, we can determine the covariance between Toxico (T) and Incubeast (I) as:

$$COV_{T,I} = \sqrt{-.15 * .10} * \sqrt{.25}$$
$$= -.15 * .316227766 * .5$$
$$- -.02371708245$$

Thus, given a covariance and the comprising standard deviations, we can calculate the linear correlation coefficient as:

$$R_{a,b} = COV_{a,b}/(S_a * S_b) \qquad (7.02)$$

where: $COV_{a,b}$ = The covariance between the ath security and the bth one.

          $R_{a,b}$ = The linear correlation coefficient between a and b.

          S_a — The standard deviation of the ath security.

          S_b = The standard deviation of the bth security.

Notice that the covariance of a security to itself is the variance, since the linear correlation coefficient of a security to itself is 1:

$$COV_{x,x} = 1 * S_x * S_x$$
$$= 1 * S_x^2$$
$$= S_x^2 \qquad (7.03)$$
$$= V_x$$

where: $COV_{x,x}$ = The covariance of a security to itself.

          S_x = The standard deviation of a security.

          V_x = The variance of a security.

We can now create a table of covariances for our example of four investment alternatives:

	T	I	L	S
T	.1	−.0237	.01	0
I	−.0237	.25	.079	0
L	.01	.079	.4	0
S	0	0	0	0

We now have compiled the basic parametric information, and we can begin to state the basic problem formally. First, the sum of the weights of the securities comprising the portfolio must be equal to 1, since this is being done in a cash account and each security is paid for in full:

$$\sum_{i=1}^{N} X_i = 1 \qquad (7.04)$$

where: N = The number of securities comprising the portfolio.
X_i = The percentage weighting of the ith security.

It is important to note that in Equation (7.04) each X_i must be nonnegative. That is, each X_i must be zero or positive.

The next equation defining what we are trying to do regards the expected return of the entire portfolio. This is the E in E–V theory. Essentially, what it says is that the expected return of the portfolio is the sum of the returns of its components times their respective weightings:

$$\sum_{i=1}^{N} U_i * X_i = E \qquad (7.05)$$

where: E = The expected return of the portfolio.
N = The number of securities comprising the portfolio.
X_i = The percentage weighting of the ith security.
U_i = The expected return of the ith security.

Finally, we come to the V portion of E–V theory, the variance in expected returns. This is the sum of the variances contributed by each security in the portfolio plus the sum of all the possible covariances in the portfolio:

$$V = \sum_{i=1}^{N} \sum_{j=1}^{N} X_i * X_j * COV_{i,j} \qquad (7.06a)$$

$$V = \sum_{i=1}^{N} \sum_{j=1}^{N} X_i * X_j * R_{i,j} * S_i * S_j \tag{7.06b}$$

$$V = \left(\sum_{i=1}^{N} X_i \wedge 2 * S_i \wedge 2 \right) + 2 * \sum_{i=1}^{N} \sum_{j=1+1}^{N} X_i * X_j * COV_{i,j} \tag{7.06c}$$

$$V = \left(\sum_{i=1}^{N} X_i \wedge 2 * S_i \wedge 2 \right) + 2 * \sum_{i=1}^{N} \sum_{j=1+1}^{N} X_i * X_j * R_{i,j} * S_i * S_j \tag{7.06d}$$

where: V = The variance in the expected returns of the portfolio.

N = The number of securities comprising the portfolio.

X_i = The percentage weighting of the ith security.

S_i = The standard deviation of expected returns of the ith security.

$COV_{i,j}$ = The covariance of expected returns between the ith security and the jth security.

$R_{i,j}$ = The linear correlation coefficient of expected returns between the ith security and the jth security.

All four forms of Equation (7.06) are equivalent. The final answer to Equation (7.06) is always expressed as a positive number.

We can now consider that our goal is to find those values of X_i that, when summed, equal 1, and result in the lowest value of V for a given value of E. When confronted with a problem such as trying to maximize (or minimize) a function, $H(X,Y)$, subject to another condition or constraint, such as $G(X,Y)$, one approach is to use the method of Lagrange

To do this, we must form the Lagrangian function, $F(X,Y,L)$:

$$F(X,Y,L) = H(X,Y) + L * G(X,Y) \tag{7.07}$$

Note the form of Equation (7.07). It states that the new function we have created, $F(X,Y,L)$, is equal to the Lagrangian multiplier, L—a slack variable whose value is as yet undetermined—multiplied by the constraint function $G(X,Y)$. This result is added to the original function $H(X,Y)$, whose extreme we seek to find.

Now, the simultaneous solution to the three equations will yield those points (X_1,Y_1) of relative extreme:

$$F_X(X,Y,L) = 0$$
$$F_Y(X,Y,L) = 0$$
$$F_L(X,Y,L) = 0$$

For example, suppose we seek to maximize the product of two numbers, given that their sum is 20. We will let the variables X and Y be the two numbers. Therefore, $H(X,Y) = X * Y$ is the function to be maximized given the constraining function $G(X,Y) = X + Y - 20 = 0$. We must form the Lagrangian function:

$$F(X,Y,L) = X * Y + L * (X + Y - 20)$$
$$F_X(X,Y,L) = Y + L$$
$$F_Y(X,Y,L) = X + L$$
$$F_L(X,Y,L) = X + Y - 20$$

Now we set $F_X(X,Y,L)$ and $F_Y(X,Y,L)$ both equal to zero and solve each for L:

$$Y + L = 0$$
$$Y = -L$$

and

$$X + L = 0$$
$$X = -L$$

Now setting $F_L(X,Y,L) = 0$ we obtain $X + Y - 20 = 0$. Lastly, we replace X and Y by their equivalent expressions in terms of L:

$$(-L) + (-L) - 20 = 0$$
$$2 * -L = 20$$
$$L = -10$$

Since Y equals $-L$, we can state that Y equals 10, and likewise with X. The maximum product is $10 * 10 = 100$.

The method of Lagrangian multipliers has been demonstrated here for two variables and one constraint function. The method can also be applied when there are more than two variables and more than one constraint function. For instance, the following is the form for finding the extreme when there are three variables and two constraint functions:

$$F(X,Y,Z,L_1,L_2) = H(X,Y,Z) + L_1 * G_1(X,Y,Z) + L_2 * G_2(X,Y,Z) \qquad (7.08)$$

In this case, you would have to find the simultaneous solution for five equations in five unknowns in order to solve for the points of relative extreme. We will cover how to do that a little later on.

We can restate the problem here as one where we must minimize V, the variance of the entire portfolio, subject to the two constraints that:

$$\left(\sum_{i=1}^{N} X_i * U_i \right) - E = 0 \qquad (7.09)$$

and

$$\left(\sum_{i=1}^{N} X_i\right) - 1 = 0 \tag{7.10}$$

where: N = The number of securities comprising the portfolio.
  E = The expected return of the portfolio.
  X_i = The percentage weighting of the ith security.
  U_i = The expected return of the ith security.

The minimization of a restricted multivariable function can be handled by introducing these Lagrangian multipliers and differentiating partially with respect to each variable. Therefore, we express our problem in terms of a Lagrangian function, which we call T. Let:

$$T = V + L_1 * \left(\left(\sum_{i=1}^{N} X_i * U_i\right) - E\right) + L_2 * \left(\left(\sum_{i=1}^{N} X_i\right) - 1\right) \tag{7.11}$$

where: V = The variance in the expected returns of the portfolio, from Equation (7.06).
  N = The number of securities comprising the portfolio.
  E = The expected return of the portfolio.
  X_i = The percentage weighting of the ith security.
  U_i = The expected return of the ith security.
  L_1 = The first Lagrangian multiplier.
  L_2 = The second Lagrangian multiplier.

The minimum variance (risk) portfolio is found by setting the first-order partial derivatives of T with respect to all variables equal to zero.

Let us again assume that we are looking at four possible investment alternatives: Toxico, Incubeast Corp., LA Garb, and a savings account. If we take the first-order partial derivative of T with respect to X_1 we obtain:

$$\partial T / \partial X_1 = 2 * X_1 * COV_{1,1} + 2 * X_2 * COV_{1,2} + 2 * X_3 * COV_{1,3}$$
$$+ 2 * X_4 * COV_{1,4} + L_1 * U_1 + L_2 \tag{7.12}$$

Setting this equation equal to zero and dividing both sides by 2 yields:

$$X_1 * COV_{1,1} + X_2 * COV_{1,2} + X_3 * COV_{1,3} + X_4 * COV_{1,4} + .5 * L_1$$
$$* U_1 + .5 * L_2 = 0$$

Likewise:

$$\partial T/\partial X_2 = X_1 * COV_{2,1} + X_2 * COV_{2,2} + X_3 * COV_{2,3} + X_4 * COV_{2,4}$$
$$+.5 * L_1 * U_2 + .5 * L_2 = 0$$
$$\partial T/\partial X_3 = X_1 * COV_{3,1} + X_2 * COV_{3,2} + X_3 * COV_{3,3} + X_4 * COV_{3,4}$$
$$+.5 * L_1 * U_3 + .5 * L_2 = 0$$
$$\partial T/\partial X_4 = X_1 * COV_{4,1} + X_2 * COV_{4,2} + X_3 * COV_{4,3} + X_4 * COV_{4,4}$$
$$+.5 * L_1 * U_4 + .5 * L_2 = 0$$

And we already have $\partial T/\partial L_1$ as Equation (7.09) and $\partial T/\partial L_2$ as Equation (7.10).

Thus, the problem of minimizing V for a given E can be expressed in the N-component case as N + 2 equations involving N + 2 unknowns. For the four-component case, the generalized form is:

$$X_1 * U_1 \quad + X_2 * U_2 \quad + X_3 * U_3 \quad + X_4 * U_4 \qquad\qquad\qquad = E$$
$$X_1 \quad\quad + X_2 \quad\quad + X_3 \quad\quad + X_4 \qquad\qquad\qquad\qquad = 1$$
$$X_1 * COV_{1,1} + X_2 * COV_{1,2} + X_3 * COV_{1,3} + X_4 * COV_{1,4} + .5 * L_1 * U_1 + .5 * L_2 = 0$$
$$X_1 * COV_{2,1} + X_2 * COV_{2,2} + X_3 * COV_{2,3} + X_4 * COV_{2,4} + .5 * L_1 * U_2 + .5 * L_2 = 0$$
$$X_1 * COV_{3,1} + X_2 * COV_{3,2} + X_3 * COV_{3,3} + X_4 * COV_{3,4} + .5 * L_1 * U_3 + .5 * L_2 = 0$$
$$X_1 * COV_{4,1} + X_2 * COV_{4,2} + X_3 * COV_{4,3} + X_4 * COV_{4,4} + .5 * L_1 * U_4 + .5 * L_2 = 0$$

where: E = The expected return of the portfolio.
X_i = The percentege weighting of the ith security.
U_i = The expected return of the ith security.
$COV_{A,B}$ = The covariance between securities A and B.
L_1 = The first Lagrangian multiplier.
L_2 = The second Lagrangian multiplier.

This is the generalized form, and you use this basic form for any number of components. For example, if we were working with the case of three components (i.e., N = 3), the generalized form would be:

$$X_1 * U_1 \quad + X_2 * U_2 \quad + X_3 * U_3 \qquad\qquad\qquad = E$$
$$X_1 \quad\quad + X_2 \quad\quad + X_3 \qquad\qquad\qquad\qquad = 1$$
$$X_1 * COV_{1,1} + X_2 * COV_{1,2} + X_3 * COV_{1,3} + .5 * L_1 * U_1 + .5 * L_2 = 0$$
$$X_1 * COV_{2,1} + X_2 * COV_{2,2} + X_3 * COV_{2,3} + .5 * L_1 * U_2 + .5 * L_2 = 0$$
$$X_1 * COV_{3,1} + X_2 * COV_{3,2} + X_3 * COV_{3,3} + .5 * L_1 * U_3 + .5 * L_2 = 0$$

You need to decide on a level of expected return (E) to solve for, and your solution will be that combination of weightings which yields that E with the least variance. Once you have decided on E, you now have all of the input variables needed to construct the coefficients matrix.

The E on the right-hand side of the first equation is the E you have decided you want to solve for (i.e., it is a given by you). The first line simply states that the sum of all of the expected returns times their weightings must equal the given E. The second line simply states that the sum of the weights must equal 1. Shown here is the matrix for a three-security case, but you can use the general form when solving for N securities. However, these first two lines are always the same. The next N lines then follow the prescribed form.

Now, using our expected returns and covariances (from the covariance table we constructed earlier), we plug the coefficients into the generalized form. We thus create a matrix that represents the coefficients of the generalized form. In our four-component case ($N = 4$), we thus have six rows ($N + 2$):

X_1	X_2	X_3	X_4	L_1	L_2			Answer
.095	.13	.21	.085					E
1	1	1	1					1
.1	−.0237	.01	0	.095	1			0
−.0237	.25	.079	0	.13	1			0
.01	.079	.4	0	.21	1			0
0	0	0	0	.085	1			0

Note that the expected returns are *not* expressed in the matrix as HPRs; rather, they are expressed in their "raw" decimal state.

Notice that we also have six columns of coefficients. Adding the answer portion of each equation onto the right, and separating it from the coefficients with a | creates what is known as an *augmented matrix*, which is constructed by fusing the coefficients matrix and the answer column, which is also known as the *right-hand side vector*.

Notice that the coefficients in the matrix correspond to our generalized form of the problem:

$$
\begin{aligned}
X_1 * U_1 \quad &+ X_2 * U_2 \quad &+ X_3 * U_3 \quad &+ X_4 * U_4 \quad &= E \\
X_1 \quad &+ X_2 \quad &+ X_3 \quad &+ X_4 \quad &= 1 \\
X_1 * COV_{1,1} &+ X_2 * COV_{1,2} &+ X_3 * COV_{1,3} &+ X_4 * COV_{1,4} + .5 * L_1 * U_1 + .5 * L_2 &= 0 \\
X_1 * COV_{2,1} &+ X_2 * COV_{2,2} &+ X_3 * COV_{2,3} &+ X_4 * COV_{2,4} + .5 * L_1 * U_2 + .5 * L_2 &= 0 \\
X_1 * COV_{3,1} &+ X_2 * COV_{3,2} &+ X_3 * COV_{3,3} &+ X_4 * COV_{3,4} + .5 * L_1 * U_3 + .5 * L_2 &= 0 \\
X_1 * COV_{4,1} &+ X_2 * COV_{4,2} &+ X_3 * COV_{4,3} &+ X_4 * COV_{4,4} + .5 * L_1 * U_4 + .5 * L_2 &= 0
\end{aligned}
$$

The matrix is simply a representation of these equations. To solve for the matrix, you must decide upon a level for E that you want to solve for. Once the matrix is solved, the resultant answers will be the optimal weightings required to minimize the variance in the portfolio as a whole for our specified level of E.

Suppose we wish to solve for E = .14, which represents an expected return of 14%. Plugging .14 into the matrix for E and putting in zeros for the variables L_1 and L_2 in the first two rows to complete the matrix gives us a matrix of:

X_1	X_2	X_3	X_4	L_1	L_2		Answer
.095	.13	.21	.085	0	0	|	.14
1	1	1	1	0	0	|	1
.1	−.0237	.01	0	.095	1	|	0
−.0237	.25	.079	0	.13	1	|	0
.01	.079	.4	0	.21	1	|	0
0	0	0	0	.085	1	|	0

By solving the matrix we will solve the N + 2 unknowns in the N + 2 equations.

SOLUTIONS OF LINEAR SYSTEMS USING ROW-EQUIVALENT MATRICES

A *polynomial* is an algebraic expression that is the sum of one or more terms. A polynomial with only one term is called a *monomial*; with two terms a *binomial*; with three terms a *trinomial*. Polynomials with more than three terms are simply called polynomials. The expression $4 * A^3 + A^2 + A + 2$ is a polynomial having four terms. The terms are separated by a plus (+) sign.

Polynomials come in different *degrees*. The degree of a polynomial is the value of the highest degree of any of the terms. The degree of a term is the sum of the exponents on the variables contained in the term. Our example is a third-degree polynomial since the term $4 * A^3$ is raised to the power of 3, and that is a higher power than any of the other terms in the polynomial are raised to. If this term read $4 * A^3 * B^2 * C$, we would have a sixth-degree polynomial since the sum of the exponents of the variables (3 + 2 + 1) equals 6.

A first-degree polynomial is also called a *linear equation*, and it graphs as a straight line. A second-degree polynomial is called a *quadratic*, and it graphs as a parabola. Third-, fourth-, and fifth-degree polynomials are also called *cubics*, *quartics*, and *quintics*, respectively. Beyond that there aren't any special names for higher-degree polynomials. The graphs of polynomials greater than second degree are rather unpredictable. Polynomials can have any number of terms and can be of any degree. Fortunately, we will be working only with linear equations, first-degree polynomials here.

When we have more than one linear equation that must be solved simultaneously we can use what is called the *method of row-equivalent matrices*. This technique is also often referred to as the *Gauss-Jordan procedure* or the *Gaussian elimination method.*

To perform the technique, we first create the augmented matrix of the problem by combining the coefficients matrix with the right-hand side vector as we have done. Next, we want to use what are called *elementary transformations* to obtain what is known as the *identity matrix.* An elementary transformation is a method of processing a matrix to obtain a different but equivalent matrix. Elementary transformations are accomplished by what are called *row operations.* (We will cover row operations in a moment.)

An identity matrix is a square coefficients matrix where all of the elements are zeros except for a diagonal line of ones starting in the upper left corner. For a six-by-six coefficients matrix such as we are using in our example, the identity matrix would appear as:

1	0	0	0	0	0
0	1	0	0	0	0
0	0	1	0	0	0
0	0	0	1	0	0
0	0	0	0	1	0
0	0	0	0	0	1

This type of matrix, where the number of rows is equal to the number of columns, is called a *square matrix.* Fortunately, due to the generalized form of our problem of minimizing V for a given E, we are always dealing with a square coefficients matrix.

Once an identity matrix is obtained through row operations, it can be regarded as equivalent to the starting coefficients matrix. The answers then are read from the right-hand-side vector. That is, in the first row of the identity matrix, the 1 corresponds to the variable X_1, so the answer in the right-hand side vector for the first row is the answer for X_1. Likewise, the second row of the right-hand side vector contains the answer for X_2, since the 1 in the second row corresponds to X_2. By using row operations we can make elementary transformations to our original matrix until we obtain the identity matrix. From the identity matrix, we can discern the answers, the weights $X_1, \ldots, X_N$, for the components in a portfolio. These weights will produce the portfolio with the minimum variance, V, for a given level of expected return, E.[3]

---

[3] That is, these weights will produce the portfolio with a minimum V for a given E only to the extent that our inputs of E and V for each component and the linear correlation coefficient of every possible pair of components are accurate and variance in returns finite.

Three types of row operations can be performed:

1. Any two rows may be interchanged.
2. Any row may be multiplied by any nonzero constant.
3. Any row may be multiplied by any nonzero constant and added to the corresponding entries of any other row.

Using these three operations, we seek to transform the coefficients matrix to an identity matrix, which we do in a very prescribed manner.

The first step, of course, is to simply start out by creating the augmented matrix. Next, we perform the first elementary transformation by invoking row operations rule 2. Here, we take the value in the first row, first column, which is .095, and we want to convert it to the number 1. To do so, we multiply each value in the first row by the constant 1/.095. Since any number times 1 divided by that number yields 1, we have obtained a 1 in the first row, first column. We have also multiplied every entry in the first row by this constant, 1/.095, as specified by row operations rule 2. Thus, we have obtained elementary transformation number 1.

Our next step is to invoke row operations rule 3 for all rows except the one we have just used rule 2 on. Here, for each row, we take the value of that row corresponding to the column we just invoked rule 2 on. In elementary transformation number 2, for row 2, we will use the value of 1, since that is the value of row 2, column 1, and we just performed rule 2 on column 1. We now make this value negative (or positive if it is already negative). Since our value is 1, we make it -1. We now multiply by the corresponding entry (i.e., same column) of the row we just performed rule 2 on. Since we just performed rule 2 on row 1, we will multiply this -1 by the value of row 1, column 1, which is 1, thus obtaining -1. Now we add this value back to the value of the cell we are working on, which is 1, and obtain 0.

Now on row 2, column 2, we take the value of that row corresponding to the column we just invoked rule 2 on. Again we will use the value of 1, since that is the value of row 2, column 1, and we just performed rule 2 on column 1. We again make this value negative (or positive if it is already negative). Since our value is 1, we make it -1. Now multiply by the corresponding entry (i.e., same column) of the row we just performed rule 2 on. Since we just performed rule 2 on row 1, we will multiply this -1 by the value of row 1, column 2, which is 1.3684, thus obtaining -1.3684. Again, we add this value back to the value of the cell we are working on, row 2, column 2, which is 1, obtaining $1 + (-1.3684) = -.3684$. We proceed likewise for the value of every cell in row 2, including the value of the right-hand side vector of row 2. Then we do the same for all other rows until the column we are concerned with, column 1 here, is all zeros. Notice that we need not invoke row operations rule 3 for the last row, since that already has a value of zero for column 1.

When we are finished, we will have obtained elementary transformation number 2. Now the first column is already that of the identity matrix. Now we proceed with this pattern, and in elementary transformation 3 we invoke row operations rule 2 to convert the value in the second row, second column to a 1. In elementary transformation number 4, we invoke row operations rule 3 to convert the remainder of the rows to zeros for the column corresponding to the column we just invoked row operations rule 2 on.

We proceed likewise, converting the values along the diagonals to ones per row operations rule 2, then converting the remaining values in that column to zeros per row operations rule 3 until we have obtained the identity matrix on the left. The right-hand side vector will then be our solution set.

Starting Augmented Matrix

X_1	X_2	X_3	X_4	L_1	L_2		Answer	Explanation
0.095	0.13	0.21	0.085	0	0		0.14	
1	1	1	1	0	0		1	
0.1	−0.023	0.01	0	0.095	1		0	
−0.023	0.25	0.079	0	0.13	1		0	
0.01	0.079	0.4	0	0.21	1		0	
0	0	0	0	0.085	1		0	

Elementary Transformation Number 1

1	1.3684	2.2105	0.8947	0	0		1.47368	row 1 ∗ (1/.095)
1	1	1	1	0	0		1	
0.1	−0.023	0.01	0	0.095	1		0	
−0.023	0.25	0.079	0	0.13	1		0	
0.01	0.079	0.4	0	0.21	1		0	
0	0	0	0	0.085	1		0	

Elementary Transformation Number 2

X_1	X_2	X_3	X_4	L_1	L_2		Answer	Explanation
1	1.3684	2.2105	0.8947	0	0		1.47368	
0	−0.368	−1.210	0.1052	0	0		−0.4736	row 2 + (−1 ∗ row 1)
0	−0.160	−0.211	−0.089	0.095	1		−0.1473	row 3 + (−.1 ∗ row 1)
0	0.2824	0.1313	0.0212	0.13	1		.03492	row 4 + (.0237 ∗ row 1)
0	0.0653	0.3778	−0.008	0.21	1		−0.0147	row 5 + (−.01 ∗ row 1)
0	0	0	0	0.085	1		0	

Elementary Transformation Number 3

1	1.3684	2.2105	0.8947	0	0	|	1.47368	
0	1	3.2857	−0.285	0	0	|	1.28571	row 2 ∗ (1/−.36842)
0	−0.160	−0.211	−0.089	0.095	1	|	−0.1473	
0	0.2824	0.1313	0.0212	0.13	1	|	0.03492	
0	0.0653	0.3778	−0.008	0.21	1	|	−0.0147	
0	0	0	0	0.085	1	|	0	

Elementary Transformation Number 4

1	0	−2.285	1.2857	0	0	|	−0.2857	row 1 + (−1.368421 ∗ row 2)
0	1	3.2857	−0.285	0	0	|	1.28571	
0	0	0.3164	−0.135	0.095	1	|	0.05904	row 3 + (.16054 ∗ row 2)
0	0	−0.796	0.1019	0.13	1	|	−0.3282	row 4 + (−.282431 ∗ row 2)
0	0	0.1632	0.0097	0.21	1	|	−0.0987	row 5 + (−.065315 ∗ row 2)
0	0	0	0	0.085	1	|	0	

Elementary Transformation Number 5

X_1	X_2	X_3	X_4	L_1	L_2	|	Answer	Explanation
1	0	−2.285	1.2857	0	0	|	−0.2857	
0	1	3.2857	−0.285	0	0	|	1.28571	
0	0	1	−0.427	0.3002	3.1602	|	0.18658	row 3 ∗ (1/.31643)
0	0	−0.796	0.1019	0.13	1	|	−0.3282	
0	0	0.1632	0.0097	0.21	1	|	−0.0987	
0	0	0	0	0.085	1	|	0	

Elementary Transformation Number 6

1	0	0	0.3080	0.6862	7.2233	|	0.14075	row 1 + (2.2857 ∗ row 3)
0	1	0	1.1196	−0.986	−10.38	|	0.67265	row 2 + (−3.28571 ∗ row 3)
0	0	1	−0.427	0.3002	3.1602	|	0.18658	
0	0	0	−0.238	0.3691	3.5174	|	−0.1795	row 4 + (.7966 ∗ row 3)
0	0	0	0.0795	0.1609	0.4839	|	−0.1291	row 5 + (−.16328 ∗ row 3)
0	0	0	0	0.085	1	|	0	

Elementary Transformation Number 7

1	0	0	0.3080	0.6862	7.2233	|	0.14075	
0	1	0	1.1196	−0.986	−10.38	|	0.67265	
0	0	1	−0.427	0.3002	3.1602	|	0.18658	
0	0	0	1	−1.545	−14.72	|	0.75192	row 4 ∗ (1/−.23881)
0	0	0	0.0795	0.1609	0.4839	|	−0.1291	
0	0	0	0	0.085	1	|	0	

Elementary Transformation Number 8

X_1	X_2	X_3	X_4	L_1	L_2	|	Answer	Explanation
1	0	0	0	1.1624	11.760	|	−0.0908	row 1 + (−.30806 ∗ row 4)
0	1	0	0	0.7443	6.1080	|	−0.1692	row 2 + (−1.119669 ∗ row 4)
0	0	1	0	−0.360	−3.139	|	0.50819	row 3 + (.42772 ∗ row 4)
0	0	0	1	−1.545	−14.72	|	0.75192	
0	0	0	0	0.2839	1.6557	|	−0.1889	row 5 + (−.079551 ∗ row 4)
0	0	0	0	0.085	1	|	0	

Elementary Transformation Number 9

1	0	0	0	1.1624	11.761	|	−0.0909	
0	1	0	0	0.7445	6.1098	|	−0.1693	
0	0	1	0	−0.361	−3.140	|	0.50823	
0	0	0	1	−1.545	−14.72	|	0.75192	
0	0	0	0	1	5.8307	|	−0.6655	row 5 ∗ (1/.28396)
0	0	0	0	0.085	1	|	0	

Elementary Transformation Number 10

1	0	0	0	0	4.9831	|	0.68280	row 1 + (−1.16248 ∗ row 5)
0	1	0	0	0	1.7685	|	0.32620	row 2 + (−.74455 ∗ row 5)
0	0	1	0	0	−1.035	|	0.26796	row 3 + (.3610 ∗ row 5)
0	0.0000	−0.000	1.0000	−0.000	−5.715	|	−0.2769	row 4 + (1.5458 ∗ row 5)
0	0	0	0	1	5.8312	|	−0.6655	
0	0	0	0	0	0.5043	|	0.05657	row 6 + (−.085 ∗ row 5)

Elementary Transformation Number 11

X_1	X_2	X_3	X_4	L_1	L_2		Answer	Explanation
1	0	0	0	0	4.9826		0.68283	
0	1	0	0	0	1.7682		0.32622	
0	0	1	0	0	−1.035		0.26795	
0	0.0000	−0.000	1.0000	−0.000	−5.715		−0.2769	
0	0	0	0	1	5.8312		−0.6655	
0	0	0	0	0	1		0.11217	row 6 ∗ (1/.50434)

Elementary Transformation Number 12

1	0	0	0	0	0		0.12391	row 1 + (−4.98265 ∗ row 6)
0	1	0	0	0	0		0.12787	row 2 + (−1.76821 ∗ row 6)
0	0	1	0	0	0		0.38407	row 3 + (1.0352 ∗ row 6)
0	0	0	1	0	0		0.36424	row 4 + (5.7158 ∗ row 6)
0	0	0	0	1	0		−1.3197	row 5 + (−5.83123 ∗ row 6)
0	0	0	0	0	1		0.11217	

Identity Matrix Obtained

1	0	0	0	0	0		0.12391	$= X_1$
0	1	0	0	0	0		0.12787	$= X_2$
0	0	1	0	0	0		0.38407	$= X_3$
0	0	0	1	0	0		0.36424	$= X_4$
0	0	0	0	1	0		−1.3197/.5	$= -2.6394 = L_1$
0	0	0	0	0	1		0.11217/.5	$= .22434 = L_2$

INTERPRETING THE RESULTS

Once we have obtained the identity matrix, we can interpret its meaning. Here, given the inputs of expected returns and expected variance in returns for all of the components under consideration, and given the linear correlation coefficients of each possible pair of components, for an expected yield of 14% this solution set is optimal. *Optimal*, as used here, means that this solution set will yield the lowest variance for a 14% yield. In a moment, we will determine the variance, but first we must interpret the results.

The first four values, the values for X_1 through X_4, tell us the weights (the percentages of investable funds) that should be allocated to these investments to achieve this optimal portfolio with a 14% expected return. Hence, we should invest 12.391% in Toxico, 12.787% in Incubeast, 38.407% in LA Garb, and 36.424% in the savings account. If we are looking at investing $50,000 per this portfolio mix:

Stock	Percentage	($*50,000 =$) Dollars to Invest
Toxico	.12391	$6,195.50
Incubeast	.12787	$6,393.50
LA Garb	.38407	$19,203.50
Savings	.36424	$18,212.00

Thus, for Incubeast, we would invest $6,393.50. Now assume that Incubeast sells for $20 a share. We would *optimally* buy 319.675 shares (6393.5/20). However, in the real world we cannot run out and buy fractional shares, so we would say that optimally we would buy either 319 or 320 shares. Now, the odd lot, the 19 or 20 shares remaining after we purchased the first 300, we would have to pay up for. Odd lots are usually marked up a small fraction of a point, so we would have to pay extra for those 19 or 20 shares, which in turn would affect the expected return on our Incubeast holdings, which in turn would affect the optimal portfolio mix. We are often better off to just buy the round lot—in this case, 300 shares. As you can see, more slop creeps into the mechanics of this. Whereas we can identify what the optimal portfolio is down to the fraction of a share, the real-life implementation requires again that we allow for slop.

Furthermore, the larger the equity you are employing, the more closely the real-life implementation of the approach will resemble the theoretical optimal. Suppose, rather than looking at $50,000 to invest, you were running a fund of $5 million. You would be looking to invest 12.787% in Incubeast (if we were only considering these four investment alternatives), and would therefore be investing $5,000,000 * .12787 = $639,350$. Therefore, at $20 a share, you would buy 639,350/20 = 31,967.5 shares. Again, if you restricted it down to the round lot, you would buy 31,900 shares, deviating from the optimal number of shares by about 0.2%. Contrast this to the case where you have $50,000 to invest and buy 300 shares versus the optimal of 319.675. There you are deviating from the optimal by about 6.5%.

The Lagrangian multipliers have an interesting interpretation. To begin with, the Lagrangians we are using here must be divided by .5 after the identity matrix is obtained before we can interpret them. This is in accordance with the generalized form of our problem. The L_1 variable equals $-\delta V/\delta E$. This means that L_1 represents the marginal variance in expected returns. In the case of our example, where $L_1 = -2.6394$, we can state that V is changing at a rate of $-L_1$, or $-(-2.6394)$, or 2.6394 units for every unit in E instantaneously at $E = .14$.

To interpret the L_2 variable requires that the problem first be restated. Rather than having $\Sigma_i = 1$, we will state that $\Sigma_i = M$, where M equals the dollar amount of funds to be invested. Then $L_2 = \delta V/\delta M$. In other words, L_2 represents the marginal risk of increased or decreased investment.

Returning now to what the variance of the entire portfolio is, we can use Equation (7.06) to discern the variance. Although we could use any variation of Equation (7.06a) through (7.06d), here we will use variation a:

$$V = \sum_{i=1}^{N} \sum_{j=1}^{N} X_i * X_j * COV_{i,j}$$

Plugging in the values and performing Equation (7.06a) gives:

X_i		X_j		$COV_{i,j}$		
0.12391	*	0.12391	*	0.1	=	0.0015353688
0.12391	*	0.12787	*	−0.0237	=	−0.0003755116
0.12391	*	0.38407	*	0.01	=	0.0004759011
0.12391	*	0.36424	*	0	=	0
0.12787	*	0.12391	*	−0.0237	=	−0.0003755116
0.12787	*	0.12787	*	0.25	=	0.0040876842
0.12787	*	0.38407	*	0.079	=	0.0038797714
0.12787	*	0.36424	*	0	=	0
0.38407	*	0.12391	*	0.01	=	0.0004759011
0.38407	*	0.12787	*	0.079	=	0.0038797714
0.38407	*	0.38407	*	0.4	=	0.059003906
0.38407	*	0.36424	*	0	=	0
0.36424	*	0.12391	*	0	=	0
0.36424	*	0.12787	*	0	=	0
0.36424	*	0.38407	*	0	=	0
0.36424	*	0.36424	*	0	=	0
						.0725872809

Thus, we see that at the value of $E = .14$, the lowest value for V is obtained at $V = .0725872809$.

Now suppose we decided to input a value of $E = .18$. Again, we begin with the augmented matrix, which is exactly the same as in the last example of $E = .14$, only the upper rightmost cell, that is the first cell in the right-hand-side vector, is changed to reflect this new E of .18:

Starting Augmented Matrix

X_1	X_2	X_3	X_4	L_1	L_2		Answer
0.095	0.13	0.21	0.085	0	0		0.18
1	1	1	1	0	0		1
0.1	−0.023	0.01	0	0.095	1		0
−0.023	0.25	0.079	0	0.13	1		0
0.01	0.079	0.4	0	0.21	1		0
0	0	0	0	0.085	1		0

Through the use of row operations . . . the identity matrix is obtained:

1	0	0	0	0	0		$0.21401 = X_1$
0	1	0	0	0	0		$0.22106 = X_2$
0	0	1	0	0	0		$0.66334 = X_3$
0	0	0	1	0	0		$-.0981 = X_4$
0	0	0	0	1	0		$-1.3197/.5 = -2.639 = L_1$
0	0	0	0	0	1		$0.11217/.5 = .22434 = L_2$

We then go about solving the matrix exactly as before, only this time we get a negative answer in the fourth cell down of the right-hand-side vector. Meaning, we should allocate a negative proportion, a disinvestment of 9.81% in the savings account.

To account for this, whenever we get a negative answer for any of the X_i's—which means if any of the first N rows of the right-hand-side vector is less than or equal to zero—we must pull that row $+ 2$ and that column out of the starting augmented matrix, and solve for the new augmented matrix. If either of the last two rows of the right-hand-side vector are less than or equal to zero, we don't need to do this. These last two entries in the right-hand-side vector always pertain to the Lagrangians, no matter how many or how few components there are in total in the matrix. The Lagrangians are allowed to be negative.

Since the variable returning with the negative answer corresponds to the weighting of the fourth component, we pull out the fourth column and the sixth row from the starting augmented matrix. We then use row operations to perform elementary transformations until, again, the identity matrix is obtained:

Starting Augmented Matrix

X_1	X_2	X_3	L_1	L_2		Answer
0.095	0.13	0.21	0	0	|	0.18
1	1	1	0	0	|	1
0.1	−0.023	0.01	0.095	1	|	0
−0.023	0.25	0.079	0.13	1	|	0
0.01	0.079	0.4	0.21	1	|	0

Through the use of row operations . . . the identity matrix is obtained:

1	0	0	0	0		0.1283688	$= X_1$
0	1	0	0	0	|	0.1904699	$= X_2$
0	0	1	0	0	|	0.6811613	$= X_3$
0	0	0	1	0	|	$-2.38/.5 = -4.76$	$= L_1$
0	0	0	0	1	|	$0.210944/.5 = .4219$	$= L_2$

When you must pull out a row and column like this, it is important that you remember what rows correspond to what variables, especially when you have more than one row and column to pull. Again, using an example to illustrate, suppose we want to solve for $E = .1965$. The first identity matrix we arrive at will show negative values for the weighting of Toxico, X_1, and the savings account, X_4. Therefore, we return to our starting augmented matrix:

Starting Augmented Matrix

X_1	X_2	X_3	X_4	L_1	L_2		Answer	Pertains to
0.095	0.13	0.21	0.085	0	0	|	0.1965	Toxico
1	1	1	1	0	0	|	1	Incubeast
0.1	−0.023	0.01	0	0.095	1	|	0	LA Garb
−0.023	0.25	0.079	0	0.13	1	|	0	Savings
0.01	0.079	0.4	0	0.21	1	|	0	L_1
0	0	0	0	0.085	1	|	0	L_2

Now we pull out row three and column one, the ones that pertain to Toxico, and also pull row six and column four, the ones that pertain to the savings account:

Starting Augmented Matrix

X_2	X_3	L_1	L_2		Answer	Pertains to
0.13	0.21	0	0	|	0.1965	Incubeast
1	1	0	0	|	1	LA Garb
0.25	0.079	0.13	1	|	0	L_1
0.079	0.4	0.21	1	|	0	L_2

So we will be working with the following matrix:

Starting Augmented Matrix

X_2	X_3	L_1	L_2		Answer	Pertains to
0.13	0.21	0	0	|	0.1965	Incubeast
1	1	0	0	|	1	LA Garb
0.25	0.079	0.13	1	|	0	L_1
0.079	0.4	0.21	1	|	0	L_2

Through the use of row operations . . . the identity matrix is obtained:

1	0	0	0	|	.169		Incubeast	
1	1	0	0	|	.831		LA Garb	
0	0	1	0	|	−2.97/.5		= −5.94	L_1
0	0	0	1	|	.2779695/.5		= .555939	L_2

Another method we can use to solve for the matrix is to use the *inverse* of the coefficients matrix. An inverse matrix is a matrix that, when multiplied by the original matrix, yields the identity matrix. This technique will be explained without discussing the details of matrix multiplication.

In matrix algebra, a matrix is often denoted with a boldface capital letter. For example, we can denote our coefficients matrix as **C**. The inverse to a matrix is denoted as superscripting −1 to it. The inverse matrix to **C** then is $\mathbf{C}^{-1}$.

To use this method, we need to first discern the inverse matrix to our co-efficients matrix. To do this, rather than start by augmenting the right-hand-side vector onto the coefficients matrix, we augment the identity matrix itself onto the coefficients matrix. For our four-stock example:

Starting Augmented Matrix

X_1	X_2	X_3	X_4	L_1	L_2		Identity Matrix					
0.095	0.13	0.21	0.085	0	0	|	1	0	0	0	0	0
1	1	1	1	0	0	|	0	1	0	0	0	0
0.1	−0.023	0.01	0	0.095	1	|	0	0	1	0	0	0
−0.023	0.25	0.079	0	0.13	1	|	0	0	0	1	0	0
0.01	0.079	0.4	0	0.21	1	|	0	0	0	0	1	0
0	0	0	0	0.085	1	|	0.	0	0	0	0	1

Now we proceed using row operations to transform the coefficients matrix to an identity matrix. In the process, since every row operation performed on the left is also performed on the right, we will have transformed the identity matrix on the right-hand side into the inverse matrix C^{-1}, of the coefficients matrix C. In our example, the result of the row operations yields:

C				C^{-1}				
1 0 0 0 0 0 |	2.2527	−0.1915	10.1049	0.9127	−1.1370	−9.8806
0 1 0 0 0 0 |	2.3248	−0.1976	0.9127	4.1654	−1.5726	−3.5056
0 0 1 0 0 0 |	6.9829	−0.5935	−1.1370	−1.5726	0.6571	2.0524
0 0 0 1 0 0 |	−11.5603	1.9826	−9.8806	−3.5056	2.0524	11.3337
0 0 0 0 1 0 |	−23.9957	2.0396	2.2526	2.3248	6.9829	−11.5603
0 0 0 0 0 1 |	2.0396	−0.1734	−0.1915	−0.1976	−0.5935	1.9826

Now we can take the inverse matrix, C^{-1}, and multiply it by our original right-hand-side vector. Recall that our right-hand-side vector is:

E
S
0
0
0
0

Whenever we multiply a matrix by a columnar vector (such as this) we multiply all elements in the first column of the matrix by the first element in the vector, all elements in the second column of the matrix by the second element in the vector, and so on. If our vector were a row vector, we would multiply all elements in the first row of the matrix by the first element in the vector, all elements in the second row of the matrix by the second element in the vector, and so on. Since our vector is columnar, and since the last four elements are zeros, we need only multiply the first column of the inverse matrix by E (the expected return for the portfolio) and the second column of the inverse matrix by S, the sum of the weights. This yields the following set of equations, which we can plug values for E and S into and obtain the optimal weightings. In our example, this yields:

$$
\begin{aligned}
E * 2.2527 + S * -0.1915 &= \text{Optimal weight for first stock} \\
E * 2.3248 + S * -0.1976 &= \text{Optimal weight for second stock} \\
E * 6.9829 + S * -0.5935 &= \text{Optimal weight for third stock} \\
E * -11.5603 + S * 1.9826 &= \text{Optimal weight for fourth stock} \\
E * -23.9957 + S * 2.0396 &= .5 \text{ of first Lagrangian} \\
E * 2.0396 + S * -0.1734 &= .5 \text{ of second Lagrangian}
\end{aligned}
$$

Thus, to solve for an expected return of 14% (E = .14) with the sum of the weights equal to 1:

$$
\begin{aligned}
.14 * 2.2527 + 1 * -0.1915 &= .315378 - .1915 &= .1239 \ \text{Toxico} \\
.14 * 2.3248 + 1 * -0.1976 &= .325472 - .1976 &= .1279 \ \text{Incubeast} \\
.14 * 6.9829 + 1 * -0.5935 &= .977606 - .5935 &= .3841 \ \text{LA Garb} \\
.14 * -11.5603 + 1 * 1.9826 &= -1.618442 + 1.9826 &= .3641 \ \text{Savings} \\
.14 * -23.9957 + 1 * 2.0396 &= -3.359398 + 2.0396 &= -1.319798 * 2 \\
&= -2.6395 \ L_1 \\
.14 * 2.0396 + 1 * -0.1734 &= .285544 - .1734 &= .1121144 * 2 \\
&= .2243 \ L_2
\end{aligned}
$$

Once you have obtained the inverse to the coefficients matrix, you can quickly solve for any value of E provided that your answers, the optimal weights, are all positive. If not, again you must create the coefficients matrix without that item, and obtain a new inverse matrix.

Thus far we have looked at investing in stocks from the long side only. How can we consider short sale candidates in our analysis?

To begin with, you would be looking to sell short a stock if you expected it would decline. Recall that the term "returns" means not only the dividends in the underlying security, but any gains in the value of the security as well. This figure is then specified as a percentage. Thus, in determining the returns of a short position, you would have to estimate what percentage gain you would expect to make on the declining stock, and from that you would then need to *subtract* the dividend (however many dividends go ex-date over the holding period you are calculating your E and V on) as a percentage.[4] Lastly, any linear correlation coefficients of which the stock you are looking to short is a member must be multiplied by −1. Therefore, since the linear correlation coefficient between Toxico and Incubeast is −.15, if you were looking to short Toxico, you would multiply this by −1. In such a case you would use $-.15 * -1 = .15$ as the linear correlation coefficient. If you were looking to short both of these stocks, the linear correlation coefficient between the two would be $-.15 * -1 * -1 = -.15$. In other words, if you are looking to short both stocks, the linear correlation coefficient between them remains unchanged, as it would if you were looking to go long both stocks.

Thus far we have sought to obtain the optimal portfolio, and its variance, V, when we know the expected return, E, that we seek. We can also solve for E when we know V. The simplest way to do this is by iteration using the techniques discussed thus far in this chapter.

There is much more to matrix algebra than is presented in this chapter. There are other matrix algebra techniques to solve systems of linear equations. Often, you will encounter reference to techniques such as Cramer's Rule, *the* Simplex Method, *or the* Simplex Tableau. *These are techniques similar to the ones described in this chapter, although more involved. There are a multitude of applications in business and science for matrix algebra, and the topic is considerably involved. We have only etched the surface, just enough for what we need to accomplish. For a more detailed discussion of matrix algebra and its applications in business and science, the reader is referred to* Sets, Matrices, and Linear Programming, *by Robert L. Childress.*

---

[4]In this chapter we are assuming that all transactions are performed in a cash account. Thus, even though a short position is required to be performed in a margin account as opposed to a cash account, we will not calculate interest on the margin.

CHAPTER 8

The Geometry of Mean Variance Portfolios

*W*e have now covered how to find the optimal fs for a given market system from a number of different standpoints. Also, we have seen how to derive the efficient frontier. In this chapter we show how to combine the two notions of optimal f and classical portfolio theory. Furthermore, we will delve into an analytical study of the geometry of portfolio construction.

THE CAPITAL MARKET LINES (CMLs)

We can improve upon the performance of any given portfolio by combining a certain percentage of the portfolio with cash. Figure 8.1 shows this relationship graphically.

In Figure 8.1, point A represents the return on the risk-free asset. This would usually be the return on 91-day Treasury bills. Since the risk, the standard deviation in returns, is regarded as nonexistent, point A is at zero on the horizontal axis.

Point B represents the tangent portfolio. It is the only portfolio lying upon the efficient frontier that would be touched by a line drawn from the risk-free rate of return on the vertical axis and zero on the horizontal axis. Any point along line segment AB will be composed of the portfolio at Point B and the risk-free asset. At point B, all of the assets would be in the portfolio, and at point A all of the assets would be in the risk-free asset. Anywhere in between points A and B represents having a portion of the assets in both the portfolio and the risk-free asset. Notice that any portfolio along line segment

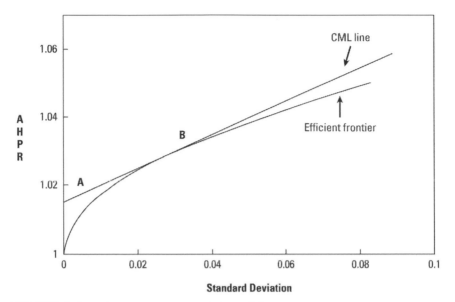

FIGURE 8.1   Enhancing returns with the risk-free asset

AB dominates any portfolio on the efficient frontier at the same risk level, since being on the line segment AB has a higher return for the same risk. Thus, an investor who wanted a portfolio less risky than portfolio B would be better off to put a portion of his or her investable funds in portfolio B and a portion in the risk-free asset, as opposed to owning 100% of a portfolio on the efficient frontier at a point less risky than portfolio B.

The line emanating from point A, the risk-free rate on the vertical axis and zero on the horizontal axis, and emanating to the right, tangent to one point on the efficient frontier, is called the *capital market line* (CML). To the right of point B, the CML line represents portfolios where the investor has gone out and borrowed more money to invest further in portfolio B. Notice that an investor who wanted a portfolio with a greater return than portfolio B would be better off to do this, as being on the CML line right of point B dominates (has higher return than) those portfolios on the efficient frontier with the same level of risk.

Usually, point B will be a very well-diversified portfolio. Most portfolios high up and to the right and low down and to the left on the efficient frontier have very few components. Those in the middle of the efficient frontier, where the tangent point to the risk-free rate is, usually are very well diversified.

It has traditionally been assumed that all rational investors will want to get the greatest return for a given risk and take on the lowest risk for a given return. Thus, all investors would want to be somewhere on the CML line. In other words, all investors would want to own the same portfolio, only with differing degrees of leverage. This distinction between the investment decision and the financing decision is known as the *Separation Theorem.*[1]

We assume now that the vertical scale, the E in E–V theory, represents the arithmetic average HPR (AHPR) for the portfolios and the horizontal, or V, scale represents the standard deviation in the HPRs. For a given risk-free rate, we can determine where this tangent point portfolio on our efficient frontier is, as the coordinates (AHPR, V) that maximize the following function are:

$$\text{Tangent Portfolio} = \text{MAX}\{(\text{AHPR} - (1 + \text{RFR}))/\text{SD}\} \qquad (8.01)$$

where: MAX{} = The maximum value.

AHPR = The arithmetic average HPR. This is the E coordinate of a given portfolio on the efficient frontier.

SD = The standard deviation in HPRs. This is the V coordinate of a given portfolio on the efficient frontier.

RFR = The risk-free rate.

In Equation (8.01), the formula inside the braces ({}) is known as the Sharpe ratio, a measurement of risk-adjusted returns. Expressed literally, the Sharpe ratio for a portfolio is a measure of the ratio of the expected excess returns to the standard deviation. The portfolio with the highest Sharpe ratio, therefore, is the portfolio where the CML line is tangent to the efficient frontier for a given RFR.

The Sharpe ratio, when multiplied by the square root of the number of periods over which it was derived, equals the t statistic. From the resulting t statistic it is possible to obtain a confidence level that the AHPR exceeds the RFR by more than chance alone, assuming finite variance in the returns.

The following table shows how to use Equation (8.01) and demonstrates the entire process discussed thus far. The first two columns represent the coordinates of different portfolios on the efficient frontier. The coordinates are given in (AHPR, SD) format, which corresponds to the Y and X axes of Figure 8.1. The third column is the answer obtained for Equation (8.01) assuming a 1.5% risk-free rate (equating to an AHPR of 1.015. We assume that the HPRs here are quarterly HPRs; thus, a 1.5% risk-free rate for the

---

[1]See Tobin, James, "Liquidity Preference as Behavior Towards Risk," *Review of Economic Studies* 25, pp. 65–85, February 1958.

quarter equates to roughly a 6% risk-free rate for the year.). Thus, to work out (8.01a) for the third set of coordinates (.00013, 1.002):

$$(\text{AHPR} - (1 + \text{RFR}))/\text{SD} = (1.002 - (1 + .015))/.00013$$
$$= (1.002 - 1.015)/.00013$$
$$= -.013/.00013$$
$$= -100$$

The process is completed for each point along the efficient frontier. Equation (8.01) peaks out at .502265, which is at the coordinates (02986, 1.03). These coordinates are the point where the CML line is tangent to the efficient frontier, corresponding to point B in Figure 8.1. This tangent point is a certain portfolio along the efficient frontier. The Sharpe ratio is the slope of the CML, with the steepest slope being the tangent line to the efficient frontier.

Efficient Frontier			CML line	
AHPR	SD	Eq. (8.1a)	Percentage	AHPR
		RFR = .015		
1.00000	0.00000	0	0.00%	1.0150
1.00100	0.00003	−421.902	0.11%	1.0150
1.00200	0.00013	−100.000	0.44%	1.0151
1.00300	0.00030	−40.1812	1.00%	1.0152
1.00400	0.00053	−20.7184	1.78%	1.0153
1.00500	0.00083	−12.0543	2.78%	1.0154
1.00600	0.00119	−7.53397	4.00%	1.0156
1.00700	0.00163	−4.92014	5.45%	1.0158
1.00800	0.00212	−3.29611	7.11%	1.0161
1.00900	0.00269	−2.23228	9.00%	1.0164
1.01000	0.00332	−1.50679	11.11%	1.0167
1.01100	0.00402	−0.99622	13.45%	1.0170
1.01200	0.00478	−0.62783	16.00%	1.0174
1.01300	0.00561	−0.35663	18.78%	1.0178
1.01400	0.00650	−0.15375	21.78%	1.0183
1.01500	0.00747	0	25.00%	1.0188
1.01600	0.00849	0.117718	28.45%	1.0193
1.01700	0.00959	0.208552	32.12%	1.0198
1.01800	0.01075	0.279036	36.01%	1.0204
1.01900	0.01198	0.333916	40.12%	1.0210
1.02000	0.01327	0.376698	44.45%	1.0217
1.02100	0.01463	0.410012	49.01%	1.0224
1.02200	0.01606	0.435850	53.79%	1.0231
1.02300	0.01755	0.455741	58.79%	1.0238
1.02400	0.01911	0.470873	64.01%	1.0246
1.02500	0.02074	0.482174	69.46%	1.0254

Efficient Frontier			CML line	
AHPR	SD	Eq. (8.1a)	Percentage	AHPR
1.02600	0.02243	0.490377	75.12%	1.0263
1.02700	0.02419	0.496064	81.01%	1.0272
1.02800	0.02602	0.499702	87.12%	1.0281
1.02900	0.02791	0.501667	93.46%	1.0290
1.03000	0.02986	0.502265 (peak)	100.02%	1.0300
1.03100	0.03189	0.501742	106.79%	1.0310
1.03200	0.03398	0.500303	113.80%	1.0321
1.03300	0.03614	0.498114	121.02%	1.0332
1.03400	0.03836	0.495313	128.46%	1.0343
1.03500	0.04065	0.492014	136.13%	1.0354
1.03600	0.04301	0.488313	144.02%	1.0366
1.03700	0.04543	0.484287	152.13%	1.0378
1.03800	0.04792	0.480004	160.47%	1.0391
1.03900	0.05047	0.475517	169.03%	1.0404
1.04000	0.05309	0.470873	177.81%	1.0417
1.04100	0.05578	0.466111	186.81%	1.0430
1.04200	0.05853	0.461264	196.03%	1.0444
1.04300	0.06136	0.456357	205.48%	1.0458
1.04400	0.06424	0.451416	215.14%	1.0473
1.04500	0.06720	0.446458	225.04%	1.0488
1.04600	0.07022	0.441499	235.15%	1.0503
1.04700	0.07330	0.436554	245.48%	1.0518
1.04800	0.07645	0.431634	256.04%	1.0534
1.04900	0.07967	0.426747	266.82%	1.0550
1.05000	0.08296	0.421902	277.82%	1.0567

The next column over, "percentage," represents what percentage of your assets must be invested in the tangent portfolio if you are at the CML line for that standard deviation coordinate. In other words, for the last entry in the table to be on the CML line at the .08296 standard deviation level corresponds to having 277.82% of your assets in the tangent portfolio (i.e., being fully invested and borrowing another $1.7782 for every dollar already invested to invest further). This percentage value is calculated from the standard deviation of the tangent portfolio as:

$$P = SX/ST \tag{8.02}$$

where: SX = The standard deviation coordinate for a particular point on the CML line.

   ST = The standard deviation coordinate of the tangent portfolio.

   P = The percentage of your assets that must be invested in the tangent portfolio to be on the CML line for a given SX.

Thus, the CML line at the standard deviation coordinate .08296, the last entry in the table, is divided by the standard deviation coordinate of the tangent portfolio, .02986, yielding 2.7782, or 277.82%.

The last column in the table, the CML line AHPR, is the AHPR of the CML line at the given standard deviation coordinate. This is figured as:

$$\text{ACML} = (\text{AT} * \text{P}) + ((1 + \text{RFR}) * (1 - \text{P})) \qquad (8.03)$$

where: ACML = The AHPR of the CML line at a given risk coordinate, or a corresponding percentage figured from (8.02).
AT = The AHPR at the tangent point, figured from (8.01a).
P = The percentage in the tangent portfolio, figured from (8.02).
RFR = The risk-free rate.

On occasion you may want to know the standard deviation of a certain point on the CML line for a given AHPR. This linear relationship can be obtained as:

$$\text{SD} = \text{P} * \text{ST} \qquad (8.04)$$

where: SD = The standard deviation at a given point on the CML line corresponding to a certain percentage, P, corresponding to a certain AHPR.
P = The percentage in the tangent portfolio, figured from (8.02).
ST = The standard deviation coordinate of the tangent portfolio.

THE GEOMETRIC EFFICIENT FRONTIER

The problem with Figure 8.1 is that it shows the arithmetic average HPR. When we are reinvesting profits back into the program we must look at the geometric average HPR for the vertical axis of the efficient frontier. This changes things considerably. The formula to convert a point on the efficient frontier from an arithmetic HPR to a geometric is:

$$\text{GHPR} = \sqrt{\text{AHPR}^2 - \text{V})}$$

where: GHPR = The geometric average HPR.
AHPR = The arithmetic average HPR.
V = The variance coordinate. (This is equal to the standard deviation coordinate squared.)

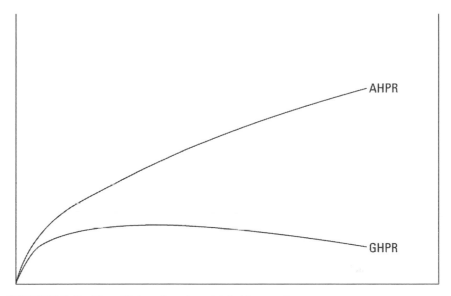

FIGURE 8.2 The efficient frontier with/without reinvestment

In Figure 8.2 you can see the efficient frontier corresponding to the arithmetic average HPRs as well as that corresponding to the geometric average HPRs. You can see what happens to the efficient frontier when reinvestment is involved.

By graphing your GHPR line, you can see which portfolio is the geometric optimal (the highest point on the GHPR line). You could also determine this portfolio by converting the AHPRs and Vs of each portfolio along the AHPR efficient frontier into GHPRs per Equation (3.04) and see which had the highest GHPR. Again, that would be the geometric optimal. However, given the AHPRs and the Vs of the portfolios lying along the AHPR efficient frontier, we can readily discern which portfolio would be geometric optimal—the one that solves the following equality:

$$AHPR - 1 - V = 0 \qquad (8.05a)$$

where: AHPR = The arithmetic average HPRs. This is the E
coordinate of a given portfolio on the efficient
frontier.

V = The variance in HPR. This is the V coordinate of a
given portfolio on the efficient frontier. This is equal
to the standard deviation squared.

Equation (8.06a) can also be written as any one of the following three forms:

$$AHPR - 1 = V \qquad (8.05b)$$

$$AHPR - V = 1 \qquad (8.05c)$$

$$AHPR = V + 1 \qquad (8.05d)$$

A brief note on the geometric optimal portfolio is in order here. Variance in a portfolio is generally directly and positively correlated to drawdown in that higher variance is generally indicative of a portfolio with higher drawdown. Since the geometric optimal portfolio is that portfolio for which E and V are equal (with $E = AHPR - 1$), then we can assume that the geometric optimal portfolio will see high drawdowns. In fact, the greater the GHPR of the geometric optimal portfolio—that is, the more the portfolio makes—the greater will be its drawdown in terms of equity retracements, since the GHPR is directly positively correlated with the AHPR. Here again is a paradox. We want to be at the geometric optimal portfolio. Yet, the higher the geometric mean of a portfolio, the greater will be the drawdowns in terms of percentage equity retracements generally. Hence, when we perform the exercise of diversification, we should view it as an exercise to obtain the highest geometric mean rather than the lowest drawdown, as the two tend to pull in opposite directions! The geometrical optimal portfolio is one where a line drawn from (0,0), with slope 1, intersects the AHPR efficient frontier.

Figure 8.2 demonstrates the efficient frontiers on a one-trade basis. That is, it shows what you can expect on a one-trade basis. We can convert the geometric average HPR to a TWR by the equation:

$$GTWR = GHPR^N$$

where: GTWR = The vertical axis corresponding to a given GHPR after N trades.

          GHPR = The geometric average HPR.

          N = The number of trades we desire to observe.

Thus, after 50 trades a GHPR of 1.0154 would be a GTWR of $1.0154^{50} = 2.15$. In other words, after 50 trades we would expect our stake to have grown by a multiple of 2.15.

We can likewise project the efficient frontier of the arithmetic average HPRs into ATWRs as:

$$ATWR = 1 + N * (AHPR - 1)$$

where: ATWR = The vertical axis corresponding to a given AHPR after N trades.

          AHPR = The arithmetic average HPR.

          N = The number of trades we desire to observe.

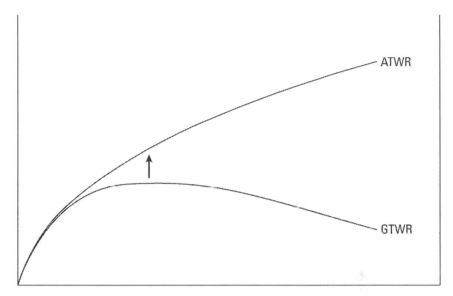

FIGURE 8.3 The efficient frontier with/without reinvestment

Thus, after 50 trades, an arithmetic average HPR of 1.03 would have made $1 + 50 * (1.03 - 1) = 1 + 50 * .03 = 1 + 1.5 = 2.5$ times our starting stake. Note that this shows what happens when we do not reinvest our winnings back into the trading program. Equation (3.06) is the TWR you can expect when constant-contract trading.

Just as Figure 8.2 shows the TWRs, both arithmetic and geometric, for one trade, Figure 8.3 shows them for a few trades later. Notice that the GTWR line is approaching the ATWR line. At some point for N, the geometric TWR will overtake the arithmetic TWR. Figure 8.4 shows the arithmetic and geometric TWRs after more trades have elapsed. Notice that the geometric has overtaken the arithmetic. If we were to continue with more and more trades, the geometric TWR would continue to outpace the arithmetic. Eventually, the geometric TWR becomes infinitely greater than the arithmetic.

The logical question is, "How many trades must elapse until the geometric TWR surpasses the arithmetic?" See the following equation, which tells us the number of trades required to reach a specific goal:

$$T = \ln(\text{Goal})/\ln(\text{Geometric Mean})$$

where: T = The expected number of trades to reach a specific goal.
           Goal = The goal in terms of a multiple on our starting stake, a TWR.
           ln( ) = The natural logarithm function.

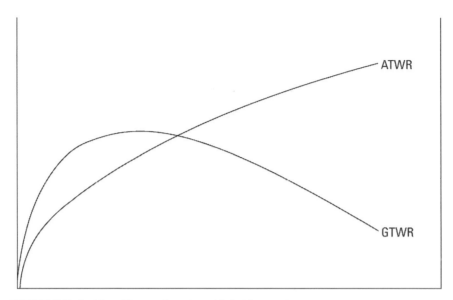

FIGURE 8.4 The efficient frontier with/without reinvestment

We let the AHPR at the same V as our geometric optimal portfolio be our goal and use the geometric mean of our geometric optimal portfolio in the denominator of the equation just mentioned. We can now discern how many trades are required to make our geometric optimal portfolio match one trade in the corresponding arithmetic portfolio. Thus:

$$T = \ln(1.031)/\ln(1.01542) \qquad (8.06)$$
$$= .035294/.0153023$$
$$= 1.995075$$

We would thus expect 1.995075, or roughly 2, trades for the optimal GHPR to be as high up as the corresponding (same V) AHPR after one trade.

The problem is that the ATWR needs to reflect the fact that two trades have elapsed. In other words, as the GTWR approaches the ATWR, the ATWR is also moving upward, albeit at a constant rate (compared to the GTWR, which is accelerating). We can relate this problem to Equations (8.07) and (8.06), the geometric and arithmetic TWRs respectively, and express it mathematically:

$$GHPR^{N} =>1 + N * (AHPR - 1) \qquad (8.07)$$

Since we know that when $N = 1$, G will be less than A, we can rephrase the question to "At how many N will G equal A?" Mathematically this is:

$$GHPR^N = 1 + N * (AHPR - 1) \qquad (8.08a)$$

which can be written as:

$$1 + N * (AHPR - 1) - GHPR^N = 0 \qquad (8.08b)$$

or

$$1 + N * AHPR - N - GHPR^N = 0 \qquad (8.08c)$$

or

$$N = (GHPR * N - 1)/(AHPR - 1) \qquad (8.08d)$$

The N that solves (8.08a) through (8.08d) is the N that is required for the geometric HPR to equal the arithmetic. All three equations are equivalent. The solution must be arrived at by iteration. Taking our geometric optimal portfolio of a GHPR of 1.01542 and a corresponding AHPR of 1.031, if we were to solve for any of Equations (8.10a) through (8.10d), we would find the solution to these equations at $N = 83.49894$. That is, at 83.49894 elapsed trades, the geometric TWR will overtake the arithmetic TWR for those TWRs corresponding to a variance coordinate of the geometric optimal portfolio.

Just as the AHPR has a CML line, so too does the GHPR. Figure 8.5 shows both the AHPR and the GHPR with a CML line for both calculated from the same risk-free rate.

The CML for the GHPR is calculated from the CML for the AHPR by the following equation:

$$CMLG = \sqrt{CMLA^2 - VT * P} \qquad (8.09)$$

where: CMLG = The E coordinate (vertical) to the CML line to the GHPR for a given V coordinate corresponding to P.
CMLA = The E coordinate (vertical) to the CML line to the AHPR for a given V coordinate corresponding to P.
P = The percentage in the tangent portfolio, figured from (8.02).
VT = The variance coordinate of the tangent portfolio.

You should know that, for any given risk-free rate, the tangent portfolio and the geometric optimal portfolio are not necessarily (and usually are

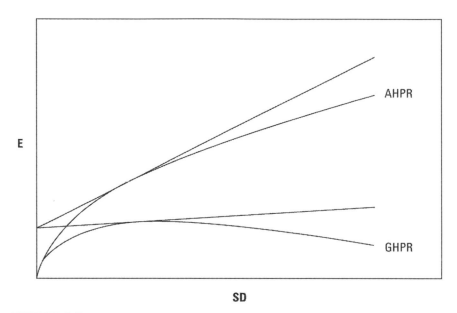

FIGURE 8.5 AHPR, GHPR, and their CML lines

not) the same. The only time that these portfolios will be the same is when the following equation is satisfied:

$$RFR = GHPROPT - 1 \qquad (8.10)$$

where: RFR = The risk-free rate.
GHPROPT = The geometric average HPR of the geometric optimal portfolio. This is the E coordinate of the portfolio on the efficient frontier.

Only when the GHPR of the geometric optimal portfolio minus 1 is equal to the risk-free rate will the geometric optimal portfolio and the portfolio tangent to the CML line be the same. If RFR > GHPROPT − 1, then the geometric optimal portfolio will be to the left of (have less variance than) the tangent portfolio. If RFR < GHPROPT − 1, then the tangent portfolio will be to the left of (have less variance than) the geometric optimal portfolio. In all cases, though, the tangent portfolio will, of course, never have a higher GHPR than the geometric optimal portfolio.

Note also that the point of tangency for the CML to the GHPR and for the CML to the AHPR is at the same SD coordinate. We could use Equation

(8.01) to find the tangent portfolio of the GHPR line by substituting the AHPR in (8.11) with GHPR. The resultant equation is:

$$\text{Tangent Portfolio} = \text{MAX}\{(\text{GHPR} - (1 + \text{RFR}))/\text{SD}\} \qquad (8.11)$$

where: MAX{} = The maximum value.
   GHPR = The geometric average HPRs. This is the E coordinate of a given portfolio on the efficient frontier.
   SD = The standard deviation in HPRs. This is the SD coordinate of a given portfolio on the efficient frontier.
   RFR = The risk-free rate.

UNCONSTRAINED PORTFOLIOS

Now we will see how to enhance returns beyond the GCML line by lifting the sum of the weights constraint. Let us return to geometric optimal portfolios. If we look for the geometric optimal portfolio among our four market systems—Toxico, Incubeast, LA Garb, and a savings account—we find it at E equal to .1688965 and V equal to .1688965, thus conforming with Equations (8.05a) through (8.05d). The geometric mean of such a portfolio would therefore be 1.094268, and the portfolio's composition would be:

Toxico	18.89891%
Incubeast	19.50386%
LA Garb	58.58387%
Savings Account	.03014%

In using Equations (8.05a) through (8.05d), you must iterate to the solution. That is, you try a test value for E (halfway between the highest and the lowest AHPRs; −1 is a good starting point) and solve the matrix for that E. If your variance is higher than E, it means the tested for value of E was too high, and you should lower it for the next attempt. Conversely, if your variance is less than E, you should raise E for the next pass. You keep on repeating the process until whichever of Equations (8.05a) through (8.05d) you choose to use, is solved. Then you will have arrived at your geometric optimal portfolio. (Note that all of the portfolios discussed thus far, whether on the AHPR efficient frontier or the GHPR efficient frontier, are determined by constraining the sum of the percentages, the weights, to 100% or 1.00.)

See the equation used in the starting augmented matrix to find the optimal weights in a portfolio. This equation dictates that the sum of the weights equal 1:

$$\left(\sum_{i=1}^{N} X_i\right) - 1 = 0$$

where: N = The number of securities comprising the portfolio.
         X_i = The percentage weighting of the ith security.

The equation can also be written as:

$$\left(\sum_{i=1}^{N} X_i\right) - 1$$

By allowing the left side of this equation to be greater than 1, we can find the unconstrained optimal portfolio. The easiest way to do this is to add another market system, called *non-interest-bearing cash* (NIC), into the starting augmented matrix. This market system, NIC, will have an arithmetic average daily HPR of 1.0 and a population standard deviation (as well as variance and covariances) in those daily HPRs of 0. What this means is that each day the HPR for NIC will be 1.0. The correlation coefficients for NIC to any other market system are always 0.

Now we set the sum of the weights constraint to some arbitrarily high number, greater than 1. A good initial value is three times the number of market systems (without NIC) that you are using. Since we have four market systems (when not counting NIC) we should set this sum of the weights constraint to $4 * 3 = 12$. Note that we are not really lifting the constraint that the sum of the weights be below some number, we are just setting this constraint at an arbitrarily high value. The difference between this arbitrarily high value and what the sum of the weights actually comes out to be will be the weight assigned to NIC.

We are not going to really invest in NIC, though. It's just a null entry that we are pumping through the matrix to arrive at the unconstrained weights of our market systems. Now, let's take the parameters of our four market systems from Chapter 7 and add NIC as well:

Investment	Expected Return as an HPR	Expected Standard Deviation of Return
Toxico	1.095	.316227766
Incubeast Corp.	1.13	.5
LA Garb	1.21	.632455532
Savings Account	1.085	0
NIC	1.00	0

The covariances among the market systems, with NIC included, are as follows:

	T	I	L	S	N
T	.1	−.0237	.01	0	0
I	−.0237	.25	.079	0	0
L	.01	.079	.4	0	0
S	0	0	0	0	0
N	0	0	0	0	0

Thus, when we include NIC, we are now dealing with five market systems; therefore, the generalized form of the starting augmented matrix is:

$$X_1 * U_1 \quad +X_2 * U_2 \quad +X_3 * U_3 \quad +X_4 * U_4 \quad X_5 * U_5 = E$$
$$X_1 \quad +X_2 \quad +X_3 \quad | X_4 \quad X_5 \quad = S$$

$$X_1 * COV_{1,1} + X_2 * COV_{1,2} + X_3 * COV_{1,3} + X_4 * COV_{1,4} + X_5$$
$$* COV_{1,5} + .5 * L_1 * U_1 + .5 * L_2 \qquad = 0$$

$$X_1 * COV_{2,1} + X_2 * COV_{2,2} + X_3 * COV_{2,3} + X_4 * COV_{2,4} + X_5$$
$$* COV_{2,5} + .5 * L_1 * U_2 + .5 * L_2 \qquad = 0$$

$$X_1 * COV_{3,1} + X_2 * COV_{3,2} + X_3 * COV_{3,3} + X_4 * COV_{3,4} + X_5$$
$$* COV_{3,5} + .5 * L_1 * U_3 + .5 * L_2 \qquad = 0$$

$$X_1 * COV_{4,1} + X_2 * COV_{4,2} + X_3 * COV_{4,3} + X_4 * COV_{4,4} + X_5$$
$$* COV_{4,5} + .5 * L_1 * U_4 + .5 * L_2 \qquad = 0$$

$$X_1 * COV_{5,1} + X_2 * COV_{5,2} + X_3 * COV_{5,3} + X_4 * COV_{5,4} + X_5$$
$$* COV_{5,5} + .5 * L_1 * U_5 + .5 * L_2 \qquad = 0$$

where: E = The expected return of the portfolio.
S = The sum of the weights constraint.
$COV_{A,B}$ = The covariance between securities A and B.
X_i = The percentage weighting of the ith security.
U_i = The expected return of the ith security.
L_1 = The first Lagrangian multiplier.
L_2 = The second Lagrangian multiplier.

Thus, once we have included NIC, our starting augmented matrix appears as follows:

X_1	X_2	X_3	X_4	X_5	L_1	L_2	Answer
.095	.13	.21	.085	0			E
1	1	1	1	0			12
.1	−.0237	.01	0	0	.095	1	0
−.0237	.25	.079	0	0	.13	1	0
.01	.079	.4	0	0	.21	1	0
0	0	0	0	0	.085	1	0
0	0	0	0	0	0	1	0

Note that the answer column of the second row, the sum of the weights constraint, is 12, as we determined it to be by multiplying the number of market systems (not including NIC) by 3.

When you are using NIC, it is important that you include it as the last, the Nth market system of N market systems, in the starting augmented matrix.

Now, the object is to obtain the identity matrix by using row operations to produce elementary transformations, as was detailed in Chapter 7. You can now create an unconstrained AHPR efficient frontier and an unconstrained GHPR efficient frontier. The unconstrained AHPR efficient frontier represents using leverage but not reinvesting.

The GHPR efficient frontier represents using leverage and reinvesting the profits. Ideally, we want to find the unconstrained geometric optimal portfolio. This is the portfolio that will result in the greatest geometric growth for us. We can use Equations (8.05a) through (8.05d) to solve for which of the portfolios along the efficient frontier is geometric optimal. In so doing, we find that no matter what value we try to solve E for (the value in the answer column of the first row), we get the same portfolio—comprised of only the savings account levered up to give us whatever value for E we want. This results in giving us our answer; we get the lowest V (in this case zero) for any given E.

What we must do, then, is take the savings account out of the matrix and start over. This time we will try to solve for only four market systems—Toxico, Incubeast, LA Garb, and NIC—and we set our sum of the weights constraint to nine. Whenever you have a component in the matrix with zero variance and an AHPR greater than one, you'll end up with the optimal portfolio as that component levered up to meet the required E.

Now, solving the matrix, we find Equations (8.05a) through (8.05d) satisfied at E equals .2457. Since this is the geometric optimal portfolio, V is also equal to .2457. The resultant geometric mean is 1.142833. The portfolio is:

Toxico	102.5982%
Incubeast	49.00558%
LA Garb	40.24979%
NIC	708.14643%

"Wait," you say. "How can you invest over 100% in certain components?" We will return to this in a moment.

If NIC is not one of the components in the geometric optimal portfolio, then you must make your sum of the weights constraint, S, higher. You must keep on making it higher until NIC becomes one of the components of the geometric optimal portfolio. Recall that if there are only two components in a portfolio, if the correlation coefficient between them is -1, and if both have positive mathematical expectation, you will be required to finance an infinite number of contracts. This is so because such a portfolio would never have a losing period. Now, the lower the correlation coefficients are between the components in the portfolio, the higher the percentage required to be invested in those components is going to be. The difference between the percentages invested and the sum of the weights constraint, S, must be filled by NIC. If NIC doesn't show up in the percentage allocations for the geometric optimal portfolio, it means that the portfolio is running into a constraint at S and is therefore not the unconstrained geometric optimal. Since you are not going to be actually investing in NIC, it doesn't matter how high a percentage it commands, as long as it is listed as part of the geometric optimal portfolio.

HOW OPTIMAL *f* FITS IN

In Chapter 7 we saw that we must determine an expected return (as a percentage) and an expected variance in returns for each component in a portfolio. Generally, the expected returns (and the variances) are determined from the current price of the stock. An optimal percentage (weighting) is then determined for each component. The equity of the account is then multiplied by a components weighting to determine the number of dollars to allocate to that component, and this dollar allocation is then divided by the current price per share to determine how many shares to have on.

That generally is how portfolio strategies are currently practiced. But it is *not* optimal. Rather than determining the expected return and variance in expected return from the current price of the component, the expected return and variance in returns should be determined from the optimal *f*, in dollars, for the component. In other words, as input you should use the arithmetic average HPR and the variance in the HPRs. Here, the HPRs used should be not of trades, but of a fixed time length such as days, weeks, months, quarters, or years—as we did in Equation (4.14):

$$\text{Daily HPR} = (A/B) + 1$$

where: A = Dollars made or lost that day.
         B = Optimal f in dollars.

We need not necessarily use days. We can use any time length we like so long as it is the same time length for all components in the portfolio (and the same time length is used for determining the correlation coefficients between these HPRs of the different components). Say the market system with an optimal f of \$2,000 made \$100 on a given day. Then the HPR for that market system for that day is 1.05.

If you are figuring your optimal f based on equalized data, you must use the following equation in order to obtain your daily HPRs:

$$\text{Daily HPR} = D\$/f\$ + 1$$

where: D\$ = The dollar gain or loss on 1 unit from the previous day.
         This is equal to
         (Tonight's Close − Last Night's Close) * Dollars per Point
         $f\$$ = The current optimal f in dollars. Here, however, the
         current price variable is last night's close.

In other words, once you have determined the optimal f in dollars for one unit of a component, you then take the daily equity changes on a one-unit basis and convert them to HPRs just mentioned—or, if you are using equalized data, you can use the equation just mentioned. When you are combining market systems in a portfolio, all the market systems should be the same in terms of whether their data, and hence their optimal fs and by-products, has been equalized or not.

Then we take the arithmetic average of the HPRs. Subtracting 1 from the arithmetic average will give us the expected return to use for that component. Taking the variance of the daily (weekly, monthly, etc.) HPRs will give the variance input into the matrix. Lastly, we determine the correlation coefficients between the daily HPRs for each pair of market systems under consideration.

Now here is the critical point. *Portfolios whose parameters (expected returns, variance in expected returns, and correlation coefficients of the expected returns) are selected based on the current price of the component will not yield truly optimal portfolios. To discern the truly optimal portfolio you must derive the input parameters based on trading one unit at the optimal f for each component. You cannot be more at the peak of the optimal* f *curve than optimal* f *itself. To base the parameters on the current market price of the component is to base your parameters arbitrarily— and, as a consequence, not necessarily optimally.*

Now let's return to the question of how you can invest more than 100% in a certain component. One of the basic premises here is that weight and quantity are not the same thing. The weighting that you derive from solving for a geometric optimal portfolio must be reflected back into the optimal *f*s of the portfolio's components. The way to do this is to divide the optimal *f*s for each component by its corresponding weight. Assume we have the following optimal *f*s (in dollars):

Toxico	$2,500
Incubeast	$4,750
LA Garb	$5,000

(Note that, if you are equalizing your data, and hence obtaining an equalized optimal *f* and by products, then your optimal *f*s in dollars will change each day based upon the previous day's closing price and Equation [2.11].)

We now divide these *f*s by their respective weightings:

Toxico	$2,500/1.025982 =	$2,436.69
Incubeast	$4,750/.4900558 =	$9,692.77
LA Garb	$5,000/.4024979 =	$12,422.43

Thus, by trading in these new "adjusted" f *values, we will be at the geometric optimal portfolio in the classical portfolio sense.* In other words, suppose Toxico represents a certain market system. By trading one contract under this market system for every $2,436.69 in equity (and doing the same with the other market systems at their new adjusted *f* values) we will be at the geometric optimal unconstrained portfolio. Likewise, if Toxico is a stock, and we regard 100 shares as "one contract," we will trade 100 shares of Toxico for every $2,436.69 in account equity. For the moment, disregard margin completely. Later in the text we will address the potential problem of margin requirements.

"Wait a minute," you protest. "If you take an optimal portfolio and change it by using optimal *f*, you have to prove that it is still optimal. But if you treat the new values as a different portfolio, it must fall somewhere else on the return coordinate, not necessarily on the efficient frontier. In other words, if you keep reevaluating *f*, you cannot stay optimal, can you?"

We are not changing the *f* values. That is, our *f* values (the number of units put on for so many dollars in equity) are still the same. We are simply performing a shortcut through the calculations, which makes it appear as though we are "adjusting" our *f* values. We derive our optimal portfolios based on the expected returns and variance in returns of trading one unit of each of the components, as well as on the correlation coefficients. We thus derive optimal weights (optimal percentages of the account to trade each component with). Thus, if a market system had an optimal f of $2,000, and an optimal portfolio weight of .5, we would trade 50% of our account at the full optimal f level of $2,000 for this market system. This is exactly the same as if we said we will trade 100% of our account at the optimal f divided by the optimal weighting ($2,000/.5) of $4000. In other words, we are going to trade the optimal f of $2,000 per unit on 50% of our equity, which in turn is exactly the same as saying we are going to trade the adjusted f of $4,000 on 100% of our equity.

The AHPRs and SDs that you input into the matrix are determined from the optimal f values in dollars. If you are doing this on stocks, you can compute your values for AHPR, SD, and optimal f on a one-share or a 100-share basis (or any other basis you like). You dictate the size of one unit.

In a nonleveraged situation, such as a portfolio of stocks that are not on margin, weighting and quantity are synonymous. Yet in a leveraged situation, such as a portfolio of futures market systems, weighting and quantity are different indeed. You can now see the idea that optimal quantities are what we seek to know, and that this is a *function* of optimal weightings.

When we figure the correlation coefficients on the HPRs of two market systems, both with a positive arithmetic mathematical expectation, we find a slight tendency toward positive correlation. This is because the equity curves (the cumulative running sum of daily equity changes) both tend to rise up and to the right. This can be bothersome to some people. One solution is to determine a least squares regression line to each equity curve and then take the difference at each point in time on the equity curve and its regression line. Next, convert this now detrended equity curve back to simple daily equity changes (noncumulative, i.e., the daily change in the detrended equity curve). Lastly, you figure your correlations on this processed data.

This technique is valid so long as you are using the correlations of daily equity changes and not prices. If you use prices, you may do yourself more harm than good. Very often, prices and daily equity changes are linked. An example would be a long-term moving average crossover system. This detrending technique must always be used with caution. Also, the daily AHPR and standard deviation in HPRs must always be figured off of non-detrended data.

A final problem that happens when you detrend your data occurs with systems that trade infrequently. Imagine two day-trading systems that give one trade per week, both on different days. The correlation coefficient between them may be only slightly positive. Yet when we detrend their data, we get very high positive correlation. This mistakenly happens because their regression lines are rising a little each day. Yet on most days the equity change is zero. Therefore, the difference is negative. The preponderance of slightly negative days with both market systems, then, mistakenly results in high positive correlation.

COMPLETING THE LOOP

One thing you will readily notice about unconstrained portfolios (portfolios for which the sum of the weights is greater than 1 and NIC shows up as a market system in the portfolio) is that the portfolio is exactly the same for any given level of E—the only difference being the degree of leverage. (This is *not* true for portfolios lying along the efficient frontier(s) when the sum of the weights is constrained). In other words, the ratios of the weightings of the different market systems to each other are always the same for any point along the unconstrained efficient frontiers (AHPR or GHPR).

For example, the ratios of the different weightings between the different market systems in the geometric optimal portfolio can be calculated. The ratio of Toxico to Incubeast is 102.5982% divided by 49.00558%, which equals 2.0936. We can thus determine the ratios of all the components in this portfolio to one another:

$$
\begin{aligned}
\text{Toxico/Incubeast} &= 2.0936 \\
\text{Toxico/LA Garb} &= 2.5490 \\
\text{Incubeast/LA Garb} &= 1.2175
\end{aligned}
$$

Now, we can go back to the unconstrained portfolio and solve for different values for E. What follows are the weightings for the components of the unconstrained portfolios that have the lowest variances for the given values of E. You will notice that the ratios of the weightings of the components are exactly the same:

	E = .1	E = .3
Toxico	.4175733	1.252726
Incubeast	.1994545	.5983566
LA Garb	.1638171	.49145

Thus, we can state that *the unconstrained efficient frontiers are the same portfolio at different levels of leverage.* This portfolio, the one that gets levered up and down with E when the sum of the weights constraint is lifted, is the portfolio that has a value of zero for the second Lagrangian multiplier when the sum of the weights equals 1.

Therefore, we can readily determine what our unconstrained geometric optimal portfolio will be. First, we find the portfolio that has a value of zero for the second Lagrangian multiplier when the sum of the weights is constrained to 1.00. One way to find this is through iteration. The resultant portfolio will be that portfolio which gets levered up (or down) to satisfy any given E in the unconstrained portfolio. That value for E which satisfies any of Equations (8.05a) through (8.05d) will be the value for E that yields the unconstrained geometric optimal portfolio.

Another equation that we can use to solve for which portfolio along the unconstrained AHPR efficient frontier is geometric optimal is to use the first Lagrangian multiplier that results in determining a portfolio along any particular point on the unconstrained AHPR efficient frontier. Recall from the previous chapter that one of the by-products in determining the composition of a portfolio by the method of row-equivalent matrices is the first Lagrangian multiplier. The first Lagrangian multiplier represents the instantaneous rate of change in variance with respect to expected return, sign reversed. A first Lagrangian multiplier equal to -2 means that at that point the variance was changing at that rate (-2) opposite the expected return, sign reversed. This would result in a portfolio that was geometric optimal.

$$L1 = -2 \qquad (8.12)$$

where: $\quad L1 =$ The first Lagrangian multiplier of a given portfolio along the unconstrained AHPR efficient frontier.[2]

Now it gets interesting as we tie these concepts together. *The portfolio that gets levered up and down the unconstrained efficient frontiers (arithmetic or geometric) is the portfolio tangent to the CML line emanating from an RFR of 0 when the sum of the weights is constrained to 1.00 and NIC is not employed.*

Therefore, we can also find the unconstrained geometric optimal portfolio by first finding the tangent portfolio to an RFR equal to 0 where the sum of the weights is constrained to 1.00, then levering this portfolio up to the point where it is the geometric optimal. But how can we determine how much to lever this constrained portfolio up to make it the equivalent of the unconstrained geometric optimal portfolio?

---

[2]Thus, we can state that the geometric optimal portfolio is that portfolio which, when the sum of the weights is constrained to 1, has a second Lagrangian multiplier equal to 0, and when unconstrained has a first Lagrangian multiplier of -2. Such a portfolio will also have a second Lagrangian multiplier equal to 0 when unconstrained.

Recall that the tangent portfolio is found by taking the portfolio along the constrained efficient frontier (arithmetic or geometric) that has the highest Sharpe ratio, which is Equation (8.01). Now we lever this portfolio up, and we multiply the weights of each of its components by a variable named q, which can be approximated by:

$$q = (E - RFR)/V \qquad (8.13)$$

where: $E =$ The expected return (arithmetic) of the tangent portfolio.

$RFR =$ The risk-free rate at which we assume you can borrow or loan.

$V =$ The variance in the tangent portfolio.

Equation (8.13) actually is a very close approximation for the actual optimal q.

An example may help illustrate the role of optimal q. Recall that our unconstrained geometric optimal portfolio is as follows:

Component	Weight
Toxico	1.025955
Incubeast	.4900436
LA Garb	.4024874

This portfolio, we found, has an AHPR of 1.245694 and variance of .2456941. Throughout the remainder of this discussion we will assume for simplicity's sake an RFR of zero. (Incidentally, the Sharpe ratio of this portfolio, (AHPR − (1 + RFR))/SD, is .49568.)

Now, if we were to input the same returns, variances, and correlation coefficients of these components into the matrix and solve for which portfolio was tangent to an RFR of zero when the sum of the weights is constrained to 1.00 and we do not include NIC, we would obtain the following portfolio:

Component	Weight
Toxico	.5344908
Incubeast	.2552975
LA Garb	.2102117

This particular portfolio has an AHPR of 1.128, a variance of .066683, and a Sharpe ratio of .49568. It is interesting to note that *the Sharpe ratio of the tangent portfolio, a portfolio for which the sum of the weights is constrained to 1.00 and we do not include NIC, is exactly the same as the Sharpe ratio for our unconstrained geometric optimal portfolio.*

Subtracting 1 from our AHPRs gives us the arithmetic average return of the portfolio. Doing so we notice that in order to obtain the same return for the constrained tangent portfolio as for the unconstrained geometric optimal portfolio, we must multiply the former by 1.9195.

$$.245694/.128 = 1.9195$$

Now if we multiply each of the weights of the constrained tangent portfolio, the portfolio we obtain is virtually identical to the unconstrained geometric optimal portfolio:

Component	Weight	* 1.9195 = Weight
Toxico	.5344908	1.025955
Incubeast	.2552975	.4900436
LA Garb	.2102117	.4035013

The factor 1.9195 was arrived at by dividing the return on the unconstrained geometric optimal portfolio by the return on the constrained tangent portfolio. Usually, though, we will want to find the unconstrained geometric optimal portfolio knowing only the constrained tangent portfolio. This is where optimal q comes in.[3] If we assume an RFR of zero, we can determine the optimal q on our constrained tangent portfolio as:

$$q = (E - RFR)/V$$
$$= (.128 - 0)/.066683$$
$$= 1.919529715$$

A few notes on the RFR. To begin with, we should always assume an RFR of zero when we are dealing with futures contracts. Since we are not actually borrowing or lending funds to lever our portfolio up or down, there is effectively an RFR of zero. With stocks, however, it is a different story. The RFR you use should be determined with this fact in mind. Quite possibly, the leverage you employ does not require you to use an RFR other than zero.

You will often be using AHPRs and variances for portfolios that were determined by using daily HPRs of the components. In such cases, you

---

[3]Latane, Henry, and Donald Tuttle, "Criteria for Portfolio Building," *Journal of Finance* 22, September 1967, pp. 362–363.

must adjust the RFR from an annual rate to a daily one. This is quite easy to accomplish. First, you must be certain that this annual rate is what is called the *effective annual interest rate*. Interest rates are typically stated as annual percentages, but frequently these annual percentages are what is referred to as the *nominal annual interest rate*. When interest is compounded semiannually, quarterly, monthly, and so on, the interest earned during a year is greater than if compounded annually (the nominal rate is based on compounding annually). When interest is compounded more frequently than annually, an effective annual interest rate can be determined from the nominal interest rate. It is the effective annual interest rate that concerns us and that we will use in our calculations. To convert the nominal rate to an effective rate we can use:

$$E = (1 + R/M)^M - 1 \qquad (8.14)$$

where: E = The effective annual interest rate.
R = The nominal annual interest rate.
M = The number of compounding periods per year.

Assume that the nominal annual interest rate is 9%, and suppose that it is compounded monthly. Therefore, the corresponding effective annual interest rate is:

$$
\begin{aligned}
E &= (1 + .09/12)^{12} - 1 \\
&= (1 + .0075)^{12} - 1 \\
&= 1.0075^{12} - 1 \\
&= 1.093806898 - 1 \\
&= .093806898
\end{aligned}
$$

Therefore, our effective annual interest rate is a little over 9.38%. Now if we figure our HPRs on the basis of weekdays, we can state that there are $365.2425/7 * 5 = 260.8875$ weekdays, on average, in a year. Dividing .093806898 by 260.8875 gives us a daily RFR of .0003595683887.

If we determine that we are actually paying interest to lever our portfolio up, and we want to determine from the constrained tangent portfolio what the unconstrained geometric optimal portfolio is, we simply input the value for the RFR into the Sharpe ratio, Equation (8.01), and the optimal q, Equation (8.13).

Now to close the loop. Suppose you determine that the RFR for your portfolio is not zero, and you want to find the geometric optimal portfolio without first having to find the constrained portfolio tangent to your applicable RFR. Can you just go straight to the matrix, set the sum of the weights to some arbitrarily high number, include NIC, and find the unconstrained geometric optimal portfolio when the RFR is greater than zero? Yes, this is easily accomplished by subtracting the RFR from the expected returns of each of the components, but not from NIC (i.e., the expected return for NIC remains at zero, or an arithmetic average HPR of 1.00). Now, solving the matrix will yield the unconstrained geometric optimal portfolio when the RFR is greater than zero.

Since the unconstrained efficient frontier is the same portfolio at different levels of leverage, you cannot put a CML line on the unconstrained efficient frontier. You can only put CML lines on the AHPR or GHPR efficient frontiers if they are constrained (i.e., if the sum of the weights equals 1). It is not logical to put CML lines on the AHPR or GHPR unconstrained efficient frontiers.

CHAPTER 9

The Leverage
Space Model

Since the 1950s, when formal portfolio construction was put forth, people have sought to discern optimal portfolios as a function of two competing entities, risk and return. The objective was to maximize return and minimize risk. This is the old paradigm. It's how we have been taught to think.

Quoting from Kuhn,[1] "Acquisition of a paradigm and of the more esoteric type of research it permits is a sign of maturity in the development of any given scientific field."

This is precisely what happened. Portfolio construction, after the second world war, acquired a mathematical rigor that had been missing prior thereto. Earlier, it was, as in so many other fields, the fact-gathering phase where each bit of data seemed equally relevant. However, with the paradigm presented as the so-called *Modern Portfolio Theory* (a.k.a. *E-V Theory or mean-variance model*), the more esoteric type of research emerged.

Particularly troubling with this earlier paradigm was the fact that the unwanted entity, risk, was never adequately defined. Initially, it was argued that risk was the variance in returns. Later, as the arguments that the variance in returns may be infinite or undefined, and that the dispersion in returns wasn't really risk, calamitous loss was risk, the definitions of risk became ever more muddled.

Overcoming ignorance often requires a new and different way of looking at things.

---

[1]Thomas S. Kuhn, *The Structure of Scientific Reduction*, The University of Chicago Press, 1962.

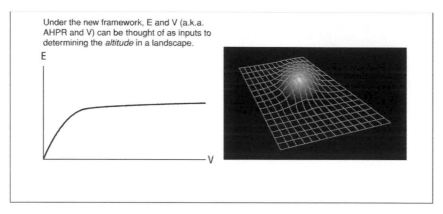

FIGURE 9.1   Conceptual view of the old framework, *left*, with the new, *right*

WHY THIS NEW FRAMEWORK IS BETTER

For nearly four decades, portfolio construction was envisioned in a two-dimensional plane, where return made up the vertical axis, and risk—actually, some surrogate measure of risk—was the horizontal axis. The basic notion was to get as great a return for a given level of risk, or as low a level of risk for a given level of return, as was possible on this two-dimensional plane (see Figure 9.1).

The new framework to be presented is an altogether new way of viewing portfolio construction, different than looking at portfolios in a two-dimensional, risk-competing-with-return sense.[2] There are a number of reasons to opt for the new framework over the old.

The new approach is superior because the inputs are no longer along the lines of expected returns and (the rather nebulous) variance in expected returns, or some other ersatz measure of risk. The inputs to this new model are different *scenarios* of different outcomes that the investments may take (a more accurate approximation for the real distribution of returns). Now, rather than estimating things like expected returns and variance in those expected returns, the inputs are much closer to what the investment manager may be thinking (e.g., a y % chance of an x % gain or loss, etc.). Now,

---

[2]In Chapter 12 we will see, however, how to take the portfolio constructed from the methods outlined in this chapter, juxtaposed to its respective drawdown and thus truly maximize return for a given level of "risk."

the investment manager can even account for the *far-out*, slim-probability scenarios as inputs to the new model.

What the investment manager uses as inputs to the new model are *spectrums* of scenarios for each market or market system (a given market traded with a given approach). The new model discerns optimal allocations to each scenario spectrum based on trading multiple, simultaneously traded scenario spectrums.

As we have seen, the old model not only used returns and variance in those returns as inputs to the model, but also used the correlation coefficients of the pairwise combinations of those streams of returns.

This last parameter, the *correlation coefficients of the pairwise combinations of the streams in returns,* is critical. Consider again the case of our two-to-one coin toss. If we are playing that particular game alone, our optimal f is .25.

If, however, we play a second and simultaneous game (and, for the sake of simplicity, we say it is the same game—a second two-to-one coin toss), the optimal f values now become a function of the correlation between those two games.

If the correlation is $+1.0$, we can show that, optimally, we bet $(.25 - x)$ on one game, and x on the other (where $x > =0$ and $x < =.25$). Thus, in total, when the correlation is 1.0, we never have more than optimal f exposure in total (i.e., in cases of perfect, positive correlation, the total exposure does not exceed the exposure of the single game).

If the correlation were -1.0, the optimal f then goes to .5 for each game, for a net exposure of 1.0 (100%) since, at such a value of the correlation coefficient, a losing sequence of such simultaneous games is impossible for even one play.

If the correlation is zero, we can determine that the optimal bet size between these two games now is .23 on each game, for a total exposure of .46 per play. Note that this exceeds the total exposure of .25 for a single game. Interestingly, when one manages the bankroll for optimal growth, *diversification* clearly does *not* reduce risk; rather, it increases it, as evident here if the one-in-four chance of both simultaneous plays were to go against the bettor, a 46% drawdown on equity would immediately occur.

Typically, correlations as they approach zero only see the optimal f buffered by a small fraction, as evidenced in this illustration of two simultaneously played two-in-one coin tosses.

Here, we are measuring the correlation of binomially distributed outcomes (heads or tails), and the outcomes are random, not generated by human emotions. In other types of environments, such as market prices, correlation coefficients begin to exhibit a very dangerous characteristic.

When a large move occurs in one component of the pairwise combination, there is a tendency for correlation to increase, often very dramatically.

Additionally, since I am speaking here of, say, market A making the large move, and its correlation to B, then too can I expect A and C to see an increase in their correlation coefficient in those time periods of the large move, and hence between B and C during those periods where I see a large move in A.

In short, when the big moves come, things tend to line up and move together (to a far greater degree than the correlation coefficient implies). In incidental time periods, which are most time periods, the correlation coefficients tend back toward zero.

To see this, consider the following study. Here, I tried to choose random and disparate markets. Surely, everyone may have picked a different basket than the random one I drew here, but this basket will illustrate the effect as well as any other. I took three commodities—crude oil (CL), gold (GC), and corn (C)—using continuous back-adjusted contracts, the use of which I devised while working with Bruce Babcock in 1985. I also put in the S&P 500 Cash Stock Index (SPX) and the prices of four individual stocks, Exxon (XOM), Ford (F), Microsoft (MSFT), and Pfizer (PFE). The data used were from the beginning of January 1986 through May 2006—nearly 20 years.

I used daily data, which required some alignment for days where some exchanges were closed and others were not. Particularly troublesome here was the mid-September 2001 period.

However, despite this unavoidable slop (which, ultimately, has little bearing on these results), the study bears out this dangerous characteristic of using correlation coefficients for market-traded pairwise price sequences.

Each market was reduced to a daily percentage of the previous day merely by converting the daily prices for each day as divided by the price of that item on the previous day. Afterward, for each market, I calculated the standard deviation in these daily price percentage changes.

Taking these eight different markets, I first ran their correlation coefficients over the daily percentage price data in question. This is shown in the "All days," section, and *is* the benchmark, as it is typically what *would* be used in constructing the classical portfolio of these components.

Next, I took each component and ran a study wherein the correlations of all components in the portfolio were looked at, but only on those days where the distinguishing component moved beyond 3 standard deviations that day. This was also done for days where the distinguishing component moved less than one standard deviation that day (the "Incidental days").

This can be seen as follows. The section titled "CL beyond 3 sigma" shows the correlation of all components in the study period on those days where crude oil had a move in excess of 3 standard deviations.

Similarly, the section that follows, where we see "CL within 1 sigma," shows the correlation of all components in the study period on those days where crude oil had a move of less than 1 standard deviation.

Consider now the correlation for crude oil and gold, which shows for "All days" as 0.18168298092886612. When crude oil has had a move in excess of 3 standard deviations, gold has moved much more lockstep in the same direction, now exhibiting a correlation of 0.6060715468257946.

On those more "Incidental days," where crude oil has moved less than 1 standard deviation, gold has moved nearly randomly with respect to it, now showing a correlation coefficient of 0.08754532513257751.

Of note on the method of calculation used in determining the means of the percentage price changes, which are used to discern standard deviations in the percentage price changes, as well as the standard deviations themselves, I did *not* calculate these simply over the entire data set. To do so would have been to have committed the error of perfect foreknowledge. Rather, at each date through the chronology of the data used, the means and standard deviations were calculated only up to that date, as a rolling 200-day window. Thus, I calculated rolling 200-day standard deviations so as to avoid the fore-knowledge trap. Thus, the actual starting date, after the 200-day required data buildup period, was (ironically) October 19, 1987 (and therefore yields a total of 4,682 trading days in this study).

This study is replete with example after example of this effect of large moves in one market portending corresponding large moves in other markets, and vice versa. As the effect of correlation is magnified, the conditions become more extreme. For example, look at Ford (F) and Pfizer (PFE). On all days, the correlation between these two stocks is 0.15208857952056634, yet, when the S&P 500 Index (SPX) moves greater than 3 standard deviations, the Ford-Pfizer correlation becomes 0.7466939906546621. On days where the S&P 500 Index moves less than 1 standard deviation, the correlation between Ford and Pfizer shrinks to a mere 0.0253249911811074.

Take a look at corn (C) and Microsoft (MSFT). On all days the correlation in the study between these two disparate, tradable items was 0.022097632770092066. Yet, when gold (GC) moved more than 3 standard deviations, the correlation between corn and Microsoft rose to 0.24606355445287773. When gold was within 1 standard deviation, this shrinks to 0.011571945077398543.

Sometimes, the exaggeration occurs in a negative sense. Consider gold and the S&P. On all days, the correlation is −0.140572093416518. On days where crude oil moves more than 3 standard deviations, this rises to −0.49033570418986916, and when crude oil's move is less than 1 standard deviation, it retracts in to −0.10905863263068859.

All days
(4682 of 4682 data points)

CL	GC	0.18168298092886612
CL	C	0.06008614529554469
CL	SPX	−0.06337343876830624
CL	XOM	0.12237528928675677
CL	F	−0.05607116699084516
CL	MSFT	−0.008336837297919815
CL	PFE	−0.03971512674407262
GC	C	0.07558861340485105
GC	SPX	−0.140572093416518
GC	XOM	−0.03185944850989464
GC	F	−0.07649165457662757
GC	MSFT	−0.06175684105762799
GC	PFE	−0.06573632473755334
C	SPX	0.03147493683616401
C	XOM	0.02623205260520187
C	F	0.030704335620653868
C	MSFT	0.022097632770092066
C	PFE	0.013735926438934488
SPX	XOM	0.4463700373245729
SPX	F	0.44747978695133384
SPX	MSFT	0.4644715701985205
SPX	PFE	0.39712431335046133
XOM	F	0.18406887477828698
XOM	MSFT	0.17555859825807965
XOM	PFE	0.17985680973424692
F	MSFT	0.19472214174383298
F	PFE	0.15208857952056634
MSFT	PFE	0.15655275607502264

CL beyond 3 sigma
(57 of 4682 data points)

CL	GC	0.6060715468257946
CL	C	0.16773966461586043
CL	SPX	−0.4889254290079874
CL	XOM	0.30834231052418093
CL	F	−0.4057990096591226
CL	MSFT	−0.04329861261414148003
CL	PFE	−0.2862619588205237
GC	C	0.2136979555796156
GC	SPX	−0.49033570418986916
GC	XOM	−0.04638590060660794
GC	F	−0.34101700944373253
GC	MSFT	−0.04792818652129692
GC	PFE	−0.23339206379967778
C	SPX	−0.13498070111097166
C	XOM	0.1282166452534864
C	F	−0.07574638268565898
C	MSFT	−0.046367278697754616
C	PFE	0.02171787217124139
SPX	XOM	0.3720220077411345
SPX	F	0.7508447148878216
SPX	MSFT	0.26583237333985554
SPX	PFE	0.5576012125272648
XOM	F	0.19597328384286486
XOM	MSFT	0.2817265916572091
XOM	PFE	0.14847216371343516
F	MSFT	0.24795671036100472
F	PFE	0.45818973137924285
MSFT	PFE	0.09703388355674258

CL within 1 sigma
(3355 of 4682 data points)

CL	GC	0.08754532513257751
CL	C	0.0257556754226136
CL	SPX	0.018864830486201915
CL	XOM	0.07275446285160611
CL	F	−0.006035919250607675
CL	MSFT	0.0039040541983706815
CL	PFE	−6.725739893499835E-4
GC	C	0.07071392644936346
GC	SPX	−0.10905863263068859
GC	XOM	−0.038050306091619565
GC	F	−0.046995783946869804
GC	MSFT	−0.035463714683264834
GC	PFE	−0.06020481387795751
C	SPX	0.028262511037748024
C	XOM	0.017421211262930312
C	F	0.027058713971227104
C	MSFT	0.023756786611237552
C	PFE	0.014823926818879715
SPX	XOM	0.41388474915130574
SPX	F	0.4175520920293062
SPX	MSFT	0.4157760485443937
SPX	PFE	0.36192135400550934
XOM	F	0.16278071355175439
XOM	MSFT	0.1319530034838986
XOM	PFE	0.147701570495354
F	MSFT	0.16753417657993877
F	PFE	0.12522622923381158
MSFT	PFE	0.12969188109495833

GC beyond 3 sigma
(49 of 4682 data points)

CL	GC	0.37610799881628454
CL	C	−0.013505453061135679
CL	SPX	−0.4663766105812081
CL	XOM	−0.1236757784439896
CL	F	−0.26893323996770363
CL	MSFT	−0.25074947066586095
CL	PFE	−0.34522609666192644
GC	C	0.12339691398398928
GC	SPX	−0.2256870226039319
GC	XOM	−0.17825193598720657
GC	F	−0.2932885892847866
GC	MSFT	−0.0942827495583651
GC	PFE	−0.08178972441698702
C	SPX	0.2589426127779489
C	XOM	0.324334753787739
C	F	0.17993600277237867
C	MSFT	0.24606355445287773
C	PFE	0.0632678902662783
SPX	XOM	0.610653892/488477
SPX	F	0.7418500480107237
SPX	MSFT	0.814073269082298
SPX	PFE	0.6333158417738232
XOM	F	0.3731941584747982
XOM	MSFT	0.29680898662233957
XOM	PFE	0.5191106683884512
F	MSFT	0.5875623837594202
F	PFE	0.35514526049741935
MSFT	PFE	0 46725966739620467

C beyond 3 sigma
(63 of 4682 data points)

CL	GC	0.09340139862063926
CL	C	0.15937424801870365
CL	SPX	−0.034836945862889324
CL	XOM	0.31262202861570143
CL	F	−0.0015035928633431528
CL	MSFT	−0.03510042846355l
CL	PFE	−0.042790208990554315
GC	C	−0.07554730971707264
GC	SPX	−0.09770624459871546
GC	XOM	−0.1178996789974603
GC	F	−0.1580599457490364
GC	MSFT	−0.017408456343824652
GC	PFE	−0.05711641234541667
C	SPX	−0.12610050901450232
C	XOM	−0.06491379177062588
C	F	0.1371318020l552985
C	MSFT	0.1184669909561641
C	PFE	0.073651177457489o/
SPX	XOM	0.6379868873961733
SPX	F	0.6386287499447472
SPX	MSFT	0.3141265015844073
SPX	PFE	0.07148466884745952
XOM	F	0.352541750183325
XOM	MSFT	0.15822517152455984
XOM	PFE	−0.01714503647656309
F	MSFT	0.2515504291514764
F	PFE	−0.l/91571598Bl66248
MSFT	PFE	4.0302517044280364E-4

GC within 1 sigma
(3413 of 4682 data points)

CL	GC	0.08685001387886367
CL	C	0.03626120508953206
CL	SPX	−0.02604251050820922З
CL	XOM	0.12444488722949365
CL	F	−0.03218089855875674
CL	MSFT	−0 0015484284736459364
CL	PFE	−0.023185426431743598
GC	C	0.036165047559364234
GC	SPX	−0.1187633862400288
GC	XOM	−4.506758967026326E-5
GC	F	−0.05680170379975439
GC	MSFT	−0.04749027255821666
GC	PFE	−0.05546821106288489
C	SPX	0.020548509330959506
C	XOM	0.0098914934447098O5
C	F	0.03164457405193553
C	MSFT	0.011571945077398543
C	PFE	0.021658621577528698
SPX	XOM	0.38127728674269895
SPX	F	0.45590091052598297
SPX	MSFT	0.4658428532832456
SPX	PFE	0.34733314433363616
XOM	F	0.15700577420431003
XOM	MSFT	0.12789055576102093
XOM	PFE	0.1226203887798495
F	MSFT	0.19737706075000538
F	PFE	0.11755272888079606
MSFT	PFE	0.13784745249948008

C within 1 sigma
(3391 of 4682 data points)

CL	GC	0.17533527416024455
CL	C	0.026858830610224073
CL	SPX	−0.0732811159519982
CL	XOM	0.1028138088787534
CL	F	−0.05102926721840804
CL	MSFT	−0.01099110090227016
CL	PFE	−0.047128710608280625
GC	C	0.05773910871663286
GC	SPX	0.1360779110437837
GC	XOM	−0.02099718827227882
GC	F	−0.06222113210658744
GC	MSFT	−0.04966940059247658
GC	PFE	−0.07413097933730392
C	SPX	−0.0088328668248l027
C	XOM	−4 4357501736777734E-4
C	F	−0.003482794137395384
C	MSFT	0.001127703028657709З
C	PFE	0.006559218632362692
SPX	XOM	0.3825048808789464
SPX	F	0.41829697072918165
SPX	MSFT	0.4395087414084105
SPX	PFE	0.49804329260547564
XOM	F	0.1475733885968429
XOM	MSFT	0.13663720618579042
XOM	PFE	0.21209220175136173
F	MSFT	0.16502841838609542
F	PFE	0.188267473055017
MSFT	PFE	0.1868337356456869

SPX beyond 3 sigma
(37 of 4682 data points)

CL	GC	0.262180235243967
CL	C	0.2282732831599413
CL	SPX	0.09510759900263809
CL	XOM	0.15585802115704978
CL	F	0.03830267479460007
CL	MSFT	0.11346892107581757
CL	PFE	0.014716269207474146
GC	C	−0.2149326327219606
GC	SPX	−0.2724333717672031
GC	XOM	−0.20973685485328555
GC	F	−0.5133205870466547
GC	MSFT	−0.2718742251789026
GC	PFE	−0.15372156278838536
C	SPX	0.27252943570443455
C	XOM	0.28696147861064464
C	F	0.28903764586090686
C	MSFT	0.2682496194114376
C	PFE	0.1575739360953595
SPX	XOM	0.8804915455367398
SPX	F	0.8854422072373676
SPX	MSFT	0.9353021184213065
SPX	PFE	0.8785677290825313
XOM	F	0.7720878305603963
XOM	MSFT	0.8107472671261666
XOM	PFE	0.8581109151100405
F	MSFT	0.867848932613579
F	PFE	0.7466939906546621
MSFT	PFE	0.8244864622745551

XOM beyond 3 sigma
(31 of 4682 data points)

CL	GC	0.08619386913767751
CL	C	0.12281769759782755
CL	SPX	0.1598136682243572
CL	XOM	0.19657554427842094
CL	F	0.20764047880440853
CL	MSFT	0.20143983941373977
CL	PFE	0.06491145921791507
GC	C	−0.3440263176542505
GC	SPX	−0.6127703828515739
GC	XOM	−0.21647163055987845
GC	F	−0.5588665697340519
GC	MSFT	−0.49757437569583096
GC	PFE	−0.6574499556463053
C	SPX	0.46950837936435447
C	XOM	0.10204725109291456
C	F	0.5528812200193067
C	MSFT	0.3962060773300878
C	PFE	0.4835629447364572
SPX	XOM	0.26560300433620926
SPX	F	0.9513940647043279
SPX	MSFT	0.951627088342409
SPX	PFE	0.939838119184664
XOM	F	0.2073529344817686
XOM	MSFT	0.23527599847538386
XOM	PFE	0.1587269337304879
F	MSFT	0.9093988443935644
F	PFE	0.8974023710639419
MSFT	PFE	0.8661556879321936

SPX within 1 sigma
(3366 of 4682 data points)

CL	GC	0.1411703426148108
CL	C	0.07065326135001565
CL	SPX	−0.04672042595452156
CL	XOM	0.1369231929185177
CL	F	−0.03833898351928496
CL	MSFT	0.008249795822319618
CL	PFE	−0.039824997750446386
GC	C	0.07487815673746215
GC	SPX	−0.098702234833124
GC	XOM	0.0126749627781548
GC	F	−0.025504778030182328
GC	MSFT	−0.007650115919919071
GC	PFE	−0.03409826874750128
C	SPX	−0.0037085243318329152
C	XOM	0.007681382976920977
C	F	0.012302593393623804
C	MSFT	0.023440459199345766
C	PFE	0.020051710510815043
SPX	XOM	0.24274905226797128
SPX	F	0.25706355236368167
SPX	MSFT	0.23491561078843676
SPX	PFE	0.22050509324437187
XOM	F	0.051567190213371944
XOM	MSFT	0.011930867235937883
XOM	PFE	0.03903218211997973
F	MSFT	0.049167377717242194
F	PFE	0.0253249911811074
MSFT	PFE	0.01813554953465995

XOM within 1 sigma
(3469 of 4682 data points)

CL	GC	0.1626123169907851
CL	C	0.06385666453921195
CL	SPX	−0.10197617432497605
CL	XOM	0.10671051194661867
CL	F	−0.06561037074518512
CL	MSFT	−0.03369575980606431
CL	PFE	−0.049070460132032751
GC	C	0.0699568184904768
GC	SPX	−0.14448139178331096
GC	XOM	−0.0218388080921421
GC	F	−0.07949839243937246
GC	MSFT	−0.06427915157699021
GC	PFE	−0.056426779255276956
C	SPX	0.002666843180930068
C	XOM	0.008152806548151075
C	F	0.02130372788477299
C	MSFT	0.02696846819596459
C	PFE	0.023479323154123974
SPX	XOM	0.443945645292686
SPX	F	0.410255598243555
SPX	MSFT	0.40962971140985116
SPX	PFE	0.3337542998608116
XOM	F	0.16171670346660708
XOM	MSFT	0.1522471847121916
XOM	PFE	0.14027113549516057
F	MSFT	0.15954186850809635
F	PFE	0.09692360471545824
MSFT	PFE	0.11103574324620878

CL	GC	0.27427702981787166
CL	C	-0.036710270159938795
CL	SPX	-0.05122250042406012
CL	XOM	0.019879344178947128
CL	F	-0.1619398623288661
CL	MSFT	0.06113040620102775
CL	PFE	-0.03052373880511025
GC	C	-0.2105245502328284
GC	SPX	-0.39275282180603993
GC	XOM	-0.2660521070959948
GC	F	-0.07998977703405707
GC	MSFT	-0.39045981709259187
GC	PFE	-0.15655811237828485
C	SPX	0.4394625985396639
C	XOM	0.5111084269242103
C	F	0.05517927015323412
C	MSFT	0.418713605628322
C	PFE	0.4114006944120061
SPX	XOM	0.8858315365005958
SPX	F	0.32710966702049354
SPX	MSFT	0.9438851500634157
SPX	PFE	0.842765820623699
XOM	F	0.23769276790825533
XOM	MSFT	0.8786892436047334
XOM	PFE	0.7950187695417785
F	MSFT	0.26860165851836737
F	PFE	0.2978173791782456
MSFT	PFE	0.8111631403849762

CL	GC	0.05288220924874525
CL	C	0.03238866347529909
CL	SPX	0.23409424184528582
CL	XOM	0.27655163811605127
CL	F	0.21291573296289484
CL	MSFT	0.2347395935937538
CL	PFE	0.22620918949312924
GC	C	0.17132011394477453
GC	SPX	-0.27621216630360723
GC	XOM	-0.31742556492355695
GC	F	-0.39376436665709946
GC	MSFT	0.03872797470182633
GC	PFE	-0.34653065475607997
C	SPX	0.2841344985841967
C	XOM	0.2722771622858543
C	F	0.1930254456039821
C	MSFT	0.10837798889022507
C	PFE	0.24059844829500385
SPX	XOM	0.9370778598925431
SPX	F	0.9173970725342884
SPX	MSFT	0.21910290988946773
SPX	PFE	0.8750562187811304
XOM	F	0.852903525597108
XOM	MSFT	0.28329029115636173
XOM	PFE	0.8689912705869133
F	MSFT	0.1224603844278996
F	PFE	0.7914349481572399
MSFT	PFE	0.08342580014726039

CL	GC	0.14512911921800759
CL	C	0.047640657886711776
CL	SPX	-0.038662740379307635
CL	XOM	0.13475499739302577
CL	F	-0.02779741081029594
CL	MSFT	0.002124836307259393
CI	PFE	-0.0346544213095382
GC	C	0.07406272080516503
GC	SPX	-0.08216193364828302
GC	XOM	0.0018927626451161
GC	F	-0.04189153921839398
GC	MSFT	-0.017773478113621854
GC	PFE	-0.03394532760699087
C	SPX	0.00863250682585783
C	XOM	-0.0024652908939917476
C	F	0.03824383087240428
C	MSFT	0.026328712743665918
C	PFE	-0.009582466225759407
SPX	XOM	0.3300910692705658
SPX	F	0.3879282004829515
SPX	MSFT	0.37619527832248406
SPX	PFE	0.3522133339947073
XOM	F	0.12461137390050991
XOM	MSFT	0.08511094562657419
XOM	PFE	0.11899749055724199
F	MSFT	0.1291334261723857
F	PFE	0.09432105016323611
MSFT	PFE	0.10326939903567782

CL	GC	0.1780064461248614
CL	C	0.05816017421928696
CL	SPX	0.08387058206522074
CL	XOM	0.11404112460697703
CL	F	-0.0581086900122653
CL	MSFT	-0.04785934015162996
CL	PFE	-0.04252837463155788
GC	C	0.06971353618749605
GC	SPX	-0.10854537254629587
GC	XOM	-0.02305369375053341
GC	F	-0.0433322968281354
GC	MSFT	-0.05714331580093729
GC	PFE	-0.04492680308546143
C	SPX	0.01597033368734557
C	XOM	0.01678577953312174
C	F	0.019585474298717553
C	MSFT	0.021226325810089326
C	PFE	0.01121828967048508
SPX	XOM	0.35173508501967765
SPX	F	0.3788577061068169
SPX	MSFT	0.510722761985027
SPX	PFE	0.3308252244568856
XOM	F	0.12245205070590215
XOM	MSFT	0.11855012193953615
XOM	PFE	0.112787193486031 9
F	MSFT	0.18490175993452032
F	PFE	0.1035829207843917
MSFT	PFE	0.16958846505571112

The point is evident throughout this study: Big moves in one market amplify the correlation between other markets, and vice versa. Some explanations can be offered to partially account for this tendency; for one, these markets are all USD denominated, yet, these elements can only partially account as the *cause* of this. Regardless of its cause, even the fact that this characteristic exists warns us that the correlation parameter fails us at those very times when we are counting on it the most.

What we are working with in using correlation is a composite of the incidental time periods and time periods with considerably more volatility and movement. Clearly, it is misleading to use the correlation coefficient as a single parameter for the joint movement of pairwise components.

Additionally, considering that in a normal distribution, 68.26894921371% of the data points will fall within one sigma either side of the mean. Given 4,682 data points, we would expect therefore to typically have 3196.352 data points be within one sigma. But we repeatedly see more than that. We would also expect, given the Normal distribution, for 99.73002039367% of the data points to be within three sigma, thus, $1 - .99730020393 = 0.002699797$ probability of being beyond three sigma. Given 4,682 data points, we would therefore expect $4,682 * 0.002699797 = 12.64045$ data points to be beyond three sigma. Yet again, we see far more than this in every case, in every market in this study. These findings are consistent with the "fat tails," notion of price distributions.

If more data points than expected fall within one sigma, and more than expected fall outside of three sigma, then the shortfall must be made up with fewer data points than would be expected between $|1|$ and $|2|$ sigma. What is germane to the discussion here, however, is that days when correlations tend more toward randomness occur far more frequently than would be expected if prices were normally distributed, but, in a manner fatal to the conventional models, the critical days where things move more lockstep occur far more often as well.

Consider again our simultaneous two-to-one coin toss example. We have seen that at a correlation coefficient of zero, we optimally bet .23 on each component. Yet, what if we later learned we were deluded about that correlation coefficient, that, rather than being zero, it was, instead +1.0?

In such a circumstance we would have been betting .46 per play, where the optimal was .25. In short, we would have been far to the right of the peak of the *f* curve.

By relying on the correlation coefficient alone, we delude ourselves. The new model disregards correlation as a solitary parameter of pairwise component movement. Rather, the new model addresses this principle as it *must* be addressed. We are concerned in the new model with the joint probabilities of two scenarios occurring, one from each of the pairwise components, simultaneously, as the history of price data dictates we do.

Furthermore, and perhaps far more importantly, the new model holds for any distribution of returns! The earlier portfolio models most often assumed a normal distribution in estimating the various outcomes the investments may have realized. Thus, the tails—the very positive or very negative outcomes—were much thinner than they would be in a non-normal, real-world distribution. That is, the very good and very bad outcomes that investments can witness tended to be underaccounted for in the earlier models. With the new model, various scenarios comprise the tails of the distribution of outcomes, and you can assign them any probability you wish. Even the mysterious Stable Paretian Distribution of returns can be characterized by various scenarios, and an optimal portfolio discerned from such. Any distribution can be modeled as a scenario spectrum; scenario spectrums can assume any probability density shape desired, and they are easy to do. You needn't ask yourself, "What is the probability of being x distance from the mode of this distribution?" but rather, "What is the probability of these scenarios occurring?"

So the new framework can be applied to any distribution of returns, not simply the normal. Thus, the real-world *fat-tails* distribution can be utilized, as a scenario spectrum is another way of drawing a distribution.

Most importantly, the new framework, unlike its predecessors, is not one so much of composition but rather of progression. It is about leverage, and it is also about how you progress your quantity through time, as the equity in the account changes.

Interestingly, *these are different manifestations of the same thing. That is, leverage (how much you borrow), and how you progress your quantity through time are really the same thing.*

Typically, leverage is thought of as "How much do I borrow to own a certain asset?" For example, if I want to own 100 shares of XYZ Corporation, and it costs $50 a share, then it costs $5,000 for 100 shares. Thus, if I have less than $5,000 in my account, how many shares should I put on? This is the conventional notion of leverage.

But leverage also applies to borrowing your own money. Let's suppose I have $1 million in my account. I buy 100 shares of XYZ. Now, suppose XYZ goes up, and I have a profit on my 100 shares. I now want to own 200 shares, although the profit on my 100 shares is not yet $5,000 (i.e., XYZ has not yet gotten to $100). However, I buy another 100 shares anyhow. The schedule upon which I base my future buys (or sells) of XYZ (or any other stock while I own XYZ) is leverage—whether I borrow money to perform these transactions, or whether I use my own money. It is the schedule, the progressions, that constitutes leverage in this sense. If you understand this concept, you are well down the line toward understanding the new framework in asset allocation.

So, we see that *leverage* is a term that refers to either the degree to which we borrow money to take a position in an asset, or the *schedule* upon which we take further positions in assets (whether we borrow to do this or not).

That said, since the focus of the new framework is on *leverage*, we can easily see that it applies to speculative vehicles in the sense that leverage refers to the level of borrowing to take a position in a (speculative) asset. However, the new framework, in focusing on leverage, applies to all assets, including the most conservative, in the sense that leverage also refers to the progression, the schedule upon which we take (or remove) further positions in an asset. Ultimately, leverage in both senses is every bit as important as market timing. That is, the progression of asset accumulation and removal in even a very conservative bond fund is every bit as important as the bond market timing or the bond selection process.

Thus, the entire notion of *optimal f* not only applies to futures and option traders as well, but to any asset allocation scheme, and not just allocating among investment vehicles.

The trading world is vastly different today than just a few decades ago as a result of the proliferation of derivatives trading. Most frequently, a major characteristic with many derivatives is the leverage they bring to bear on an account. The old framework, the old two-dimensional E-V framework, was ill-equipped to handle problems of this sort. The modern environment *demands* a new asset allocation framework focused on the effects of leverage. The framework presented herein addresses exactly this.

This focus on leverage, more than any other explanation, is the main reason why the new framework is superior to its predecessors. Like the old framework, the new framework tells us optimal relative allocations among assets. But the new framework does far more. The new framework is dynamic—it tells us the immense consequences and payoffs of our schedule of taking (and removing) assets through time, giving us a *framework*, a map, of what consequences and rewards we can expect by following such-and-such a schedule. Certain points on the map may be more appealing than others to different individuals with different needs and desires. What may be optimal to one person may not be optimal to another. Yet this *map* allows us to see what we get and give up by progressing according to a certain schedule—something the earlier frameworks did not. This feature, this map of leverage space (and remember, leverage has two meanings here), distinguishes the new framework from its predecessors in many ways, and it alone makes the new framework superior.

Lastly, the new framework is superior to the old in that the user of the new framework can more readily see the consequences of his or her actions. Under the old framework, "So what if I have a little more V for a given E?" Under the new framework, you can see exactly what altitude that puts you at on the landscape, that is, exactly what multiple you make on your starting stake (versus the peak of the landscape) for operating at different levels of leverage (remember, leverage has two meanings throughout this book), or exactly what kind of a *minimum* drawdown to expect for operating at different levels of *leverage*. Under the new framework, you can more readily see how important the asset allocation function is to your bottom line and your pain threshold.

To summarize, the new framework is superior to the older, two-dimensional, risk-competing-with-return frameworks primarily because the focus is on the *dynamics* of leverage. Secondarily, it is superior because the input is more straightforward, using scenarios (i.e., actual distributions that are "binned") unperverted by the misuse of the delusional correlation coefficient parameter, and because it will work on any distribution of returns. Lastly, users of the new framework will more readily be able to see the rewards and consequences of their actions.

MULTIPLE SIMULTANEOUS PLAYS

Refer to Figure 9.2 for our two-to-one coin-toss game. Now suppose you are going to play two of these very same games simultaneously. Each coin will be used in a separate game similar to the first game. Now what quantity should be bet? The answer depends upon the relationship of the two games. If the two games are not correlated to each other, then optimally you would bet 23% on each game (Figure 9.3). However, if there is perfect positive correlation, then you would bet 12.5% on each game. If you bet 25% or more

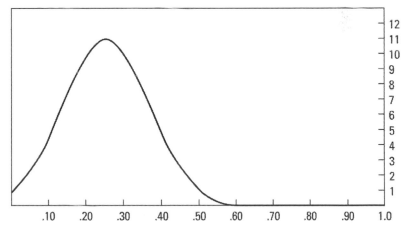

FIGURE 9.2 Two-to-one coin toss game, 40 plays. Ending multiple of starting stake betting different percentages of stake on each play

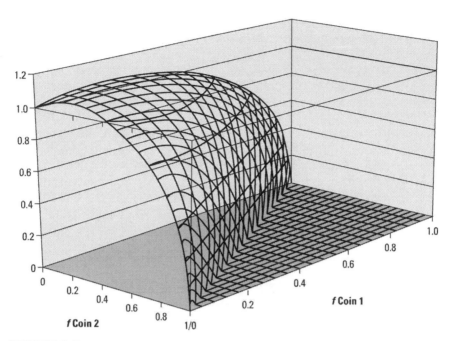

FIGURE 9.3 Two-to-one coin toss—one play

on each game, you will now go broke, with a probability that approaches certainty as the length of the game increases.

When you begin trading more than one market system, you no longer reside on a line that has a peak; instead, you reside in an $n + 1$ (where n = the number of market systems you are trading) dimensional terrain that has a single peak! In our single-coin-toss example, we had a peak on the line at 25%. Here we have one game ($n = 1$) and thus a two (i.e., $n + 1$) dimensional landscape (the line) with a single peak. When we play two of these games simultaneously, we now have a three-dimensional landscape (i.e., $n + 1$) within leverage space with a single peak. If the correlation coefficient between the coins is zero, then the peak is at 23% for the first game and 23% for the second as well. Notice that there is still only one peak, even though the dimensions of the landscape have increased!

When we are playing two games simultaneously, we are faced with a three-dimensional landscape, where we must find the highest point. If we were playing three games simultaneously, we would be looking for the peak in a four-dimensional landscape. The dimensions of the topography within which we must find a peak are equal to the number of games (markets and systems) we are playing plus one.

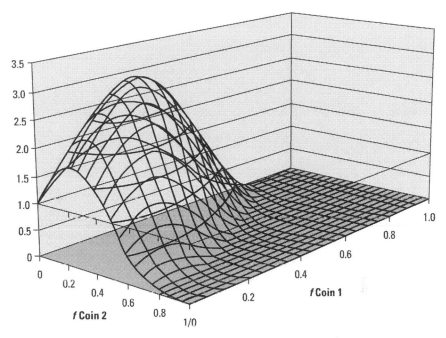

FIGURE 9.4 Two-to-one coin toss—10 plays

Notice, that as the number of plays increases, the peak gets higher and higher, and the difference between the peak and any other point on the landscape gets greater and greater (see Figures 9.3, 9.4, and 9.5). Thus, as more plays elapse, the difference between being at the peak and any other point increases. This is true regardless of how many markets or systems we are trading, even if we are trading only one.

To miss the peak is to pay a steep price. Recall in the simple single-coin-toss game the consequences of missing the peak. These consequences are no less when multiple simultaneous plays are involved. In fact, when you miss the peak in the $n + 1$-dimensional landscape, you will go broke faster than you would in the single game!

Whether or not we acknowledge these concepts, it does not affect the fact that *they are at work on us*. Remember, we can assign an *f* value to any trader in any market with any method at any time. If we are trading a single market system and we miss the peak of the *f* curve for that market system, we might, if we are lucky, make a fraction of the profits we should have made, while we will very likely endure greater drawdowns than we should have. If we are unlucky, we will go broke with certainty *even with an extremely profitable system!*

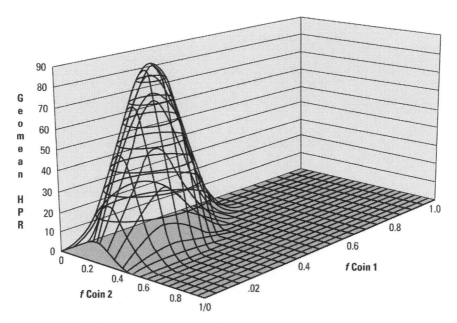

FIGURE 9.5 Two-to-one coin toss—40 plays

When we trade a portfolio of markets and/or systems, we simply magnify the effect of missing the peak of the curve in $n + 1$ space.

A COMPARISON TO THE OLD FRAMEWORKS

Let's take a look at a simple comparison of the results generated by this new framework versus those of the old E-V framework.

Suppose, for the sake of simplicity, we are going to play two simultaneous games. Each game will be the now-familiar two-to-one coin toss. Further assume that all of the pairwise correlations are zero. The new framework tells us that the optimal point, the peak in the three-dimensional $(n + 1)$ landscape is at 23% for both games.

The old framework, in addition to the zero values for the pairwise correlations, has .5 as the E value, the mean, and 2.25 as the V value, the variance. The result of this, through the old framework, generates .5 for both games.

This means that one-half of your account should be allocated toward each game. But what does this mean in terms of leverage? How much is a game? If a game is $1, the most I can lose, then .5 is way beyond the optimal of .23. How do I progress my stake as I go on? The correct answer, the mathematically optimal answer with respect to leverage (including how I progress my stake as I go on), would be .5 of .46 of the account. But the old mean variance models do not tell me that. They are not attuned to the use of leverage (with both of its meanings). The answers tell me nothing of where I am in the $n + 1$ dimensional landscape. Also, there are important points within the $n + 1$ dimensional landscape other than the peak. For instance, as we will see in the next chapter, the points of inflection in the landscape are also very important. The old E-V models tell us nothing about any of this.

In fact, the old models simply tell us that allocating one-half of our stake to each of these games will be *optimal* in that you will get the greatest return for a given level of variance, or the lowest variance for a given level of return. How much you want to lever it is a matter of your utility—your personal preference.

In reality, though, there is an optimal point of leverage, an optimal place in the $n + 1$ dimensional landscape. There are also other important points in this landscape. When you trade, you automatically reside somewhere in this landscape (again, just because you do not acknowledge it does not mean it does not apply to you). The old models were oblivious to this. This new framework addresses this problem and has the users aware of the use/misuse of leverage within an optimal portfolio in a foremost sense. In short, the new framework simply yields more and more useful information than its predecessors.

Again, if a trader is utilizing two market systems simultaneously, then where he resides on the three-dimensional landscape is everything. Where he resides on it is every bit as important as his market systems, his timing, or his trading ability.

MATHEMATICAL OPTIMIZATION

Mathematical optimization is an exercise in finding a maximum or minimum value of an objective function for a given parameter(s). The objective function is, thus, something that can be solved only through an iterative procedure.

For example, the process of finding the optimal f for a single market system, or a single scenario spectrum, is an exercise in mathematical optimization. Here, the mathematical optimization technique can be something quite brutish like trying all f values from 0 to 1.0 by .01. The objective function can be one of the functions presented in Chapter 4 for finding the geometric mean HPR for a given value of f under different conditions. The parameter is that value for f being tried between 0 and 1.

The answer returned by the objective function, along with the parameters pumped into the objective function, gives us our coordinates at a certain point in $n + 1$ space. In the case of simply finding the optimal f for a single market system or a single scenario spectrum, n is 1, so we are getting coordinates in two-dimensional space. One of the coordinates is the f value sent to the objective function, and the other coordinate is the value returned by the objective function for the f value passed to it.

Since it is a little difficult for us to mentally picture any more than three dimensions, we will think in terms of a value of 2 for n (thus, we are dealing with the three-dimensional, i.e., $n + 1$, landscape). Since, for simplicity's sake, we are using a value of 2 for n, the objective function gives us the height or *altitude* in a three-dimensional landscape. We can think of the north-south coordinates as corresponding to the f value associated with one scenario spectrum, and the east-west coordinates as the f value associated with another scenario spectrum. Each scenario spectrum pertains to the possible outcomes for a given market system. Thus, we could say, for example, that the north-south coordinates pertain to the f value for such-and-such a market under such-and-such a system, and the east-west coordinates pertain to the f values of trading a different market and/or a different system, when both market systems are traded simultaneously.

The objective function gives us the altitude for a given set of f values. That is, the objective function gives us the altitude corresponding to a single east-west coordinate and a single north-south coordinate. That is, a single point where the length and depth are given by the f values we are pumping into the objective function, and the height at that point is the value returned by the objective function.

Once we have the coordinates for a single point (its length, depth, and height), we need a search procedure, a mathematical optimization technique, to alter the f values being pumped into the objective function in such a way so as to get us to the peak of the landscape as quickly and easily as possible.

What we are doing is trying to map out the terrain in the $n + 1$-dimensional landscape, because the coordinates corresponding to the peak in that landscape give us the optimal f values to use for each market system.

Many mathematical optimization techniques have been worked out over the years and many are quite elaborate and efficient. We have a number of these techniques to choose from. The burning question for us is, "Upon what objective function shall we apply these mathematical optimization techniques?" under this new framework. The objective function is the heart of this new framework in asset allocation, and we will discuss it and show examples of how to use it before looking at optimization techniques.

THE OBJECTIVE FUNCTION

The objective function we wish to maximize is the geometric mean HPR, simply called G:

$$G(f_1 \ldots f_n) = \left(\prod_{k=1}^{m} \mathrm{HPR}_k \right)^{\left(1 / \sum_{k=1}^{m} \mathrm{Prob}_k \right)} \tag{9.01}$$

where: $n =$ The number of scenario spectrums (market systems or portfolio components).

$m =$ The possible number of combinations of outcomes between the various scenario spectrums (market systems) based on how many scenarios are in each set.

$m =$ The number of scenarios in the first spectrum $*$ the number of scenarios in the second spectrum $* \ldots *$ the number of scenarios in the nth spectrum.

Prob $=$ The sum of probabilities of all m of the HPRs for a given set of f values. Prob_k is the sum of the values in brackets $\{\}$ in Equation (9.02) for all m values of a given set of f values.

HPR $=$ The holding period return of each k. This is given as:

$$\mathrm{HPR}_k = \left(1 + \left(\sum_{i=1}^{n} (f_i * (-\mathrm{PI}_{k,i} / \mathrm{BL}_i)) \right) \right)^{\mathrm{Prob}_k} \tag{9.02}$$

where: $n =$ The number of components (scenario spectrums, i.e., market systems) in the portfolio.

$f_i =$ The f value being used for component i.

f_i must be > 0, and can be infinitely high (i.e., can be greater than 1.0).

$\mathrm{PL}_{k,i} =$ The outcome profit or loss for the ith component (i.e., scenario spectrum or market system) associated with the kth combination of scenarios.

$\mathrm{BL}_i =$ The worst outcome of scenario spectrum (market system) i.

We can estimate Prob_k in the earlier equation for G as:

$$\mathrm{Prob}_k = \left(\prod_{i=1}^{n-1} \left(\prod_{j=i+1}^{n} P(i_k | j_k) \right) \right)^{(1/(n-l))} \tag{9.03}$$

The expression $P(i_k | j_k)$ is simply the joint probability of the scenario in the ith spectrum and the jth spectrum, corresponding to the kth combination of scenarios. For example, if we have three coins, each coin represents a scenario spectrum, represented by the variable n, and each spectrum contains two scenarios: heads and tails. Thus, there are eight $(2 * 2 * 2)$ possible combinations, represented by the variable m.

In Equation (9.01), the variable k proceeds from 1 to m, in *odometric* fashion:

Coin 1	Coin 2	Coin 3	k
t	t	t	1
t	t	h	2
t	h	t	3
t	h	h	4
h	t	t	5
h	t	h	6
h	h	t	7
h	h	h	8

That is, initially all spectrums are set to their worst (leftmost) values. Then, the rightmost scenario spectrum cycles through all of its values, after which the second rightmost scenario spectrum increments to the next (next right) scenario. You proceed as such again, with the rightmost scenario spectrum cycling through all of its scenarios, and when the second rightmost scenario spectrum has cycled through all of its values, the third rightmost scenario spectrum increments to its next scenario. The process is exactly the same as an odometer on a car, hence the term *odometrically*.

So in the expression $P(i_k | j_k)$, if k were at the value 3 above (i.e., $k = 3$), and i was 1 and j was 3, we would be looking for the joint probability of coin 1 (coming up tails and coin 3) coming up tails. Equation (9.03) helps us in estimating the joint probabilities of particular individual scenarios occurring in n spectrums simultaneously. To put it simply, if I have two scenario spectrums, at any given k I will have only one joint probability to incorporate. If I have three scenario spectrums, I will have three joint probabilities to incorporate (spectrums 1 and 2, spectrums 1 and 3, and spectrums 2 and 3). If four scenario spectrums, I will have six joint probabilities to compute using (9.03); if five scenario spectrums, then I have 10 joint probabilities to compute using (9.03). Quite simply, in (9.03) the number of joint probabilities you will have to incorporate at any $P(i)$ is:

$$n! / (n - 2)! / 2 = \text{number of joint probabilities required as input}$$

to (9.03)

To demonstrate (9.03) in a simple manner, if I have three scenario spectrums (called A, B, and C), and each has two possible outcomes, H and T, then I want to find the multiplicative product of the probabilities of a given outcome of all three at each i, across all values of i (of which there are q).

So, if I have $n = 3$, then, at $k = 1$, I have the tails scenario (with a probability of .5) in all three scenario spectrums. Thus, to find the probability of this spectrum, I need to multiply the probability of $((A_T|B_T) \times (A_T|C_T) \times (B_T|C_T))^{\wedge}(1/(n-1))$

$$= (.25 \times .25 \times .25))^{\wedge} (1/(3-1))$$
$$= .015625^{\wedge} (1/2) = .125$$

Note that this is a simple example. Our joint probabilities between any two scenarios from any of the three different scenario spectrums was always .25 in this case. In the real world, however, such conveniences are rare coincidences.

Equation (9.03) is merely an *estimate*, which makes a major assumption (that all elements move randomly with respect to all other elements, i.e. if we were to take a correlation coefficient of any pairwise elements, it would be zero). Note that we are constructing a $Prob_k$ here using (9.03); we are attempting to actually composite a joint probability of n events occurring simultaneously, knowing only the probabilities of pairwise occurrences of those events (at two scenario spectrums, this assumption is, in fact, *not* an assumption). In truth, this is an attempt to approximate the actual joint probability

For example, say I have three conditions called A, B, and C. A and B occur with .5 probability. A and C occur with .6 probability. B and C occur with .1 probability.

However, that probability of .1 of B and C's occurring may be 0 if A and B occur. It may be any value between 0 and 1 in fact. In order to determine then, what the probability of A, B, and C's occurring simultaneously is, I would have to look at when those three conditions actually *did* occur. I cannot infer the probability of all three events occurring simultaneously given the probabilities of their pairwise joint probabilities unless I am dealing with less than three elements or their pairwise correlations were all zero.

We need to derive the actual joint probability via empirical data, or accurately approximate the joint probabilities of occurrence among three or more simultaneous events. Equation (9.03) is invalid if there are more than two joint probabilities or the correlation coefficients between any pairwise elements is not 0. However, in the examples that follow in this chapter, we *will* use (9.03) merely as a proxy for whatever the actual joint probabilities may be, for sake of illustration. We can create one complete objective function. Thus, we wish to maximize G as:

$$G(f_i \ldots f_n) = \left(\prod_{k=1}^{m} \left(\left(1 + \sum_{i=1}^{n} \left(f_i * \left(\frac{-PL_{k,i}}{BL_i} \right) \right) \right)^{Prob_k} \right) \right)^{\left(1/ \sum_{k=1}^{m} Prob_k \right)}$$

(9.04)

This is the objective function, the equation we wish to maximize. It is the equation or mathematical expression of this new framework in asset allocation. It gives you the *altitude*, the geometric mean HPR, in $n + 1$ space for the coordinates, the values of f used. It is *exact*, regardless of how many scenarios or scenario spectrums are used as input. *It is the objective function of the leverage space model.*

Although Equation (9.04) may look a little daunting, there isn't any reason to fear it. As you can see, Equation (9.04) is a lot easier to work with in the compressed form, expressed earlier in Equation (9.01).

Returning to our three coin example, suppose we win \$2 on heads and lose \$1 on tails. We have three scenario spectrums, three market systems, named Coin 1, Coin 2, and Coin 3. Two scenarios, heads and tails, comprise each coin, each scenario spectrum. We will assume, for the sake of simplicity, that the correlation coefficients of all three scenario spectrums (coins) to each other are zero.

We must therefore find three different f values. We are seeking an optimal f value for Coin 1, Coin 2, and Coin 3, as f_1, f_2, and f_3, respectively, that results in the greatest growth—that is, the combination of the three f values that results in the greatest geometric mean HPR [Equation (9.01) or (9.04)].

For the moment, we are not paying any attention to the optimization technique selected. The purpose here is to show how to perform the objective function. Since optimization techniques usually assign an initial value to the parameters, we will arbitrarily select .1 as the initial value for all three values of f.

We will use Equation (9.01) in lieu of (9.04) for the sake of simplicity. Equation (9.01) has us begin by cycling through all scenario set combinations, all values of k between 1 and m, compute the HPR of the scenario set combination per Equation (9.02), and multiply all of these HPRs together. When we perform Equation (9.02) each time, we must keep track of the $Prob_k$ values, because we will need the sum of these values later.

Thus, we start at $k = 1$, where scenario spectrum 1 (Coin 1) is tails, as are the other two scenario spectrums (coins).

We can rewrite Equation (9.02) as:

$$\mathrm{HPR}_k = (1 + C)^x$$

$$C = \sum_{i=1}^{n} (f_i * (-\mathrm{PL}_{k,i}/\mathrm{BL}_i))$$

$$x = \left(\prod_{k=1}^{n-1} \left(\prod_{j=i+1}^{n} P(i_k|j_k) \right) \right)^{(1/(n-1))}$$

Notice that the exponent in Equation (9.02), which we must keep track of, is expressed as the variable x in Equation (9.02a). This is also expressed in Equation (9.03).

So, to obtain C, we simply go through each scenario spectrum, taking the outcome of the scenario currently being used in that spectrum as dictated by k, dividing its negative by the scenario in that spectrum with the worst outcome, and multiplying this quotient by the f value being used with that scenario spectrum. As we go through all of the scenario spectrums, we total these values.

The variable i is the scenario spectrum we are looking at. The biggest loss in scenario spectrum 1 is tails, which sees a loss of one dollar (i.e., 1). Thus, BL_1 is -1 (as will be BL_2 and BL_3 since the biggest loss in each of the other two scenario spectrums—the other two coins—is -1). The associated PL, that is, the outcome of the scenario in spectrum i corresponding to the scenario in that spectrum that k points to, is -1 in scenario spectrum 1 (as it is in the other two spectrums). The f value is currently .1 (as it also is now in the other two spectrums). Thus:

$$C = \sum_{i=1}^{n} \left(f_i * \left(\frac{-\mathrm{PL}_{k,i}}{\mathrm{BL}_i} \right) \right)$$

$$C = \left(.1 * \left(\frac{--1}{-1} \right) \right) + \left(.1 * \left(\frac{--1}{-1} \right) \right) + \left(-1 * \left(\frac{--1}{-1} \right) \right)$$

$$C = (.1 * -1) + (.1 * -1) + (.1 * -1)$$

$$C = -.1 + -.1 + -.1 = -.3$$

Notice that the PLs are negative and, since PL has a minus sign in front of it, that makes them positive.

Now we take the value for C in Equation (9.02) above and add 1 to it, obtaining .7 (since $1 + -.3 = .7$). Now we must figure the exponent, the variable x in Equation (9.02) above.

$P(i_k|j_k)$ means, simply, the joint probability of the scenario in spectrum i pointed to by k, and the scenario in spectrum j pointed to by k. Since k is presently 1, it points to tails in all three scenario spectrums. To find x, we simply take the sum of the joint probabilities of the scenarios in spectrum 1 and 2 times the joint probability of the scenarios in spectrum 1 and 3, times the joint probabilities of the scenarios in spectrums 2 and 3. Expressed differently:

i	j
1	2
1	3
2	3

If there were four spectrums, we would take the product of all the joint probabilities as:

i	j
1	2
1	3
1	4
2	3
2	4
3	4

Since all of our joint probabilities are .25, we get for x:

$$x = \left(\prod_{i=1}^{n-1} \left(\prod_{j=i+1}^{n} P(i_k | j_k) \right) \right)^{(1/(n-1))}$$

$$x = (.25 * .25)^{(1/(n-1))}$$

$$x = (.015625)^{1/(3-1)}$$

$$x = (.015625)^{1/2}$$

$$x = .125$$

Thus, x equals .125, which represents the joint probability of the kth combination of scenarios. (Note that we are going to determine a joint probability of three random variables by using joint probabilities of two random variables!)

Thus, $HPR_k = .7^{.125} = .9563949076$ when $k = 1$. Per Equation (9.02), we must figure this for all values of k from 1 through m (in this case, m equals 8). Doing this, we obtain:

k	HPR$_k$	Prob$_k$
1	0.956395	0.125
2	1	0.125
3	1	0.125
4	1.033339	0.125
5	1	0.125
6	1.033339	0.125
7	1.033339	0.125
8	1.060511	0.125

Summing up all the Prob$_k$, given by Equation (9.03), per Equation (9.04), we get 1. Now, taking the product of all of the HPRs, per Equations (9.01) and (9.04), we obtain 1.119131. Performing Equation (9.01), then, we get a value of G of 1.119131 which corresponds to the f values .1, .1, .1 for f_1, f_2, and f_3, respectively.

$$G(.1, .1, .1) = \left(\prod_{k=1}^{m} \mathrm{HPR}_k \right)^{\left(1 / \sum_{k=1}^{m} \mathrm{Prob}_k \right)}$$

$$G(.1, .1, .1) = (.956395 * 1 * .1 * 1.033339 * 1 * 1.033339$$
$$* 1.033339 * 1.0605011)^{(1/(.125 + .125 + .125 + .125 + .125 + .125 + .125 + .125))}$$

$$G(.1, .1, .1) = (1.119131)^{(1/1)}$$

$$G(.1, .1, .1) = 1.119131$$

Now, depending upon what mathematical optimization method we were using, we would alter our f values. Eventually, we would find our optimal f values at .21, .21, .21 for f_1, f_2, and f_3, respectively. This would give us:

k	HPR$_k$	Prob$_k$
1	0.883131	0.125
2	1	0.125
3	1	0.125
4	1.062976	0.125
5	1	0.125
6	1.062976	0.125
7	1.062976	0.125
8	1.107296	0.125

Thus, Equation (9.01) gives us:

$$G(.21, .21, .21) = \left(\prod_{k=1}^{m} \text{HPR}_k\right)^{\left(1 / \sum_{k=1}^{m} \text{Prob}_k\right)}$$

$$G(.21, .21, .21) = (.883131 * 1 * .1 * 1.062976 * 1 * 1.062976 * 1.062976$$
$$* 1.107296)^{(1/(.125+.125+.125+.125+.125+.125+.125+.125))}$$

$$G(.21, .21, .21) = 1.174516^{(1/1)}$$

$$G(.21, .21, .21) = 1.174516$$

This is the f value combination that results in the greatest G for these scenario spectrums. Since this is a very simplified case, that is, all scenario spectrums were identical, and all had correlation of zero between them, we ended up with the same f value for all three scenario spectrums of .21. Usually, this will not be the case, and you will have a different f value for each scenario spectrum.

Now that we know the optimal f values for each scenario spectrum, we can determine how much those decimal f values are, in currency, by dividing the largest loss scenario in each of the spectrums by the negative optimal f for each of those spectrums. For example, for the first scenario spectrum, Coin 1, we had a largest loss of -1. Dividing -1 by the negative optimal f, $-.21$, we obtain 4.761904762 as $f\$$ for Coin 1.

To summarize the procedure, then:

1. Start with an f value set for $f_1 \ldots f_n$ where n is the number of components in the portfolio, that is, market systems or scenario spectrums. This initial f value set is given by the optimization technique selected.

2. Go through the combinations of scenario sets k from 1 to m, odometrically, and calculate an HPR for each k, multiplying them all together. While doing so, keep a running sum of the exponents of the HPRs.

3. When k equals m, and you have computed the last HPR, the final product must be taken to the power of 1, divided by the sum of the exponents (probabilities) of all the HPRs, to get G, the geometric mean HPR.

4. This geometric mean HPR gives us one *altitude* in $n + 1$ space. We wish to find the peak in this space, so we must now select a new set of f values to test to help us find the peak. This is the mathematical optimization process.

MATHEMATICAL OPTIMIZATION VERSUS ROOT FINDING

Equations have a left and a right side. Subtracting the two makes the equation equal to 0. In *root finding*, you want to know what values of the independent variable(s) make the answer of this equation equal to 0 (these are the *roots*). There are traditional root-finding techniques, such as the *Newton-Rapheson* method, to do this.

It would seem that root finding is related to mathematical optimization in that the first derivative of an optimized function (i.e., extremum located) will equal 0. Thus, you would assume that traditional root-finding techniques, such as the Newton-Rapheson method, could be used to solve optimization problems (careful to use what is regarded as an optimization technique to solve for the roots of an equation can lead to a Pandora's box of problems).

However, our discussion will concern only optimization techniques and not root finding techniques per se. The single best source for a listing of these techniques is *Numerical Recipes* and much of the following section on optimization techniques is referenced therefrom.[3]

OPTIMIZATION TECHNIQUES

Mathematical optimization, in short, can be described as follows: You have a function (we call it G), the objective function, which depends on one or more independent variables (which we call $f_l \ldots f_n$). You want to find the value(s) of the independent variable(s) that results in a minimum (or sometimes, as in our case, a maximum) of the objective function. Maximization or minimization is essentially the same thing (that is one person's G is another person's $-G$).

In the crudest case, you can optimize as follows: Take every combination of parameters, run them through the objective function, and see which produce the best results. For example, suppose we want to find the optimal f for two coins tossed simultaneously, and we want the answer to be precise to .01. We could, therefore, test Coin 1 at the 0.0 level, while testing Coin 2 at the 0.01 level, then .01, .02, and proceed until we have tested Coin 2 at the 1.0 level. Then, we could go back and test with Coin 1 at the .01 level, and cycle Coin 2 through all of its possible values while holding Coin 1 at the .01 level. We proceed until both levels are at their maximum, that is, both values equal 1.0. Since each variable in this case has 101 possible values (0 through 1.0 by .01 inclusive), there are $101 * 101$ combinations which must be tried, or 10,201 times the objective function must be evaluated.

We could, if we wanted, demand precision greater than .01. Suppose we wanted precision to the .001 level. Then we would have $1,001 * 1,001$

---

[3] William H. Press, Brian P. Flannery, Saul A. Teukolsky, and William T. Vetterling, *Numerical Recipes: The Art of Scientific Computing*, New York: Cambridge University Press, 1986.

combinations that we would need to try, or 1,002,001 times the objective function would have to be calculated. If we were then to include three variables rather than just two, and demand .001 precision this way, we would then have to evaluate the objective function 1001 * 1001 * 1001, or 1,003,003,001; that is, we would have to evaluate the objective function in excess of one billion times. We are using only three variables and we are demanding precision to only .001!

Although this crude case of optimizing has the advantage of being the most robust of all optimization techniques, it is also has the dubious distinction of being too slow to apply to most problems.

Why not cycle through all variables for the first variable and get its optimal; then cycle through all variables for the second while holding the first at its optimal; get the second variable's optimal, so that you now have the optimal for the first two parameters; go find the optimal for the third while setting the first two to their optimal, and so on, until you have solved the problem?

The problem with this second approach is that it is often impossible to find the optimum parameter set this way. Notice that by the time we get to the third variable, the first two variables equal their optimum as if there were no other variables. Thus, when the third variable is optimized, with the first two variables set to their optimums, they interfere with the solution of the third optimum. What you would end up with is not the optimum parameter set of the three variables, but, rather, an optimum value for the first parameter, an optimum for the second when the first is set to its optimum, an optimum for the third when the first is set to its optimum, and the second set to a suboptimum, but optimum given the interference of the first, and so on. It may be possible to keep cycling through the variables and eventually resolve to the optimum parameter set, but with more than three variables, it becomes more and more lengthy, if at all possible, given the interference of the other variables.

There exist superior techniques that have been devised, rather than the two crude methods described, for mathematical optimization. This is a fascinating branch of modern mathematics, and I strongly urge you to study it, simply in the hope that you derive a fraction of the satisfaction from the study as I have.

An extremum, that is the maximum or minimum, can be either *global* (truly the highest or lowest value) or *local* (the highest or lowest value in the immediate neighborhood). To truly know a global extremum is nearly impossible, since you do not know the range of values of the independent variables. If you do not know the range, then you have simply found a local extremum. Therefore, oftentimes, when people speak of a global extremum, they are really referring to a local extremum over a very wide range of values for the independent variables.

There are a number of techniques for finding the maximum or minimum in such cases. Usually, in any type of mathematical optimization, there are *constraints* placed on the variables, which must be met with respect to the extremum. For example, in our case, there are the constraints that all independent variables (the *f* values) must be greater than or equal to zero. Oftentimes, there are constraining functions that must be met [i.e., other functions involving the variable(s) used which must be above/below or equal to certain values]. *Linear programming*, including the *simplex algorithm*, is one very well developed area of this type of constrained optimization, but will work only where the function to be optimized and the constraint functions are linear functions (first-degree polynomials).

Generally, the different methods for mathematical optimization can be broken down by the following categories, and the appropriate technique selected:

1. Single-variable (two-dimensional) vs. multivariable (three- or more dimensional) objective functions.
2. Linear methods vs. nonlinear methods. That is, as previously mentioned, if the function to be optimized and the constraint functions are linear functions (i.e., do not have exponents greater than one to any of the terms in the functions), there are a number of very well developed techniques for solving for extrema.
3. Derivatives. Some methods require computation of the first derivative of the objective function. In the multivariable case, the first derivative is a vector quantity called the *gradient*.
4. Computational efficiency. That is, you want to find the extremum as quickly (i.e., with as few computations) and easily (something to consider with those techniques which require calculation of the derivative) as possible, using as little computer storage as possible.
5. Robustness. Remember, you want to find the extremum that is local to a very wide range of parameter values, to act as a surrogate global extremum. Therefore, if there is more than one extremum in this range, you do not want to get hung up on the less extreme extremum.

In our discussion, we are concerned only with the multidimensional case. That is, we concern ourselves only with those optimization algorithms that pertain to two or more variables (i.e., more than one scenario set). *In searching for a single f value, that is, in finding the f of one market system or one scenario set, parabolic interpolation, as detailed in Chapter 4,* Portfolio Management Formulas, *will generally be the quickest and most efficient technique.*

In the multidimensional case, there are many good algorithms, yet there is no perfect algorithm. Some methods work better than others for certain types of problems. Generally, personal preference is the main determinant in selecting a multidimensional optimization technique (provided one has the computer hardware necessary for the chosen technique).

Multidimensional techniques can be classified according to five broad categories.

First are the *hill-climbing simplex methods*. These are perhaps the least efficient of all, if the computational burden gets a little heavy. However, they are often easy to implement and do not require the calculation of partial first derivatives. Unfortunately, they tend to be slow and their storage requirements are on the order of n^2.

The second family are the *direction set methods*, also known as the *line minimization methods* or *conjugate direction methods*. Most notable among these are the various methods of Powell. These are more efficient, in terms of speed, than the hill-climbing simplex methods (not to be confused with the simplex algorithm for linear functions mentioned earlier), do not require the calculation of partial first derivatives, yet the storage requirements are still on the order of n^2.

The third family is the *conjugate gradient methods*. Notable among these are the Fletcher-Reeves method and the closely related Polak-Ribiere method. These tend to be among the most efficient of all methods in terms of speed and storage (requiring storage on the order of n times x), yet they do require calculations of partial first derivatives.

The fourth family of multidimensional optimization techniques are the *quasi-Newton*, or *variable metric methods*. These include the Davidson-Fletcher-Powell (DFP) and the Broyden-Fletcher-Goldfarb-Shanno (BFGS) algorithms. Like the conjugate gradient methods, these require calculation of partial first derivatives, tend to rapidly converge to an extremum, yet these require greater storage, on the order of n^2. However, the tradeoff to the conjugate gradient methods is that these have been around longer, are in more widespread use, and have greater documentation.

The fifth family is the *natural simulation* family of multidimensional optimization techniques. These are by far the most fascinating, as they seek extrema by simulating processes found in nature, where nature herself is thought to seek extrema. Among these techniques are the *genetic algorithm* method, which seeks extrema through a survival-of-the-fittest process, and *simulated annealing*, a technique which simulates crystallization, a process whereby a system finds its minimum energy state. These techniques tend to be the most robust of all methods, nearly immune to local extrema, and can solve problems of gigantic complexity. However, they are not necessarily the quickest, and, in most cases, will not be. These techniques are still so new that very little is known about them yet.

Although you can use any of the aforementioned multidimensional optimization algorithms, I have opted for the genetic algorithm because it is perhaps the single most robust mathematical optimization technique, aside from the very crude technique of attempting every variable combination.

It is a *general* optimization and search method that has been applied to many problems. Often it is used in neural networks, since it has the characteristic of scaling well to noisy or large nonlinear problems. Since the technique does not require gradient information, it can also be applied to discontinuous functions, as well as empirical functions, just as it is applied to analytic functions.

The algorithm, although frequently used in neural networks, is not limited solely to them. Here, we can use it as a technique for finding the optimal point in the $n + 1$ dimensional landscape.

THE GENETIC ALGORITHM

In a nutshell, the algorithm works by examining many possible candidate solutions and ranking them on how well their value output, by whatever objective function, is used. Then, like the theory of natural selection, the most fit survive and reproduce a new generation of candidate solutions, which inherit characteristics of both *parent* solutions of the earlier generation. The average fitness of the population will increase over many generations and approach an optimum.

The main drawback to the algorithm is the large amount of processing overhead required to evaluate and maintain the candidate solutions. However, due to its robust nature and effective implementation to the gamut of optimization problems, however large, nonlinear, or noisy, it is this author's contention that it will become the de facto optimization technique of choice in the future (excepting the emergence of a better algorithm which possesses these desirable characteristics). As computers become ever more powerful and inexpensive, the processing overhead required of the genetic algorithm becomes less of a concern. Truly, if processing speed were zero, if speed were not a factor, the genetic algorithm would be the optimization method of choice for nearly all mathematical optimization problems.

The basic steps involved in the algorithm are as follows:

1. *Gene length.* You must determine the length of a *gene*. A gene is the binary representation of one member of the population of candidate solutions, and each member of this population carries a value for each variable (i.e., an f value for each scenario spectrum). Thus, if we allow a gene length of 12 times the number of scenario spectrums, we have 12 bits assigned to each variable (i.e., f value). Twelve bits allows for values in the range of 0 to 4095. This is figured as:

$$2^0 + 2^1 + 2^2 + \ldots + 2^{11} = 4095$$

Simply take 2 to the 0th power plus 2 to the next power, until you reach the power of the number of bits minus 1 (i.e., 11 in this case). If there are, say, three scenario spectrums, and we are using a length of 12 bits per scenario spectrum, then the length of a gene for each candidate solution is $12 * 3 = 36$ bits. That is, the gene in this case is a string of 36 bits of 1s and 0s.

Notice that this method of encoding the bit strings only allows for integer values. We can have it allow for floating-point values as well by using a uniform divisor. Thus, if we select a uniform divisor of, say, 1,000, then we can store values of 0/1000 to 4095/1000, or 0 to 4.095, and get precision down to .001.

What we need then is a routine to convert the candidate solutions to encoded binary strings and back again.

2. *Initialization.* A starting population is required—that is, a population of candidate solutions. The bit strings of this first generation are encoded randomly. Larger population sizes make it more likely that we will find a good solution, but they require more processing time.

3. *Objective function evaluation.* The bit strings are decoded to their decimal equivalents, and are used to evaluate the objective function. (The objective function, for example, if we are looking at two scenario spectrums, gives us the Z coordinate value, the altitude of the three-dimensional terrain, assuming the f values of the respective scenario spectrums are the X and Y coordinates.) This is performed for all candidate solutions, and their objective functions are saved. (*Important:* Objective function values must be non-negative!)

4. *Reproduction*

    a. *Scaling based upon fitness.* The objective functions are now scaled. This is accomplished by first determining the lowest objective function of all the candidate solutions, and subtracting this value from all candidate solutions. The results of this are summed up. Then, each objective function has the smallest objective function subtracted from it, and the result is divided by the sum of these, to obtain a fitness score between 0 and 1. The sums of the fitness scores of all candidate solutions will then be 1.0.

    b. *Random selection based upon fitness.* The scaled objective functions are now aligned as follows. If, say, the first objective function has a scaled fitness score of .05, the second has one of .1, and the third .08, then they are set up in a selection scheme as follows:

First candidate   0 to .05
Second candidate   .05 to .15
Third candidate   .15 to .23

This continues until the last candidate has its upper limit at 1.0.

Now, two random numbers are generated between 0 and 1, with the random numbers determining from the preceding selection scheme who the two parents will be. Two parents must now be selected for each candidate solution of the next generation.

c. *Crossover.* Go through each bit of the *child*, the new population candidate. Start by copying the first bit of the first parent to the first bit of the child. At each bit carryover, you must also generate a random number. If the random number is less than or equal to (probability of crossover/gene length), then switch to copying the bits over from the other parent. Thus, if we have three scenario spectrums and 12 bits per each variable, then the gene length is 36. If we use a probability of crossover of .6, then the random number generated at any bit must be less than .6/36, or less than .01667, in order to switch to copying the other parent's code for subsequent bits. Continue until all the bits are copied to the child. This must be performed for all new population candidates.

Typically, probabilities of crossover are in the range .6 to .9. Thus, a .9 probability of crossover means there is a 90% chance, on average, that there will be crossover to the child, that is, a 10% chance the child will be an exact replicant of one of the parents.

d. *Mutation.* While copying over each bit from parent to child, generate a second random number. If this random number is less than or equal to the probability of mutation, then toggle that bit. Thus, a bit which is 0 in the parent becomes 1 in the child and vice versa. Mutation helps maintain diversity in the population. The probability of mutation should generally be some small value (i.e., $< =.001$); otherwise the algorithm tends to deteriorate into a random search. As the algorithm approaches an optimum, however, mutation becomes more and more important, since crossover cannot maintain genetic diversity in such a localized space in the $n + 1$ terrain.

Now you can go back to step three and perform the process for the next generation. Along the way, you must keep track of the highest objective function returned and its corresponding gene. Keep repeating the process until you have reached X unimproved generations, that is, X generations where the best objective function value has not been exceeded. You then quit, at that point, and use the gene corresponding to that best objective function value as your solution set.

For an example of implementing the genetic algorithm, suppose our objective function is one of the form:

$$Y = 1500 - (X - 15)^2$$

For the sake of simplicity in illustration, we will have only a single variable; thus, each population member carries only the binary code for that one variable.

Upon inspection, we can see that the optimal value for X is 15, which would result in a Y value of 1500. However, rarely will we know what the optimal values for the variables are, but for the sake of this simple illustration, it will help if we know the optimal so that we can see how the algorithm takes us there.

Assume a starting population of three members, each with the variable values encoded in five-bit strings, and each initially random:

		First Generation		
Individual #	X	Binary X	Y	Fitness Score
1	10	01010	1475	.4751
2	0	00000	1275	0
3	13	01101	1496	.5249

Now, through random selection based on fitness, Individual 1 for the second generation draws Parents 1 and 3 from the first generation (note that Parent 2, with a fitness of 0, has died and will not pass on its genetic characteristics). Assume that random crossover occurs after the fourth bit, so that Individual 1 in the second generation inherits the first four bits from Individual 1 of the first generation, and the last bit from Individual 3 of the first generation, producing 01011 for Individual 1 of the second generation.

Assume Individual 2 for the second generation also draws the same parents; crossover occurs only after the first and third bits. Thus, it inherits bit 0 from Individual 1 in the first generation, bit 11 as the second and third bits from the third individual in the first generation, and the last two bits from the first individual of the first generation, producing 01110 as the genetic code for the second individual in the second generation.

Now, assume that the third individual of the second generation draws Individual 1 as its first parent as well as its second. Thus, the third individual in the second generation ends up with exactly the same genetic material as the first individual in the first generation, or 01010.

Second Generation		
Individual #	X	Binary X
1	11	01011
2	14	01110
3	10	01010

Now, through random mutation, the third bit of the first individual is flipped, and the resulting values are used to evaluate the objective function:

Second Generation				
Individual #	X	Binary X	Y	Fitness Score
1	15	01111	1500	.5102
2	14	01110	1499	.4898
3	10	01010	1475	0

Notice how the average Y score has gone up, or evolved, after two generations.

IMPORTANT NOTES

It is often advantageous to carry the strongest individual's code to the next generation in its entirety. By so doing, good solution sets are certain to be maintained, and this has the effect of expediting the algorithm. Then, you can work to aggressively maintain genetic diversity by increasing the values used for the probability of crossover and the probability of mutation. I have found that you can work with a probability of crossover of 2, a probability of mutation of .05, and converge to solutions quicker, provided you retain the code of the most fit individual from one generation to the next, which keeps the algorithm from deteriorating to a random search.

As population size approaches infinity, that is, as you use a larger and larger value for the population size, the answer converged upon is exact. Likewise, with the unimproved generations parameter, as it approaches infinity—that is, as you use a larger and larger value for unimproved generations—the answer converged upon is exact. However, both of these parameter increases are at the expense of extra computing time.

The algorithm can be time intensive. As the number of scenario sets increases, and the number of scenarios increases, the processing time grows geometrically. Depending upon your time constraints, you may wish to keep your scenario sets and the quantity of scenarios to a manageable number. The genetic algorithm is particularly appropriate as we shall see by Chapter 12, where we find the landscape of leverage space to be discontinuous for our purposes.

Once you have found the optimal portfolio, that is, once you have f values, you simply divide those f values by the largest loss scenario of the respective scenario spectrums to determine the $f\$$ for that particular scenario spectrum. This is exactly as we did in the previous chapter for determining how many contracts to trade in an optimal portfolio.

CHAPTER 10

The Geometry of Leverage Space Portfolios

J ust as everyone is at a value for f whether they acknowledge it or not, so too therefore is everyone in leverage space, at some point on the terrain therein, whether they acknowledge it or not. The consequences they must pay for this are not exorcised by their ignorance to this.

DILUTION

If we are trading a portfolio at the full optimal allocations, we can expect tremendous drawdowns on the entire portfolio in terms of equity retracement.

Even a portfolio of blue chip stocks, if traded at their geometric optimal portfolio levels, will show tremendous drawdowns. Yet, these blue chip stocks must be traded at these levels, as these levels maximize potential geometric gain relative to dispersion (risk), and also provide for attaining a goal in the least possible time. When viewed from such a perspective, trading blue chip stocks is no more risky than trading pork bellies, and pork bellies are no less conservative than blue chip stocks. The same can be said of a portfolio of commodity trading systems and a portfolio of bonds.

Typically, investors practice dilution, whether inadvertent or not. That is, if, optimally, one should trade a certain component in a portfolio at the $f\$$ level of, say, $2,500, they may be trading it consciously at an $f\$$ level of, say, $5,000, in a conscious effort to smooth out the equity curve and buffer drawdowns, or, unconsciously, at such a half-optimal f level, since all positions can be assigned an f value as detailed in earlier chapters. Often, people practice asset allocation is by splitting their equity into two subaccounts, an active subaccount and an inactive subaccount. These are not two separate accounts; rather, in theory, they are a way of splitting a single account.

The technique works as follows. First, you must decide upon an initial fractional level. Let's suppose that, initially, you want to emulate an account at the half f level. Therefore, your initial fractional level is .5 (the initial fractional level must be greater than 0 and less than 1). This means you will split your account, with .5 of the equity in your account going into the inactive subaccount and .5 going into the active subaccount. Let's assume we are starting out with a $100,000 account. Therefore, $50,000 is initially in the inactive subaccount and $50,000 is in the active subaccount. It is the equity in the active subaccount that you use to determine how many units to trade. These subaccounts are not real; they are a hypothetical construct you are creating in order to manage your money more effectively. You always use the full optimal fs with this technique. Any equity changes are reflected in the active portion of the account. Therefore, each day, you must look at the account's total equity (closed equity plus open equity, marking open positions to the market) and subtract the inactive amount (which will remain constant from day to day). The difference is your active equity, and it is on this difference that you will calculate how many units to trade at the full f levels.

Let's suppose that the optimal f for market system A is to trade one contract for every $2,500 in account equity. You come into the first day with $50,000 in active equity and, therefore, you will look to trade 20 units. If you were using the straight half f strategy, you would end up with the same number of units on day one. At half f, you would trade one contract for every $5,000 in account equity ($2,500/.5) and you would use the full $100,000 account equity to figure how many units to trade. Therefore, under the half f strategy, you would trade 20 units on this day as well.

However, as soon as the equity in the account changes, the number of units you will trade changes as well. Let's assume that you make $5,000 this next day, thus pushing the total equity in the account up to $105,000. Under the half f strategy, you will now be trading 21 units. However, under the split equity technique, you must subtract the now-constant inactive amount of $50,000 from your total equity of $105,000. This leaves an active equity portion of $55,000, from which you will figure your contract size at the optimal f level of one contract for every $2,500 in equity. Therefore, under the split equity technique, you will now look to trade 22 units.

The procedure works the same on the downside of the equity curve as well, with the split equity technique peeling off units at a faster rate than the fractional f strategy. Suppose we lost $5,000 on the first day of trading, putting the total account equity at $95,000. Under the fractional f strategy, you would now look to trade 19 units ($95,000/$5,000). However, under the split equity technique you are now left with $45,000 of active equity and, thus, you will look to trade 18 units ($45,000/$2,500).

Notice that with the split equity technique, the exact fraction of optimal *f* that we are using changes with the equity changes. We specify the fraction we want to start with. In our example, we used an initial fraction of .5. When the equity increases, this fraction of the optimal *f* increases, too, approaching 1 as a limit as the account equity approaches infinity. On the downside, this fraction approaches 0 as a limit at the level where the total equity in the account equals the inactive portion. This fact, that there is built-in portfolio insurance with the split equity technique, is a tremendous benefit and will be discussed at length later in this chapter.

Because the split equity technique has a fraction for *f* that moves, we will refer to it as a dynamic fractional *f* strategy, as opposed to the straight fractional *f* (which we will call a *static* fractional *f*) strategy.

Using the dynamic fractional *f* technique is analogous to trading an account full out at the optimal *f* levels, where the initial size of the account is the active equity portion.

So, we see that there are two ways to dilute an account down from the full geometric optimal portfolio. We can trade a static fractional or a dynamic fractional *f*. Although the two techniques are related, they also differ. Which is best?

To begin with, we need to be able to determine the arithmetic average HPR for trading n given scenario spectrums simultaneously, as well as the variance in those HPRs for those n simultaneously traded scenario spectrums, for given *f* values $(f_1 \ldots f_n)$ operating on those scenario spectrums. These are given now as:

$$\text{AHPR}\,(f_1 \ldots f_n) = \frac{\sum_{k=1}^{m}\left[\left(1 + \sum_{i=1}^{n}\left(f_i^* \left(\frac{-PL_{k,i}}{BL_i}\right)\right)\right) * \text{Prob}_k\right]}{\sum_{k-1}^{m}\text{Prob}_k} \quad (10.01)$$

where: n = The number of scenario spectrums (market systems or portfolio components).

m = The possible number of combinations of outcomes between the various scenario spectrums (market systems) based on how many scenarios are in each set. m = The number of scenarios in the first spectrum * the number of scenarios in the second spectrum $* \ldots *$ the number of scenarios in the nth spectrum.

Prob = The sum of probabilities of all m of the HPRs for a given set of f values. $Prob_k$ is the sum of the values in brackets {} in the numerator, for all m values of a given set of f values.

f_i = The f value being used for component i. f_i must be greater than 0, and can be infinitely high (i.e., it can be greater than 1.0).

$PL_{k,j}$ = The outcome profit or loss for the ith component (i.e., scenario spectrum or market system) associated with the kth combination of scenarios.

BL_i = The worst outcome of scenario spectrum (market system) i.

Thus, $Prob_k$ in the equation is equal to Equation (9.03)

Equation (10.01) simply takes the coefficient of each HPR *times* its probability and sums these. The resultant sum is then divided by the sum of the probabilities.

The variance in the HPRs for a given set of multiple simultaneous scenario spectrums being traded at given f values can be determined by first taking the *raw coefficient* of the HPRs, the rawcoef:

$$\text{rawcoef}_k = 1 + \sum_{i=1}^{n} \left(f_i * \left(\frac{-PL_{k,i}}{BL_i} \right) \right) \tag{10.02}$$

Then, these raw coefficients are averaged for all values of k between 1 and m, to obtain arimeanrawcoef:

$$\text{arimeanrawcoef} = \frac{\left(\sum_{k=1}^{m} \text{rawcoef}_k \right)}{m} \tag{10.03}$$

Now, the variance V can be determined as:

$$V = \frac{\sum_{k=1}^{m} (\text{rawcoef}_k - \text{arimeanrawcoef})^2 * Prob_k}{\sum_{k=1}^{m} Prob_k} \tag{10.04}$$

Where again, $Prob_k$ is determined by Equation (9.03).

If we know what the AHPR is, and the variance at a given f level (say the optimal f level for argument's sake), we can convert these numbers into what they would be trading at a level of dilution we'll call FRAC. And, since we are able to figure out the two legs of the right triangle, we can also figure the estimated geometric mean HPR at the diluted level. The formulas are now given for the diluted AHPR, called FAHPR, the diluted standard deviation (which is simply the square root of variance), called FSD, and the diluted geometric mean HPR, called FGHPR here:

$$FAHPR = (AHPR - 1) * FRAC + 1$$
$$FSD = SD * FRAC$$
$$FGHPR = \sqrt{FAHPR^2 - FSD^2}$$

where: FRAC = The fraction of optimal f we are solving for.
          AHPR = The arithmetic average HPR at the optimal f.
          SD = The standard deviation in HPRs at the optimal f.
          FAHPR = The arithmetic average HPR at the fractional f.
          FSD = The standard deviation in HPRs at the fractional f.
          FGHPR = The geometric average HPR at the fractional f.

Let's assume we have a system where the AHPR is 1.0265. The standard deviation in these HPRs is .1211 (i.e., this is the square root of the variance given by Equation (10.04)); therefore, the estimated geometric mean is 1.019. Now, we will look at the numbers for a .2 static fractional f and a .1 static fractional f. The results, then, are:

	Full f	.2 f	.1 f
AHPR	1.0265	1.0053	1.00265
SD	.1211	.02422	.01211
GHPR	1.01933	1.005	1.002577

Here is what will also prove to be a useful equation, the time expected to reach a specific goal:

$$T = \frac{\ln(goal)}{\ln(geometric\ mean)}$$

where: T = The expected number of holding periods to reach a specific goal.
          goal = The goal in terms of a multiple on our starting stake, a TWR.
          ln () = The natural logarithm function.

Now, we will compare trading at the .2 static fractional f strategy, with a geometric mean of 1.005, to the .2 dynamic fractional f strategy (20% as initial active equity) with a daily geometric mean of 1.01933. The time (number of days, since the geometric means are daily) required to double the static fractional f is given by Equation (5.07) as:

$$\frac{\ln(2)}{\ln(1.005)} = 138.9751$$

To double the dynamic fractional f requires setting the goal to 6. This is because, if you initially have 20% of the equity at work, and you start out with a $100,000 account, then you initially have $20,000 at work. The goal is to make the active equity equal $120,000. Since the inactive equity remains at $80,000, you will have a total of $200,000 on your account that started at $100,000. Thus, to make a $20,000 account grow to $120,000 means you need to achieve a TWR of 6. Therefore, the goal is 6 in order to double a .2 dynamic fractional f:

$$\frac{\ln(6)}{\ln(1.01933)} = 93.58634$$

Notice how it took 93 days for the dynamic fractional f versus 138 days for the static fractional f.

Now let's look at the .1 fraction. The number of days expected in order for the static technique to double is expected as:

$$\frac{\ln(2)}{\ln(1.002577)} = 269.3404$$

If we compare this to doubling a dynamic fractional f that is initially set to .1 active, you need to achieve a TWR of 11. Hence, the number of days required for the comparative dynamic fractional f strategy is:

$$\frac{\ln(11)}{\ln(1.01933)} = 125.2458$$

To double the account equity, at the .1 level of fractional f is, therefore, 269 days for our static example, compared to 125 days for the dynamic. The lower the fraction for f, the faster the dynamic will outperform the static technique.

Let's take a look at tripling the .2 fractional f. The number of days expected by static technique to triple is:

$$\frac{\ln(3)}{\ln(1.005)} = 220.2704$$

This compares to its dynamic counterpart, which requires:

$$\frac{\ln(11)}{\ln(1.01933)} = 125.2458$$

To make 400% profit (i.e., a goal or TWR, of 5) requires of the .2 static technique:

$$\frac{\ln(5)}{\ln(1.005)} = 322.6902$$

Which compares to its dynamic counterpart:

$$\frac{\ln(21)}{\ln(1.01933)} = 1590201$$

It takes the dynamic almost half the time it takes the static to reach the goal of 400% in this example. However, if you look out in time 322.6902 days to where the static technique doubled, the dynamic technique would be at a TWR of:

$$= .8 + 1.01933^{322.6902} * .2$$
$$= .8 + 482.0659576 * .2$$
$$= 97.21319$$

This represents making over 9,600% in the time it took the static to make 400%.

We can now amend Equation (5.07) to accommodate both the static and fractional dynamic f strategies to determine the expected length required to achieve a specific goal as a TWR. To begin with, for the static fractional f, we can create Equation (5.07b):

$$T = \frac{\ln(\text{goal})}{\ln(\text{FGHPR})}$$

where: T = The expected number of holding periods to reach a specific goal.

goal = The goal in terms of a multiple on our starting stake, a TWR.

FGHPR = The adjusted geometric mean. This is the geometric mean, run through Equation (5.06) to determine the geometric mean for a given static fractional f.

$\ln(\ )$ = The natural logarithm function.

For a dynamic fractional f, we have Equation (5.07c):

$$T = \frac{\ln\left(\left(\frac{(\text{goal}-1)}{\text{FRAC}}\right) + 1\right)}{\ln(\text{geometric mean})}$$

where: $T =$ The expected number of holding periods
                               to reach a specific goal.
                  goal $=$ The goal in terms of a multiple on our starting
                               stake, a TWR.
          FRAC $=$ The initial active equity percentage.
  geometric mean $=$ the raw geometric mean HPR at the
                               optimal f; there is no adjustment performed
                               on it as there is in Equation (5.07b)
              $\ln(\ ) =$ The natural logarithm function.

Thus, to illustrate the use of Equation (5.07c), suppose we want to determine how long it will take an account to double (i.e., TWR $= 2$) at .1 active equity and a geometric mean of 1.01933:

$$
\begin{aligned}
T &= \frac{\ln\left(\left(\frac{(\text{goal}-1)}{\text{FRAC}}\right)+1\right)}{\ln(\text{geometric mean})} \\[2mm]
&= \frac{\ln\left(\left(\frac{(2-1)}{.1}\right)+1\right)}{\ln(1.01933)} \\[2mm]
&= \frac{\ln\left(\frac{(1)}{.1}+1\right)}{\ln(1.01933)} \\[2mm]
&= \frac{\ln(10+1)}{\ln(1.01933)} \\[2mm]
&= \frac{\ln(11)}{\ln(1.01933)} \\[2mm]
&= \frac{2.397895273}{.01914554872} \\[2mm]
&= 125.2455758
\end{aligned}
$$

Thus, if our geometric means are determined off scenarios which have a daily holding period basis, we can expect about 125¼ days to double. If our scenarios used months as holding period lengths, we would have to expect about 125¼ months to double.

As long as you are dealing with a T large enough that Equation (5.07c) is greater than Equation (5.07b), then you are benefiting from dynamic fractional f trading. This can, likewise, be expressed as Equation (10.05):

$$\text{FGHPR}^T <= \text{geometric mean}^T * \text{FRAC} + 1 - \text{FRAC} \qquad (10.05)$$

Thus, you must iterate to that value of T where the right side of the equation exceeds the left side—that is, the value for T (the number of holding periods) at which you should wait before reallocating; otherwise, you are better off to trade the static fractional f counterpart.

Figure 10.1 illustrates this graphically. The arrow is that value for T at which the left-hand side of Equation (10.05) is equal to the right-hand side.

Thus, if we are using an active equity percentage of 20% (i.e., FRAC = .2), then FGHPR must be figured on the basis of a .2f. Thus, for the case where our geometric mean at full optimal f is 1.01933, and the .2 f (FGHPR) is 1.005, we want a value for T that satisfies the following:

$$1.005^T <= 1.01933^T * .2 + 1 - .2$$

We figured our geometric mean for optimal f and, therefore, our geometric mean for the fractional f (FGHPR) on a daily basis, and we want to see if one quarter is enough time. Since there are about 63 trading days per quarter, we want to see if a T of 63 is enough time to benefit by dynamic fractional f. Therefore, we check Equation (10.05) at a value of 63 for T:

$$1.005^{63} <= 1.01933^{63} * .2 + 1 - .2$$
$$1.369184237 <= 3.340663933 * .2 \mid 1 - .2$$
$$1.369184237 <= .6681327866 + 1 - .2$$
$$1.369184237 <= 1.6681327866 - .2$$
$$1.369184237 <= 1.4681327866$$

The equation is satisfied, since the left side is less than or equal to the right side of the equation. Thus, we can reallocate on a quarterly basis under the given values and benefit from using dynamic fractional f.

Figure 10.1 demonstrates the relationship between trading at a static versus a dynamic fractional f strategy over a period of time.

This chart shows a 20% initial active equity, traded on both a static and a dynamic basis. Since they both start out trading the same number of units, that very same number of units is shown being traded straight through as a constant contract. The geometric mean HPR, at full f used in this chart, was 1.01933; therefore, the geometric mean at the .2 static fractional f was 1.005, and the arithmetic average HPR at full f was 1.0265.

All of this leads to a couple of important points, that the *dynamic fractional f will outpace the static fractional f faster, the lower the fraction and the higher the geometric mean*. That is, using an initial active equity percentage of .1 (for both dynamic and static) means that the dynamic will overtake the static faster than if you used a .5 fraction for both. Thus, generally, the dynamic fractional f will overtake its static counterpart faster, the lower the portion of initial active equity. In other words, a portfolio with an initial active equity of .1 will overcome its static counterpart faster than a portfolio with an initial active equity allocation of .2 will overtake

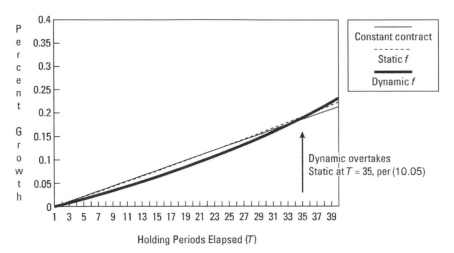

FIGURE 10.1 Percent growth per period for constant contract, static, and dynamic f

its static counterpart. At an initial active equity allocation of 100% (1.0), the dynamic never overtakes the static fractional f (rather, they grow at the same rate). Also affecting the rate at which the dynamic fractional f overtakes its static counterpart is the geometric mean of the portfolio itself. The higher the geometric mean, the sooner the dynamic will overtake its static counterpart. At a geometric mean of 1.0, the dynamic never overtakes its static counterpart.

The more time that elapses, the greater the difference between the static fractional f and the dynamic fractional f strategy. Asymptotically, the dynamic fractional f strategy has infinitely greater wealth than its static counterpart.

One last important point about Figure 10.1. The constant contract line crosses the other two lines before they cross over each other.

In the long run, you are better off to practice asset allocation with a dynamic fractional f technique. That is, you determine an initial level—a percentage—to allocate as active equity. The remainder is inactive equity. The day-to-day equity changes are reflected in the active portion only. The inactive dollar amount remains constant. Therefore, each day you subtract the constant inactive dollar amount from your total account equity. This difference is the active portion, and it is on this active portion that you will figure your quantities to trade in, based on the optimal f levels.

Now, when the margin requirement is calculated for the positions, it will not be exactly the same as your active equity. It can be more or less; it doesn't matter. Thus, unless your margin requirement is for 100% of the equity in the account, you will have some unused cash in the account on any given holding period. Thus, you are almost always inadvertently allocating something to cash (or cash equivalents). So you can see that there isn't any need for a scenario spectrum for cash or cash equivalents—they already get their proper allocation when you do the active and inactive equity split.

REALLOCATION

Notice in Figure 10.1 that, trading at a dynamic fractional f, eventually the active portion of your equity will dwarf the inactive portion, and you will be faced with a portfolio that is far too aggressive for your blood—the same situation you faced in the beginning when you looked at trading the portfolio at the full optimal f amount. Thus, at some point in time in the future, you will want to *reallocate* back to some level of initial active equity.

For instance, you start out at a 10% initial active equity on a $100,000 account. You, therefore, have $10,000 active equity—equity that you are trading full out at the optimal f level. Each day, you will subtract $90,000 from the equity on the account. The difference is the active equity, and it is on the active equity that you trade at the full optimal f levels.

Now, assume that this account got up to $1 million equity. Thus, subtracting the constant dollar inactive amount of $90,000 leaves you at an active equity of $910,000, which means you are now at 91% active equity. Thus, you face those gigantic drawdowns that you sought to avoid initially, when you diluted f and started trading at a 10% initial active equity.

Consider the case of reallocating after every trade or every day. Such is the case with static fractional f trading. Recall again Equation (10.08a), the time required to reach a specific goal.

Let's return to our system that we are trading with a .2 active portion and a geometric mean of 1.01933. We will compare this to trading at the static fractional .2 f, where the resultant geometric mean is 1.005. Now, if we are starting out with a $100,000 account, and we want to reallocate at $110,000 total equity, the number of days (since our geometric means here are on a per-day basis) required by the static fractional .2 f is:

$$\frac{\ln(1.1)}{\ln(1.005)} = 19.10956$$

This compares to using $20,000 of the $100,000 total equity at the full f amount, and trying to get the total account up to $110,000. This would represent a goal of 1.5 times the $20,000:

$$\frac{\ln(1.5)}{\ln(1.01933)} = 21.17807$$

At lower goals, the static fractional f strategy grows faster than its corresponding dynamic fractional f counterpart. As time elapses, the dynamic overtakes the static until, eventually, the dynamic is infinitely further ahead. Figure 10.1 graphically displays this relationship between the static and dynamic fractional f's.

If you reallocate too frequently, you are only shooting yourself in the foot, as the technique would be inferior to its static fractional f counterpart. Therefore, since you are better off, in the long run, to use the dynamic fractional f approach to asset allocation, you are also better off to reallocate funds between the active and inactive subaccounts as infrequently as possible. Ideally, you will only make this division between active and inactive equity once, at the outset of the program.

It is not beneficial to reallocate too frequently. Ideally, you will never reallocate. Ideally, you will let the fraction of optimal f you are using keep approaching 1 as your account equity grows. In reality, however, you most likely will reallocate at some point in time. Hopefully, you will not reallocate so frequently that it becomes a problem.

Reallocation seems to do just the opposite of what we want to do, in that reallocation trims back after a run up in equity, or adds more equity to the active portion after a period in which the equity has been run down.

Reallocation is a compromise. It is a compromise between the theoretical ideal and the real-life implementation. The techniques discussed allow us to make the most of this compromise. Ideally, you would never reallocate. Your humble little \$10,000 account, when it grew to \$10 million, would never go through reallocation. Ideally, you would sit through the drawdown which took your account down to \$50,000 from the \$10 million mark before it shot up to \$20 million. Ideally, if your active equity were depleted down to one dollar, you would still be able to trade a fractional contract (a *microcontract?*). In an ideal world, all of these things would be possible. In real life, you are going to reallocate at some point on the upside or the downside. Given that you are going to do this, you might as well do it in a systematic, beneficial way.

In reallocating—compromising—you *reset* things back to a state where you would be if you were starting the program all over again, only at a different equity level. Then, you let the outcome of the trading dictate where the fraction of f floats to by using a dynamic fractional f in between reallocations. Things can get levered up awfully fast, even when starting out with an active equity allocation of only 5%. Remember, you are using the full optimal f on this 5%, and if your program does modestly well, you'll be trading in substantial quantities relative to the total equity in the account in short order.

The first, and perhaps most important, thing to realize about reallocation, can be seen in Figure 10.1. Note the arrow in the figure, which is identified as that T where Equation (10.09) is equal. This amount of time, T, is critical. If you reallocate before T, you are doing yourself harm in trading the dynamic, rather than the static, fractional f.

The next critical thing to realize about reallocation is that you have some control over the maximum drawdown in terms of percentage equity retracements. Notice that you are trading the active portion of an account as though it were an account of exactly that size, full out at the optimal levels. Since you should expect to see nearly 100% equity retracements when trading at the full optimal f levels, you should expect to see 100% of the active equity portion wiped out at any one time.

Further, many traders who have been using the fractional dynamic f approach over the last couple of years relate what appears to be a very good rule of thumb: *Set your initial active equity at one half of the maximum drawdown you can tolerate.* Thus, if you can take up to a 20% drawdown, set your initial active equity at 10% (however, if the account is profitable and your active equity begins to exceed 20%, you are very susceptible to seeing drawdowns in excess of 20%).

There is a more accurate implementation of this very notion. Notice, that for portfolios, you must use the sum of all f in determining exposure. That is, you must sum the f values up across the components. This is important in that, suppose you have a portfolio of three components with f values determined, respectively, of .5, .7, and .69. The total of these is 1.89. That is the f you are working with in the portfolio, as a whole. Now, if each of these components saw the worst-case scenario manifest, the account would see a 189% drawdown on active equity! When working with portfolios, you should be very careful to be ever-vigilant for such an event, and to bear this in mind when determining initial active equity allocations.

The third important notion about reallocation pertains to the concept of portfolio insurance and its relationship to optimal f.

PORTFOLIO INSURANCE AND OPTIMAL f

Assume for a moment that you are managing a stock fund. Figure 10.2 depicts a typical portfolio insurance strategy, also known as dynamic hedging. The floor in this example is the current portfolio value of 100 (dollars per share). The typical portfolio will follow the equity market one for one. This is represented by the unbroken line. The insured portfolio is depicted by the dotted line. You will note that the dotted line is below the unbroken

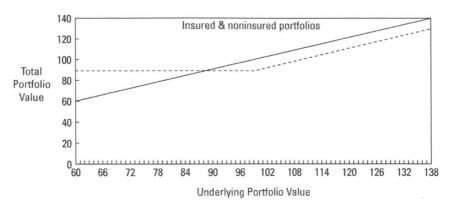

FIGURE 10.2 Portfolio insurance

line when the portfolio is at its initial value (100) or greater. This difference represents the cost of performing the portfolio insurance. Otherwise, as the portfolio falls in value, portfolio insurance provides a floor on the value of the portfolio at a desired level (in this case, the present value of 100) minus the cost of performing the strategy.

In a nutshell, portfolio insurance is akin to buying a put option on the portfolio. Let's suppose that the fund you are managing consists of only one stock, which is currently priced at $100. Buying a put option on this stock, with a strike price of $100, at a cost of $10, would replicate the dotted line in Figure 10.2. The worst that could happen to your portfolio of one stock and a put option on it is that you could exercise the put, which sells your stock at $100, and you lose the value of the put, $10. Thus, the worst that this portfolio can be worth is $90, regardless of how far down the underlying stock goes.

On the upside, your insured portfolio suffers somewhat, in that the value of the portfolio is always reduced by the cost of the put.

Now, consider that being long a call option will give you the same profile as being long the underlying and long a put option with the same strike price and expiration date as the call option. When we speak of the same profile, we mean an equivalent position in terms of the risk/reward characteristics at different values for the underlying. Thus, the dotted line in Figure 10.2 can also represent a portfolio composed of simply being long the $100 call option at expiration.

Here is how *dynamic hedging* works to provide portfolio insurance. Suppose you buy 100 shares of this single stock for your fund, at a price of $100 per share. Now, you will replicate the call option by using this underlying stock. The way you will do this is by determining an initial floor for the stock. The floor you choose is, say, 100. You also determine an expiration date for this hypothetical option which you are going to create. Let's say that the expiration date you choose is the date on which this quarter ends.

Now, you will figure the delta (the instantaneous rate of change in the price of a call option relative to the change in price of the underlying instrument) for this 100 call option with the chosen expiration date. Suppose the delta is .5. This means that you should be 50% invested in the given stock. Thus, you would have only 50 shares of stock rather than the 100 shares you would have if you were not practicing portfolio insurance. As the value of the stock increases, so, too, will the delta, and likewise the number of shares you hold. The upside limit is a delta at 1, where you would be 100% invested. In our example, at a delta of 1, you would have 100 shares.

As the stock decreases, so, too, does the delta, and likewise the size of the position in the stock decreases. The downside limit is at a delta of 0, where you wouldn't have any position in the stock.

Operationally, stock fund managers have used *noninvasive methods* of dynamic hedging. Such a technique involves not having to trade the cash portfolio. Rather, the portfolio as a whole is adjusted to what the current delta should be, as dictated by the model by using stock index futures, and, sometimes, put options. One benefit of a technique using futures is that futures have low transactions cost.

Selling short futures against the portfolio is equivalent to selling off part of the portfolio and putting it into cash. As the portfolio falls, more futures are sold, and as it rises, these short positions are covered. The loss to the portfolio, as it goes up and the short futures positions are covered, is what accounts for the portfolio insurance cost, the cost of the replicated put options. Dynamic hedging, though, has the benefit of allowing us to closely estimate this cost at the outset. To managers trying to implement such a strategy, it allows the portfolio to remain untouched, while the appropriate asset allocation shifts are performed through futures trades. This noninvasive technique of using futures permits the separation of asset allocation and active portfolio management.

To someone implementing portfolio insurance, you must continuously adjust the portfolio to the appropriate delta. This means that, say, each day, you must input into the option pricing model the current portfolio value, time until expiration, interest rate levels, and portfolio volatility, to determine the delta of the put option you are trying to replicate. Adding this delta (which is a number between 0 and -1) to 1 will give you the corresponding call's delta. This is the hedge ratio, the percentage that you should be investing in the fund.

Suppose your hedge ratio for the present moment is .46. Let's say that the size of the fund you are managing is the equivalent of 50 S&P futures units. Since you want to be only 46% invested, it means you want to be

54% disinvested. Fifty-four percent of 50 units is 27 units. Therefore, at the present price level of the fund at this point in time, for the given interest rate and volatility levels, the fund should be short 27 S&P units along with its long position in cash stocks.

Because the delta needs to be recomputed on an ongoing basis, and portfolio adjustments must be constantly monitored, the strategy is called a dynamic hedging strategy.

One problem with using futures in the strategy is that the futures market does not exactly track the cash market. Further, the portfolio you are selling futures against may not exactly follow the cash index upon which the futures market is traded. These tracking errors can add to the expense of a portfolio insurance program. Furthermore, when the option being replicated gets very near to expiration, and the portfolio value is near the strike price, the gamma of the replicated option goes up astronomically. Gamma is the instantaneous rate of change of the delta or hedge ratio. In other words, gamma is the delta of the delta. If the delta is changing very rapidly (i.e., if the replicated option has a high gamma), portfolio insurance becomes increasingly more cumbersome to perform. There are numerous ways to work around this problem, some of which are very sophisticated. One of the simplest involves the concept of a perpetual option. For instance, you can always assume that the option you are trying to replicate expires in, say, three months. Each day you will move the replicated option's expiration date ahead by a day. Again, this high gamma usually becomes a problem only when expiration draws near and the portfolio value and the replicated option's strike price are very close.

There is a very interesting relationship between optimal f and portfolio insurance. When you enter a position, you can state that f percent of your funds are invested. For example, consider a gambling game where your optimal f is .5, biggest loss -1, and bankroll is $10,000. In such a case, you would bet one dollar for every two dollars in your stake since -1, the biggest loss, divided by $-.5$, the negative optimal f, is 2. Dividing $10,000 by 2 yields $5,000. You would, therefore, bet $5,000 on the next bet, which is f percent (50%) of your bankroll. Had we multiplied our bankroll of $10,000 by f (.5), we would have arrived at the same $5,000 result. Hence, we have bet f percent of our bankroll.

Likewise, if our biggest loss were $250 and everything else the same, we would be making one bet for every $500 in our bankroll since $-\$250/-.5 = \$500$. Dividing $10,000 by $500 means that we would make twenty bets. Since the most we can lose on any one bet is $250, we have thus risked f percent, 50% of our stake in risking $5,000 ($250 $*$ 20).

Therefore, we can state that f equals the percentage of our funds at risk, or f equals the hedge ratio. Remember, when discussing portfolios, we are discussing the sum of the f values of the components. Since f is only applied on the active portion of our portfolio in a dynamic fractional f strategy, we can state that the hedge ratio of the portfolio, H, equals:

$$H = \left(\sum_{i=1}^{n} f_i \right) * \frac{\text{active\$}}{\text{total equity}} \qquad (10.06a)$$

where: H = The hedge ratio of the portfolio.
        f_i = The f value of the ith component in the portfolio.
   active\$ = The active portion of funds in an account.

Equation (10.06a) gives us the hedge ratio for a portfolio being traded on a dynamic fractional f strategy. Portfolio insurance is also at work in a static fractional f strategy, only the quotient active\$/total equity equals 1, and the value for f (the optimal f) is multiplied by whatever value we are using for the fraction of f. Thus, in a static fractional f strategy, the hedge ratio is:

$$H = \left(\sum_{i=1}^{n} f_i \right) * \text{FRAC} \qquad (10.06b)$$

We can state that in trading an account on a dynamic fractional f basis, we are performing portfolio insurance. Here, the floor is known in advance and is equal to the initial inactive equity plus the cost of performing the insurance. However, it is often simpler to refer to the floor of a dynamic fractional f strategy as the initial inactive equity of an account.

We can state that Equation (10.06a) or (10.06b) equals the delta of the call option of the terms used in portfolio insurance. Further, we find that this delta changes much the way a call option, which is deep out of the money and very far from expiration, changes. Thus, by using a constant inactive dollar amount, trading an account on a dynamic fractional f strategy is equivalent to owning a put option on the portfolio which is deep in the money and very far out in time. Equivalently, we can state that trading a dynamic fractional f strategy is the same as owning a call option on the portfolio which doesn't expire for a very long time and is very far out of the money.

However, it is also possible to use portfolio insurance as a reallocation technique to steer performance somewhat. This steering may be analogous to trying to steer a tanker with a rowboat oar, but this is a valid reallocation technique. The method initially involves setting parameters for the program. First, you must determine a floor value. Once chosen, you must decide upon an expiration date, volatility level, and other input parameters to the particular option model you intend to use. These inputs will give you the option's delta at any given point in time. Once the delta is known, you can determine what your active equity should be. Since the delta for the account, the variable H in Equation (10.06a), must equal the delta for the call option being replicated:

$$H = \left(\sum_{i=1}^{n} f_i \right) * \frac{\text{active\$}}{\text{total equity}}$$

Therefore:

$$\frac{H}{\sum_{i=1}^{n} f_i} = \frac{\text{active\$}}{\text{total equity}} \qquad \text{if } H < \sum_{i=1}^{n} f_i \qquad (10.07)$$

Otherwise:

$$H = \frac{\text{active\$}}{\text{total equity}} = 1$$

Since active\$/total equity is equal to the percentage of active equity, we can state that the percentage of funds we should have in active equity, of the total account equity, is equal to the delta on the call option divided by the sum of the f values of the components. However, you will note that if H is greater than the sum of these f values, then it is suggesting that you allocate greater than 100% of an account's equity as active. Since this is not possible, there is an upper limit of 100% of the account's equity that can be used as active equity.

Portfolio insurance is great in theory, but poor in practice. As witnessed in the 1987 stock market crash, the problem with portfolio insurance is that, when prices plunge, there isn't any liquidity at any price. This does not concern us here, however, since we are looking at the relationship between active and inactive equity, and how this is mathematically similar to portfolio insurance.

The problem with implementing portfolio insurance as a reallocation technique, as detailed here, is that reallocation is taking place constantly. This detracts from the fact that a dynamic fractional f strategy will asymptotically dominate a static fractional f strategy. As a result, trying to steer performance by way of portfolio insurance as a dynamic fractional f reallocation strategy probably isn't such a good idea. However, anytime you use fractional f, static or dynamic, you are employing a form of portfolio insurance.

UPSIDE LIMIT ON ACTIVE EQUITY AND THE MARGIN CONSTRAINT

Even if you are trading only one market system, margin considerations can often be a problem. Consider that the optimal f in dollars is very often less than the initial margin requirement for a given market. Now, depending on what fraction of f you are using at the moment, whether you are using a static or dynamic fractional f strategy, you will encounter a margin call if the fraction is too high.

When you trade a portfolio of market systems, the problem of a margin call becomes even more likely.

What is needed is a way to reconcile how to create an optimal portfolio within the bounds of the margin requirements on the components in the portfolio. This can very easily be found. The way to accomplish this is to find what fraction of f you can use as an upper limit. This upper limit, L, is given by Equation (10.08):

$$ L = \frac{\overset{n}{\underset{i=1}{MAX}}(f_i\$)}{\sum\limits_{k=1}^{n}\left(\left(\overset{n}{\underset{i-1}{MAX}}(f_i\$)/f_k\$\right) * \text{margin}_k\right)} \tag{10.08} $$

where: L = The upside fraction of f. At this particular fraction of f, you are trading the optimal portfolio as aggressively as possible without incurring an initial margin call.

$f_k\$$ = The optimal f in dollars for the kth market system.

$\text{margin}_k\$$ = The initial margin requirement of the kth market system.

n = The total number of market systems in the portfolio.

Equation (10.08) is really much simpler than it appears. For starters, in both the numerator and the denominator, we find the expression $\overset{n}{\underset{i=1}{MAX}}$, which simply means to take the greatest f \$ of all of the components in the portfolio.

Let's assume a two-component portfolio, which we'll call Spectrums A and B. We can arrange the necessary information for determining the upside limit on active equity in a table as follows:

Component	f$	Margin	Greatest f$/f$
Spectrum A	$2,500	$11,000	2500/2500 = 1
Spectrum B	$1,500	$2,000	2500/1500 = 1.67

Now we can plug these values into Equation (10.08). Notice that $\overset{n}{\underset{i=1}{MAX}}$ is $2,500, since the only other $f\$$ is $1,500, which is less. Thus:

$$L = \frac{2500}{1*11000 + 1.67*2000} = \frac{2500}{11000 + 3340} = \frac{2500}{14,340} = 17.43375174\%$$

This tells us that 17.434% should be our maximum upside percentage.

Now, suppose we had a $100,000 account. If we were at 17.434% active equity, we would have $17,434 in active equity. Thus, assuming we can trade in fractional units for the moment, we would buy 6.9736 (17,434/2,500) of Spectrum A and 11.623 (17,434/1,500) of Spectrum B. The margin requirements on this would then be:

$$6.9726 * 11,000 = 76,698.60$$
$$11.623 * 2,000 = 23,245.33$$
$$\text{Total Margin Requirement} = \$99,943.93$$

If, however, we are still employing a static fractional f strategy (despite this author's protestations), then the highest you should set that fraction is 17.434%. This will result in the same margin call as above.

Notice that using Equation (10.08) yields the highest fraction for f without incurring an initial margin call that gives you the same ratios of the different market systems to one another.

Earlier in the text we saw that adding more and more market systems (scenario spectrums) results in higher and higher geometric means for the portfolio as a whole. However, there is a trade-off in that each market system adds marginally less benefit to the geometric mean, but marginally more detriment in the way of efficiency loss due to simultaneous rather than sequential outcomes. Therefore, we have seen that you do not want to trade an infinitely high number of scenario spectrums. What's more, theoretically optimal portfolios run into the real-life application problem of margin constraints. In other words, you are usually better off to trade three scenario spectrums at the full optimal f levels than to trade 10 at dramatically reduced levels as a result of Equation (10.08). Usually, you will find that the optimal number of scenario spectrums to trade in, particularly when you have many orders to place and the potential for mistakes, is but a handful.

f SHIFT AND CONSTRUCTING A ROBUST PORTFOLIO

There is a polymorphic nature to the $n + 1$ dimensional landscape; that is, the landscape is undulating—the peak in the landscape tends to move around as the markets and techniques we use to trade them change in character.

This f shift is doubtless a problem to all traders. Oftentimes, if the f shift is toward zero for many axes—that is, as the scenario spectrums weaken—it can cause what would otherwise be a winning method on a constant unit basis to be a losing program because the trader is beyond the peak of the f curve (to the right of the peak) to an extent that he is in a losing position.

f shift exists in all markets and approaches. It frequently occurs to the point at which many scenario spectrums get allocations in one period in an optimal portfolio construction, then no allocations in the period immediately following. This tells us that the performance, out of sample, tends to greatly diminish. The reverse is also true. Markets that appear as poor candidates in one period where an optimal portfolio is determined, then come on strong in the period immediately following, since the scenarios do not measure up.

When constructing scenarios and scenario sets, you should pay particular attention to this characteristic: Markets that have been performing well will tend to underperform in the next period and vice versa. Bearing this in mind when constructing your scenarios and scenario spectrums will help you to develop more robust portfolios, and help alleviate f shift.

TAILORING A TRADING PROGRAM THROUGH REALLOCATION

Often, money managers may opt for the dynamic f, as opposed to the static, even when the number of holding periods is less than that specified by Equation (10.05) simply because the dynamic provides a better implementation of portfolio insurance.

In such cases, it is important that the money manager not reallocate until Equation (10.05) is satisfied—that is, until enough holding periods elapse that the dynamic can begin to outperform the static counterpart.

A real key to tailoring trading programs to fit the money manager's goals in these instances is by reallocating on the upside. That is, at some upside point in active equity, you should reallocate to achieve a certain goal, yet that point is beyond some minimum point in time (i.e., number of elapsed holding periods).

Returning to Figure 10.1, Equation (10.05) gives us T, or where the crossing of the static f line by the dynamic f line occurs with respect to the horizontal coordinate. That is the point, in terms of number of elapsed holding periods, at which we obtain more benefit from trading the dynamic f rather than the static f. However, once we know T from Equation (10.05), we can figure the Y, or vertical, axis where the points cross as:

$$Y = FRAC * \text{Geometric Mean}^T - FRAC \qquad (10.09)$$

where: $T =$ The variable T derived from Equation (10.05).
                        FRAC $=$ The initial active portion of funds in an account.
        Geometric Mean $=$ The raw geometric mean HPR; there is no
                                adjustment performed on it as there is in
                                Equation (5.07b).

Example:

> Initial Active Equity Percentage $= 5\%$ (i.e., .05)
>
> Geomean HPR per period $= 1.004171$
>
> $T = 316$

We know at 316 periods, on average, the dynamic will begin to outperform the corresponding static f for the same value of f, per Equation (10.05). This is the same as saying that, starting at an initial active equity of 5%, when the account is up by 13.63% ($.05 * 1.004171^{316} - .05$), the dynamic will begin to outperform the corresponding static f for the same value of f.

So, we can see that there is a minimum number of holding periods which must elapse in order for the dynamic fractional f to overtake its static counterpart (*prior to which, reallocation is harmful if implementing the dynamic fractional* f, *and, after which, it is harmful to trade the static fractional* f), which can also be converted from a horizontal point to a vertical one. That is, rather than a minimum number of holding periods, a minimum profit objective can be used.

Reallocating when the equity equals or exceeds this target of active equity will generally result in a much smoother equity curve than reallocating based on T, the horizontal axis. That is, most money managers will find it advantageous to reallocate based on upside progress rather than elapsed holding periods.

What is most interesting here is that for a given level of initial active equity, the upside target will always be the same, regardless of what values you are using for the geometric mean HPR or T *! Thus, a 5% initial active equity level will always see the dynamic overtake the static at a 13.63% profit on the account!*

Since we have an optimal upside target, we can state that there is, as well, an optimal delta on the portfolio on the upside. So, what is the formula for the optimal upside delta? This can be discerned by Equations (10.06a) and (10.06b), where FRAC equals that fraction of active equity which would be seen by satisfying Equation (10.09). This is given as:

$$\text{FRAC} = \frac{(\text{Initial Active Equity} + \text{Upside Target})}{1 + \text{Upside Target}} \qquad (10.10)$$

Thus, if we start out with an initial active equity of 5%, then 13.63% is the upside point where the dynamic would be expected to overtake the static, and we would use the following for FRAC in Equations (10.10a) and (10.10b) in determining the hedge ratio at the upside point, Y, dictated by Equation (10.13):

$$\begin{aligned} \text{FRAC} &= \frac{(0.5 + 1.363)}{(1 + .1363)} \\ &= \frac{.1863}{1.1363} \\ &= .1639531814 \end{aligned}$$

Thus, when we have an account which is up 13.63%, and we start with a 5% initial active equity, we know that the active equity is then 16.39531814%.

GRADIENT TRADING AND CONTINUOUS DOMINANCE

We have seen throughout this text, that trading at the optimal f for a given market system or scenario spectrum (or the set of optimal fs for multiple simultaneous scenario spectrums or market systems) will yield the greatest growth asymptotically, that is, in the long run, as the number of holding periods we trade for gets greater and greater. However, we have seen in Chapter 5, with "Threshold to Geometric," and in Chapter 6, that if we have a finite number of holding periods and we know how many holding periods we are going to trade for, what is truly optimal is somewhat more aggressive even than the optimal f values; that is, it is those values for f which maximize expected average compound growth (EACG).

Ultimately, each of us can only trade a finite number of holding periods—none of us will live forever. Yet, in all but the rarest cases, we do not know the exact length of that finite number of holding periods, so we use the asymptotic limit as the next best approximation.

Now you will see, however, a technique that can be used in this case of an unknown, but finite, number of holding periods over which you are going to trade at the asymptotic limit (i.e., the optimal f values), which, if you are trading any kind of a diluted f (static or dynamic), allows for dominance not only asymptotically, but for *any given holding period in the future.*

That is, we will now introduce a technique for a diluted f (which nearly all money managers must use in order to satisfy the real-world demands of clients pertaining to drawdowns) that not only en-sures that an account will be at the highest equity in the very long

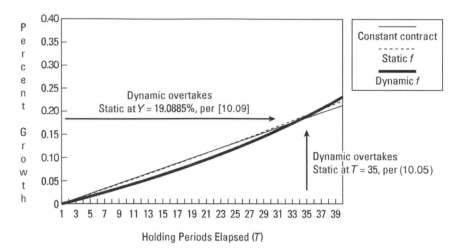

FIGURE 10.3 Points where one method overtakes another can be viewed with respect to time or return

run sense, but ensures that it will be at the highest equity at any point in time, however near or far into the future that is! No longer must someone adhering to optimal f (or, in a broader sense, this new framework) reconcile themselves with the notion that it will be dominant in the long run. Rather, the techniques about to be illustrated seek dominance at all points in time!

Everyone is at an f value whether they acknowledge it or not. Since nearly everyone is diluting what their optimal f values are—either intentionally or unintentionally through ignorance—these techniques always maximize the profitability of an account in cases of diluted f values, not just, as has always been the case with geometric mean maximization, in the very long run.

Again, we must turn our attention to growth functions and rates. Look at Figure 10.3 where growth (the growth functions) is represented as a percentage of our starting stake. Now consider Figure 10.4, which shows the growth rate as a percentage of our stake.

Again, these charts show a 20% initial active equity, traded on both a static and a dynamic basis. Since they both start out trading the same number of units, that very same number of units is shown being traded straight through as a constant contract. The geometric mean HPR (at full f) used in this chart was 1.01933; therefore, the geometric mean at the .2 static fractional f was 1.005, and the arithmetic average HPR at full f was 1.0265.

Notice that by always trading that technique which has the highest gradient at the moment, we ensure the probability of the account being at its greatest equity at any point in time. Thus, we start out trading on a

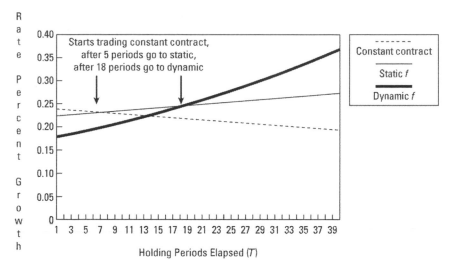

R
a
t
e

P
e
r
c
e
n
t

G
r
o
w
t
h

FIGURE 10.4   Growth rate as a percentage of stake

constant contract basis, with the number of units being determined as that number which would be traded initially if we were trading a fractional f.

Next, the static f gradient dominates, at which point in time (or on the upside in equity) we switch to trading the static f Finally, the dynamic gradient dominates, at which point we switch to trading on a dynamic f basis, Notice that by always trading that technique which has the highest gradient at that moment means you will be on the highest of the three lines in Figure 10.3.

The growth function, Y, for the constant contract technique is now given as:[*]

$$Y = 1 + (\text{AHPR} - 1) * \text{FRAC} * T \qquad (10.11)$$

---

[*]Just as Equation (10.05) gave us that point where the dynamic overtakes the static with respect to the horizontal axis T, we can determine from Equations (10.11) and (10.12) where the static overtakes a constant contract as that value of T where Equation (10.12) equals Equation (10.11).

$$1 + (\text{AHPR} - 1) * \text{FRAC} * T => \text{FGHPR}^T$$

Likewise, this can be expressed in terms of the Y coordinate, to tell us at what percentage of profit, on the total equity in the account, we should switch from a constant contract to static f trading:

$$Y = \text{FGHPR}^T - 1$$

The value for T used in the preceding equation is derived from the one above it.

The growth functions are taken from Equation (10.05). Thus, the static f growth function is the left side of (10.05) and the dynamic f is the right side. Thus, the growth function for static f is:

$$Y = \text{FGHPR}^T \tag{10.12}$$

And for dynamic f, it is:

$$Y = \text{geometric mean}^T * \text{FRAC} + 1 - \text{FRAC} \tag{10.13}$$

Equations (10.11) through (10.13) give us the growth function as a multiple of our starting stake, at a given number of elapsed holding periods, T. Thus, by subtracting 1 from Equations (10.11) through (10.13), we obtain the percent growth as depicted in Figure 10.3.

The gradients, depicted in Figure 10.4, are simply the first derivatives of Y with respect to T, for Equations (10.11) through (10.13). Thus, the gradients are given by the following.

For constant contract trading:

$$\frac{dY}{dT} = \frac{((\text{AHPR} - 1) * \text{FRAC})}{(1 + \text{AHPR}-) * \text{FRAC} * T} \tag{10.14}$$

For static fractional f:

$$\frac{dY}{dT} = \text{FGHPR}^T * \ln(\text{FGHPR}) \tag{10.15}$$

And finally for dynamic fractional f:

$$\frac{dY}{dT} = \text{geometric mean}^T * \ln(\text{geometric mean}) * \text{FRAC} \tag{10.16}$$

where:
$T =$ The number of holding periods.
$\text{FRAC} =$ The initial active equity percentage.
$\text{geometric mean} =$ The raw geometric mean HPR at the optimal f.
$\text{AHPR} =$ The arithmetic average HPR at full optimal f.
$\text{FGHPR} =$ The fractional f geometric mean HPR given by Equation (5.06).
$\ln( ) =$ The natural logarithm function.

The way to implement these equations, especially as your scenarios (scenario spectrums) and joint probabilities change from holding period to holding period, is as follows. Recall that just before each holding period we must determine the optimal allocations. In the exercise of doing that, we derive all of the necessary information to get the values for the variables listed above (for FRAC, geometric mean, AHPR, and the inputs to Equation

(5.06) to determine the FGHPR) Next we plug these values into Equations (10.14), (10.15), and (10.16). Whichever of these three equations results in the greatest value is the technique we go with.

To illustrate by way of an example, we now return to our familiar two-to-one coin toss. Let's assume that this is our only scenario set, comprising the two scenarios heads and tails. Further, suppose we are going to trade it at a .2 fraction (i.e., one-fifth optimal f). Thus, FRAC is .2, the geometric mean is 1.06066, and the AHPR is 1.125. To figure the FGHPR, from Equation (5.06) we already have FRAC and AHPR; we need only SD, the standard deviation in HPRs, which is .375. Thus, the FGHPR is:

$$1.022252415 = \left(\sqrt{((1.125 - 1) * .2 + 1)^2 - (.375 * .2)^2} \right)$$

Plugging these values into the three gradient functions, Equations (10.14) through (10.16), gives us the following table:

	Eq. (10.14)	Eq. (10.15)	Eq. (10.16)
T	**Constant Contract**	**Static *f***	**Dynamic *f***
1	0.024390244	0.022498184	0.012492741
2	0.023809524	0.022998823	0.013250551
3	0.023255814	0.023510602	0.014054329
4	0.022727273	0.02403377	0.014906865
5	0.022222222	0.024568579	0.015811115
6	0.02173913	0.025115289	0.016770217
7	0.021276596	0.025674165	0.017787499
8	0.020833333	0.026245477	0.018866489
9	0.020408163	0.026829503	0.02001093
10	0.02	0.027426524	0.021224793
11	0.019607843	0.02803683	0.022512289
12	0.019230769	0.028660717	0.023877884
13	0.018867925	0.029298488	0.025326317
14	0.018518519	0.02995045	0.026862611
15	0.018181818	0.030616919	0.028492097
16	0.017857143	0.03129822	0.030220427
17	0.01754386	0.031994681	0.032053599
18	0.017241379	0.03270664	0.03399797
19	0.016949153	0.033434441	0.036060287
20	0.016666667	0.034178439	0.038247704

We find that we are at the greatest gradient for the first two holding periods by trading on a constant contract basis, and that on the third period, we should switch to static f. On the seventeenth period, we should switch

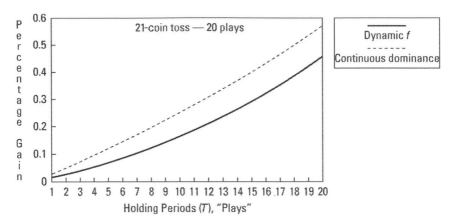

FIGURE 10.5 Continuous dominance vs. dynamic *f*

to dynamic *f*. If we were to do this, Figure 10.5 shows how much better we would have fared, on average, over the first 20 plays or holding periods, than by simply trading a dynamic fractional *f* strategy:

Notice that, at every period, an account traded this way has a higher expected value than even the dynamic fractional *f*. Further, from period 17 on, where we switched from static to dynamic, both lines are forevermore on the same gradient. That is, the dynamic line will never be able to catch up to the continuous dominance line. Thus, the principle of always trading the highest gradient to achieve continuous dominance helps a money manager maximize where an account will be at any point in the future, not just in an asymptotic sense.

To clarify by carrying the example further, suppose we play this two-to-one coin-toss game, and we start out with an account of $200. Our optimal *f* is .25, and a .2 *f*, one-fifth of this, means we are trading an *f* value of .05, or we bet $1 for every $20 in our stake. Therefore, on the first play we bet $10. Since we are trading constant contract, regardless of where the account equity is thereafter, we bet $10 on each subsequent play until we switch over to static *f*. This occurs on the third play. So, on the third bet, we take where our stake is and bet $1 for every $20 we have in equity. We proceed as such through play 16, where, going into the seventeenth play, we will switch over to dynamic. Thus, as we go into every play, from play 3 through play 16, we divide our total equity by $20 and bet that many dollars, thus performing a static fractional *f*.

So, assume that after the second play we have $210 in our stake. We would now bet $10 on the next play (since 210/20 = 10.5, and we must round down to the integer). We keep doing this going into each play through the sixteenth play.

On the seventeenth play, we can see that the dynamic f gradient overtakes the others, so we must now switch over to trading on a dynamic f basis. Here is how. When we started, we decided that we were going to trade a 20% active equity, in effect (because we decided to trade at one-fifth the full optimal f). Since our starting stake was $200, then it means we would have started out, going into play 1, with $40 active equity. We would therefore have $160 inactive equity.

So, going into play 17, where we want to switch over to dynamic, we subtract $160 from whatever is our equity. The difference we then divide by $4, the optimal f\$, and that is how many bets we make on play 17. We proceed by doing this before each play, ad infinitum.

Therefore, let's assume our stake stood at $292 after the sixteenth play. We subtract $160 from this, leaving us with $132, which we then divide by the optimal f\$, which is $4, for a result of 33. We would thus make 33 bets on the seventeenth play (i.e., bet $33).

If you prefer, you can also figure these continuous dominance breakpoints as an upside percentage gain which must be made before switching to the next level. This is the preferred way. Just as Equation (10.09) gives us the vertical, or Y, coordinate corresponding to Equation (10.05)'s horizontal coordinate, we can determine the vertical coordinates corresponding to Equations (10.14) through (10.16). Since you move from a constant contract to static f at that value of T whereby Equation (10.15) is greater than Equation (10.14), you can then plug that T into Equation (10.12) and subtract 1 from the answer. This is the percentage gain on your starting equity required to switch from a constant contract to static f.

Since you move to dynamic f from static f at that value of T whereby Equation (10.16) is greater than Equation (10.15), you can then plug that value for T into Equation (10.13), subtract 1 from the answer, and that is the percentage profit from your starting equity to move to trading on a dynamic f basis.

IMPORTANT POINTS TO THE LEFT OF THE PEAK IN THE $n + 1$-DIMENSIONAL LANDSCAPE

We continue this discussion that is directed towards most money managers, who will trade a diluted f set (whether they know it or not), that is, they will trade at less aggressive levels than optimal for the different scenario spectrums or market systems they are employing. We refer to this as being to the *left*, a term which comes from the idea that, if we were looking at trading one scenario spectrum, we would have one curve drawn out in two-dimensional space, where being to the left of the peak corresponds to having less units on a trade than is optimal. If we are trading two scenario spectrums, we have a topographical map in three-dimensional space, where such money managers would restrict themselves to that real estate which is to the left of the peak when looking from south to north at the landscape, and left of the peak when looking from east to west at the landscape. We could carry the thought into more dimensions, but the term *to the left*, is irrespective of the number of dimensions; it simply means at less than full optimal with respect to each axis (scenario spectrum).

Money managers are *not* wealth maximizers. That is, their utility function or, rather, the utility functions imposed on them by their clients and their industry, their $U''(x)$, is less than zero. They are, therefore, to the left of the peaks of their optimal *f*s.

Thus, given the real-world constraints of smoother equity curves than full optimal calls for, as well as the realization that a not-so-typical drawdown at the optimal level will surely cause a money manager's clients to flee, we are faced with the prospect of where, to the left, is an opportune point (to satisfy their $U''(x)$)? Once this opportune point is found, we can then exercise continuous dominance. In so doing, we will be ensuring that by trading at this opportune point to the left, we will have the highest expected value for the account at any point thereafter. It does not mean, however, that it will outpace an account traded at the full optimal f set. It will not.

Now we actually begin to work with this new framework. Hence, the point of this section is twofold: first, to point out that there are possible advantageous points to the left, and, second, but more importantly, to show you, by way of examples, how the new framework can be used.

There are a number of advantageous points to the left of the peak, and what follows is not exhaustive. Rather, it is a starting place for you.

The first point of interest to the left pertains to constant contract trading, that is, always trading in the same unit size regardless of where equity runs up or shrinks. This should not be dismissed as overly simplistic by candidate money managers for the following reason: *Increasing your bet size after a loss maximizes the probability of an account being profitable at any arbitrary future point. Varying the trading quantity relative to account equity attempts to maximize the profitability (yet it does not maximize the probability of being profitable).*

The problem with trading the same constant quantity is that it not only puts you to the left of the peak, but, as the account equity grows, you are actually migrating toward zero on the various f axes.

For instance, let's assume we are playing the two-to-one coin toss game. The peak is at $f = .25$, or making one bet for every $4 in account equity. Let's say we have a twenty dollar account, and we plan to always make two bets, that is, to always bet $2 regardless of where the equity goes. Thus, we start out (fortunately, this is a two-dimensional case since we are only discussing one scenario spectrum) trading at an $f\$$ of $10, which is an f of .1, since $f\$ = -BL/f$, it follows that $f = - BL/f\$$. Now, let us assume that we continue to always bet $2; that if the account were to get to $30 total equity, our f, given that we are still only betting $2, corresponding to an $f\$$ of $15, has migrated to .067. As the account continues to make money, the f we are employing would continues to migrate left. However, it also works in reverse—that, if we are losing money, the f we are employing migrates right, and at some point may actually round over the peak of the landscape. Thus, the peak represents where a constant contract trader should stop constant contract trading on the downside. Thus, the f is migrating around, passing through other points in the landscape, some of which are still to be discussed.

Another approach is to begin by defining the worst case drawdown the money manager can afford, in terms of percentage equity retracements, and use that in lieu of the optimal f in determining $f\$$.

$$f\$ = \frac{\text{abs(Biggest Loss Scenario)}}{\text{Maximum Drawdown Percent}} \qquad (10.17a)$$

Thus, if the maximum tolerable drawdown to a money manager is 20%, and the worst-case scenario calls for a loss of $-\$1,000$:

$$f\$ = \frac{\$1,000}{.2} = \$5,000$$

He should thus use $5,000 for his $f\$$. In doing so, he still does not restrict his worst-case drawdown to 20% retracement on equity. Rather, what he has accomplished is that the drawdown to be experienced with the manifestation of the single catastrophic event is defined in advance.

Note that in using this technique, the money manager must make certain that the maximum drawdown percent is not greater than the optimal f, or this technique will put him to the right of the peak. For instance, if the optimal f is actually .1, but the money manager uses this technique with a .2 value for maximum drawdown percentage, he will then be trading an $f\$$ of $5,000 when he should be trading an $f\$$ of $10,000 at the optimal level! Trouble is certain to befall him.

Further, the example given only shows for trading one scenario spectrum. If you are trading more than one scenario spectrum, you must change your denominator to reflect this, dividing the maximum drawdown percent by n, the number of scenario spectrums:

$$f\$ = \frac{\text{abs(Biggest Loss Scenarion)}}{\left(\dfrac{\text{Maximum Drawdown Percent}}{n}\right)} \qquad (10.17b)$$

where: $n =$ The number of components (scenario spectrums or market systems) in the portfolio.

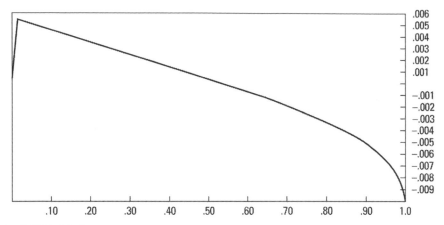

FIGURE 10.6 Two-to-one coin toss, GRR at $T = 1$

Notice that by doing this, if each scenario spectrum realizes its worst-case scenario simultaneously, you will still have defined the maximum drawdown percent for the entire portfolio.

Next, we move on to another important point to the left, which may be of importance to certain money managers: the *growth risk ratio*, or GRR (Figure 10.6). If we take the TWR as the growth, the numerator, and the f used (or the sum of the f values used for portfolios) as representing risk, since it represents the percentage of your stake you would lose if the worst case scenarios(s) manifest, then we can write the growth risk ratio as:

$$\text{GRR}_T = \frac{\text{TWR}_T}{\sum\limits_{i=1}^{n} f_i} \tag{10.18}$$

This ratio is exactly what its name implies, the ratio of growth (TWR_T, the expected multiple on our stake after T plays) to risk (sum of the f values, which represent the total percentage of our stake risked). If TWR is a function of T, then too is the GRR. That is, as T increases, the GRR moves from that point where it is an infinitesimally small value for f, towards the optimal f (see Figure 10.7). At infinite T, the GRR equals the optimal f. Much like the EACG, you can trade at the f value to maximize the GRR if you know, a priori, what value for T you are trying to maximize for.

The migration from an infinitesimally small value for f at $T = 1$ to the optimal f at $T =$ infinity happens with respect to all axes, although in Figures 10.6 and 10.7 it is shown for trading one scenario spectrum. If you were trading two scenario spectrums simultaneously, the peak of the GRR would migrate through the three-dimensional landscape as T increased,

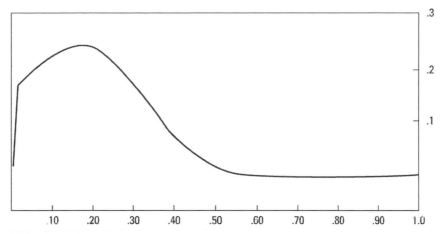

FIGURE 10.7 Two-to-one coin toss, GRR at $T - 30$

from nearly 0,0 for both values of f, to the optimal values for f (at .23,.23 in the two-to-one coin toss).

Discerning the GRR for more than one scenario spectrum traded simultaneously is simple, using Equation (10.18), regardless of how many multiple simultaneous scenario spectrums we are looking at.

The next and final point to be covered to the left, which may be quite advantageous for many money managers, is the point of inflection in the TWR with respect to f.

Refer again to Figure 9.2 in Chapter 9. Notice that as we approach the peak in the optimal f from the left, starting at 0, we gain TWR (vertical) at an ever-increasing rate, up to a point. We are thus getting greater and greater benefit for a linear increase in risk. However, at a certain point, the TWR curve gains, but at a slower and slower rate for every increase in f. This point of changeover, called *inflection*, because it represents where the function goes from concave up to concave down, is another important point to the left for the money manager. The point of inflection represents the point where the marginal increase in gain stops increasing and actually starts to diminish for every marginal increase in risk. Thus, it may be an extremely important point for a money manager, and may even, in some cases, be optimal in the eyes of the money manager in the sense of what it does in fact, maximize.

However, recall that Figure 9.2 represents the TWR after 40 plays. Let's look at the TWR after one play for the two-to-one coin loss, also simply called the geometric mean HPR, as shown in Figure 10.8.

Interestingly, there isn't any point here where the function goes from concave up to concave down, or vice versa. There aren't any points of inflection. The whole thing is concave down.

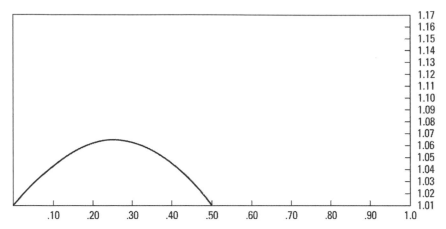

FIGURE 10.8 Geometric mean HPR two-to-one coin toss (= TWR at $T = 1$)

For a positive arithmetic expectation, the geometric mean does not have any points of inflection. However, the TWR, if $T > 1$, has two points of inflection, one to the left of the peak and one to the right. The one which concerns us is, of course, the one to the left of the peak.

The left point of inflection is nonexistent at $T = 1$ and, as T increases, it approaches the optimal f from the left (Figures 10.9 and 10.10). When T is infinite, the point of inflection converges upon optimal f.

Unfortunately, the left point of inflection migrates toward optimal f as T approaches infinity, just like with the GRR. Again, just like EACG, if you

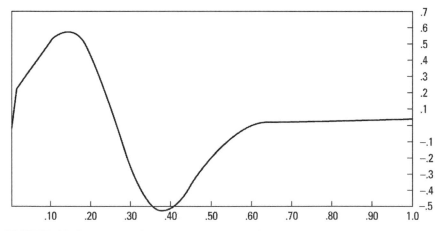

FIGURE 10.9 $d\text{TWR}/df$ for 40 plays ($T = 40$) of the two-to-one coin toss. The peak to the left and the trough to the right are the points of inflection

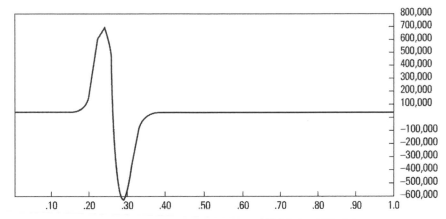

FIGURE 10.10 dTWR/df for 800 plays ($T = 800$) of the two-to-one coin toss. The peak to the left and the trough to the right are the points of inflection. The left peak is at $f = .23$

knew how many finite T you were to trade before you started trading, you could maximize the left point of inflection.[*]

To recap how the left point of inflection migrates towards optimal f, the following table depicts the numbers for the two-to-one coin-toss game:

2:1 Coin Toss

# plays (T)	f inflection left
1	0
30	.12
40	.13
80	.17
800	.23

Thus, we again see that, as more time elapses, as T increases, missing the optimal f carries with it a steep penalty. Asymptotically, nearly everything is maximized, whether it is EACG, GRR, or the left point of inflection. As T increases, they all converge on optimal f. Thus, as T

---

[*]Interestingly, though, if you were trying to maximize the EACG for a given T, you would be seeking a point to the *right* of the peak of the f curve, as the f value which maximizes EACG migrates toward the optimal f as T approaches infinity from the *right*.

increases, the distance between these advantageous points and optimal f diminishes.

Suppose a money manager uses daily HPRs and wants to be optimal (with respect to inflection or GRR) over the course of the current quarter (63 days). He would use a value of 63 for T and set himself at those coordinates to be optimal for each quarter.

When we begin working in more than two dimensions, that is, when we are dealing with more than one scenario spectrum, we enter an altogether more complicated problem.

The solution can be expressed mathematically as that point where the second partial derivatives of the TWR [Equation (9.04), raised to the power of T, the number of holding periods at which we are seeking the points of inflection] with respect to a each particular f equals zero, and each point is to the left (on its axis) of the peak. This becomes ever more complicated in that such a point, where the second partials of the TWR with respect to each f equaling zero may not, depending upon the parameters of the scenario spectrums themselves and how high or low T is, exist. If T equals one, the TWR equals the geometric mean HPR, which is upside down parabolic—it doesn't have any points of inflection! Yet as T approaches infinity, the point(s) of inflection approach the optimal f(s)! Shy of infinite T, there may not be in most cases, such a conveniently common point of inflection with respect to all axes.[*]

All of this brings us right back to the notion of the $n + 1$ dimensional terrain in leverage space, if you will, the axes of which correspond to the f values of the different scenario sets, is to act as a *framework* for analyzing portfolio construction and quantity determination through time. There is so much more to be done in working with this new framework. This chapter is not the end-all on the subject. Rather, it is a mere introduction to an altogether new and, I believe, better way of determining asset allocation. Almost certainly, portfolio strategists, applied mathematicians, asset allocators, and programmers have much new fertile ground to work. Truly, there is a great deal to be done in analyzing, working with, and adding to this new framework, the rewards of which cannot yet even be determined. More importantly, whether one attempts to actively employ the Leverage Space Model, the tenets of The New Framework, as expressed here, are at work and apply to him regardless.

---

[*]Remember that the primary thing gained by diversification, that is, trading more than one scenario spectrum, or working in more than two dimensions, is that you increase T, the number of holding periods in a given period of time—you do not reduce risk. In light of this, someone looking to maximize the marginal increase in gain to a marginal increase in *risk*, may well opt to trade only one scenario spectrum.

DRAWDOWN MANAGEMENT AND THE NEW FRAMEWORK

Drawdowns occur from one of three means. The first of these, the most common, is a cataclysmic loss on one trade. I started in this business as a margin clerk where my job was to oversee hundreds of accounts. I have worked as a programmer and consultant to many of the largest traders in the world. I have been trading and working around the trading arena for my entire adult life, often with a bird's-eye view of the way people operate in and around the markets. I have witnessed many people being obliterated over the course of a single trade. I have plenty of firsthand experience in getting destroyed on a single trade as well.

The common denominator of every single occasion when this has happened has been a lack of liquidity in the market. The importance of liquidity cannot be overemphasized. Liquidity is not something I have been able to quantify. It isn't simply a function of open interest and volume. Further, liquidity need not dry up for long periods of time in order to do tremendous harm. The U.S. Treasury Bond futures were the most liquid contract in the world in 1987. Yet, that, too, was a very arid place for a few days in October of 1987. You must be ever vigilant regarding liquidity.

The second way people experience great drawdowns is the common, yet even more tragic, means of not knowing their position until the market has moved ferociously against them. This is tragic because, in all cases, this can be avoided. Yet it is a common occurrence. You must always know your position in every market.

The third cause of drawdowns is the most feared, although the consequences are the same as with the first two causes. This type of drawdown is characterized by a protracted losing streak, maybe with some occasional winning trades interspersed between many losers. This is the type of drawdown most traders live in eternal fear of. It is this type of drawdown that makes systems traders question whether or not their systems are still working. However, this is exactly the type of drawdown that can be managed and greatly buffered through the new framework.

The new framework in asset allocation concerns itself with growth optimality. However, the money management community, as a general rule, holds growth optimality as a secondary concern. The primary concern for the money management community is *capital preservation.*

This is true not only of money managers, but of most investors as well. Capital preservation is predicated upon reducing drawdowns. The new framework presented allows us for the first time to reduce the activity of drawdown minimization to a mathematical exercise. This is one of the many fortuitous consequences of—and the great irony of—the new framework.

Everything I have written of in the past and in this book pertains to growth optimality. Yet, in constructing a framework for viewing things in a growth optimal sense, we are able to view things in a drawdown optimal sense within the same framework. The conclusions derived therefrom are conclusions which otherwise would not have been arrived at.

The notion of optimal f, which has evolved into this new framework in asset allocation, can now go beyond the theoretical formulations and concepts and into the real-world implementation to achieve the goals of money managers and individual investors alike.

The older mean-variance models were ill-equipped to handle the notion of drawdown management. The first reason for this is that risk is reduced to the simplified notion that dispersion in returns constitutes risk. It is possible, in fact quite common, to reduce dispersion in returns yet not reduce drawdowns.

Imagine two components that have a correlation to each other that is negative. Component 1 is up on Monday and Wednesday, but down on Tuesday and Thursday. Component 2 is exactly the opposite, being down on Monday and Wednesday, but up on Tuesday and Thursday. On Friday, both components are down. Trading both components together reduces the dispersion in returns, yet on Friday the drawdown experienced can actually be worse than just trading one of the two components alone. *Ultimately, all correlations reduce to one.* The mean variance model does not address drawdowns, and simply minimizing the dispersion in returns, although it may buffer many of the drawdowns, still leaves you open to severe drawdowns.

To view drawdowns under the new framework, however, will give us some very useful information. Consider for a moment that drawdown is minimized by not trading (i.e., at $f = 0$). Thus, if we are considering two simultaneous coin-toss games, each paying two-to-one, growth is maximized at an f value of .23 for each game, while drawdown is minimized at an f value of zero for both games.

The first important point to recognize about drawdown optimality (i.e., minimizing drawdowns) is that it can be *approached* in trading. The optimal point, unlike the optimal growth point, cannot be achieved unless we do not trade; however, it can be approached. Thus, to minimize drawdowns, that is, to approach drawdown optimality, requires that we use as small a value for f, for each component, as possible. In other words, to approach drawdown optimality, you must hunker down in the corner of the landscape where all f values are near zero.

In something like the two-to-one coin-toss games, depicted in Figure 10.11, the peak does not move around. It is a theoretical ideal, and, in itself, can be used as a superior portfolio model to the conventional models.

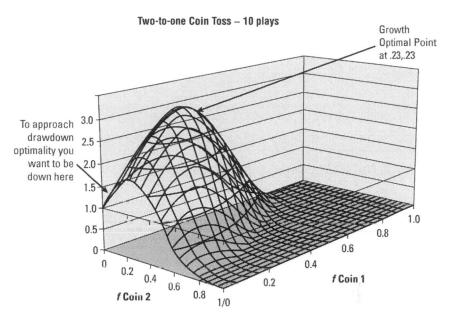

FIGURE 10.11 Drawdown optimality is approached at a different point on the landscape than the growth optimal point

However, as was mentioned earlier in this text, in the real world of trading, the markets do not conform so neatly to the theoretical ideal. The problem is that, unlike the two-to-one coin-toss games shown, the distribution of returns changes through time as market conditions change. The landscape is polymorphic, moving around as market conditions change. The closer you are to where the peak is, the more dramatic the negative effects will be on you when it moves, simply because the landscape is the steepest in the areas nearest the peak. If we were to draw a landscape map, such as the one in Figure 10.11, but only incorporating data over a period when both systems were losing, the landscape (the altitude or TWR) would be at 1.0 at the f coordinates 0,0, and then it would slide off, parabolically, from there.

We approach drawdown optimality by hunkering down in those f values near zero for all components. In Figure 10.11 we would want to be tucked down in the upper-left corner, near zero for all f values. The reason for this is that, as the landscape undulates, as the peak moves around, the negative effects on those in that corner are very minimal. In other words, as market conditions *change*, the effect on a trader down in this corner is minimized.

The seeming problem, then, is that growth is sacrificed and this sacrifice in growth occurs with an exponential difference. However, the solution to this problem can be found by the fundamental equation for trading. Since growth—that is, TWR—is the geometric mean holding period return to the power T, the number of plays is given by:

$$\text{TWR} = G^T \tag{10.19}$$

By hiding out in the corner, we have a much smaller G. However, by increasing T, that is, the number of trades, the effect of an exponential decrease in growth is countered, by itself an exponential function.

In short, if a trader must minimize drawdowns, he or she is far better off to trade at a very low f value and get off many more holding periods in the same span of time.

For example, consider playing only one of the two-to-one coin-toss games. After 40 holding periods, at the optimal f value of .25, the geometric mean HPR is 1.060660172, and the TWR is 10.55. If we were to play this same game with an f value of .01, our geometric mean HPR would be 1.004888053, which crosses 10.55 when raised to the power of 484. Thus, if you can get off 484 plays (holding periods) in the same time it would take you to get off 40 plays, you would see equivalent growth, with a dramatic reduction in drawdowns. Further, you would have insulated yourself tremendously from changes in the landscape. That is, you would also have insulated yourself a great deal from changing market conditions.

It may appear that you want to trade more than one component (i.e., scenario spectrum) simultaneously. That is, to increase T, you want to trade many more components simultaneously. This is counter to the idea presented earlier in discussing the points of inflection that you may be better off to trade only one component. However, by increasing the number of components traded simultaneously, you increase the composite f of the portfolio. For example, if you were to trade 20 scenario spectrums simultaneously, each with a .005 value of f, you would have a composite f of the entire portfolio of 0.1. At such a level, if the worstcase scenarios were to manifest simultaneously, you would experience a 10% drawdown on equity. By contrast, you are better off to trade only one scenario spectrum whereby you can get off the equivalent of 20 holding periods in the same spame of time. This may not be possible, but it is the direction you want to be working in to minimize drawdowns.

Finally, when a trader seeks to approach drawdown minimization, he or she can use the continuous dominance notion in doing so. Continuous dominance is great in the theoretical ideal model. However, it is extremely sensitive to changes in the landscape. That is, as the scenarios used as input change to accommodate changing market characteristics, continuous dominance begins to run into trouble. In a gambling game where the conditions do not change from one period to the next, continuous dominance is ideal.

In the real world of trading, you must insulate yourself from the undulations in the landscape. Thus, drawdown minimization under the new framework lends itself very well to implementing continuous dominance.

So we have now gone full circle, from discerning the landscape of leverage space and finding the growth optimal point on it to retreating away from that point to approach the real-world primary constraint of drawdown minimization and capital preservation. By simply increasing the exponent, by whatever means available to us, we achieve growth. We can possibly achieve equivalent growth if we can get a high enough T, a high enough exponent. Since the exponent is the number of holding periods in a given span of time, we want to get as many holding periods in a given span of time as possible. This does not necessarily mean, however, to trade as many components as possible. All correlations revert to one. Further, we must always assume that worst-case scenarios will manifest simultaneously for all components traded. We must consider that the composite f, the sum of the f values for all components being simultaneously traded, is a drawdown that we will, therefore, experience. This suggests that, in seeking to approach drawdown optimality, yet still striving for equivalent growth as at the growth optimal point, we trade as few components as possible, with as small an f for each component as possible, while managing to get as many holding periods in a given span of time as possible.

The growth optimal point is a dangerous place to be. However, if we hit it just right, that is, if we are at the place where the peak will be, we can see tremendous growth. Even so, we will endure severe drawdowns. However, the leverage space framework allows us to formulate a plan, a place to be on the map of leverage space, to achieve drawdown minimization. It further allows us an alternate avenue to achieve growth, by increasing T, the exponent, by whatever means necessary. This strategy is not so mathematically obvious when viewed under the earlier frameworks.

This is but one means, and a very crude one at that, for mitigating drawdowns in the Leverage Space Model. In Chapter 12, we will see how the terrain of leverage space is "pock-marked," by "holes," where the probability of a given drawdown is too high for an investor's utility preference.

When viewed in the sense to be presented in Chapter 12, the drawdown mitigation technique just mentioned, virtually insures the investor will not be within a pock-marked-out area of the terrain. However, he pays a steep price here for being way to the left of the peak of all curves in leverage space. In Chapter 12, we will see that, although the $0, \ldots 0$ edge point in leverage space is never pock-marked out, we can determine other, far more favorable areas in the terrain.

PART II

Practice

CHAPTER 11

What the Professionals Have Done

I n this chapter we examine those common denominators, in terms of portfolio and systems management, that seem to be shared among the more successful commodities funds.

In looking at the real world now, versus the theoretical one in which we have been mired thus far in the text, we will now consider those fund managers regarded as the long-term trend followers, and not some of the more novelty acts.

We're looking at the *successful long-term trend followers* here. That is, we are focusing on those who manage the most money in the CTA business, have for a number of years, and, over those years, have had considerable success both in the markets with these funds as well as raising investment money in the funds. These are all quite recognizable names as of this writing.

Why pick them? First, they manage the largest amounts that are invested in futures speculation. In fact, I would venture to say that as of this writing, and for decades preceding it, well over half of the money in managed futures has been under the control of what might be termed *long-term trend following*.

Second, the larger investors—that is, the institutions that now allocate a small portion of their enormous funds to futures and alternative-type investments—have been attracted by these funds. Therefore, from a business standpoint, the funds have been unquestionably successful, and it is exactly that type of success that so many fund managers covet.

Lastly, exactly because there *is* this enormous chasm between what these successful funds and individuals do, versus the optimal *f*/Leverage Space Model framework, as discussed to this point, it bears discussion.

Here are the main, common tactics that most of these successful long-term trend-following funds are following:

COMMONALITIES

1. Most everyone is risking x percent per trade on a given market system. Typically, this is in the neighborhood of 0 to 2 % per trade. This risk is essentially determined by where the stop out is on the given trade, and the money at risk in the account. Thus, if the risk is 1% per trade, and there is $10 million in the account, the percentage of risk on the trade is $100,000. If the stop-out point on this trade was $1,000 from the entry, there would be 100 contracts put on.

2. The stop-out points are almost always a function of recent volatility in one fashion or another—often, the stop-out being a multiplier of the previous X bar's average range (times something, usually a constant value like 3) , or something along the lines of "the lowest low in the past X bars" (which, too, is a function of recent volatility). There is always seemingly a *recent volatility* metric that is employed in the quantity calculation. Thus, the more volatile the market, the less the quantity traded will come out to be, and vice versa.

3. Trend-following funds have typically shown virtually no concern for correlation, though stock traders do. In other words, a manager who trades, among other things, dollar yen, sterling, and dollar euro may have a 1% position concurrently in all three markets, lined up on the same side of the dollar, while at other times will have only one such position on, for a net risk of 1% versus the dollar. This is not at all uncommon to see in the real world of successful fund managers, the rationale being that if, say, in this example, all three are making a good run, and you are trading all three as separate market systems, then that dictates you should take this 3% risk versus the dollar at this time.

Of course, if you were risking 20% per position, you might not follow this rule and have 60% of the equity on the line against the dollar! The luxury of being able to nearly disregard correlation is a function of not being anywhere near what would otherwise be the optimal f on these positions. Again, this is a major divergence between theory and practice.

DIFFERENCES

The main differences between these funds, then, aside from where their stops are, is the markets they trade (this is the biggest difference between most of them). Essentially, long-term trend-following systems will be long a raging bull market in whatever tradable you're looking at. However, the major differences are the stops. The different philosophies are:

1. Always in a market. This is a two-phase approach ("long/short") versus a three-phase ("long/flat/short") approach. Often, these two approaches are combined and netted out. However, a two-phase approach in a long-term trend-following system will typically have stops much farther away than a three-phase system. Since the distance the stop is away from the market will dictate the quantity, very often the two-phased types of systems will have on considerably less quantity.

2. Markets traded. Typically, most managers of the successful long-term trend-following systems (I'm speaking of the larger fund managers here) trade about 20 markets, give or take half a dozen. These typically are the markets that are liquid enough to facilitate the quantity they are trading in. This is where a lot of guys are fooling themselves by selecting a handful of lucky markets. However, though they may trade client money in only those markets, they will often trade their own money in ALL markets.

Some managers *do* trade all markets—rough rice, rapeseed, etc.—with the thought that there are going to be giant trends somewhere, and the only way to participate in them is to be in those markets.

Of course, there are the novelty acts that trade only grains or only currencies, for example, but these are of no real interest to us in this discussion.

FURTHER CHARACTERISTICS OF LONG-TERM TREND FOLLOWERS

There is also the decision of how frequently the size of a position is altered. The answer to this could have fallen under either the "Commonalities" section or the "Differences" section.

Typically, when you speak with these fund managers, they will almost unanimously tell you that if there were no costs to doing so, they would alter their position sizes as frequently as possible—in fact, if it were possible, they would adjust them continuously. This is indicative of someone practicing portfolio insurance of some sort—that is, replicating an option, where their size is the option's delta percent of what it would be if it were very profitable (i.e., the systems equity curve were deep in the money). In other words, they replicate an option on that market system's equity curve.

Yet, in application, managers differ wildly from this. One of the most successful (who always has a position in a given market) will alter his size only on every rollover occurrence.

There is also the practice of staggering entries and exits. That is, most of these funds are so large that if they are required to execute a trade at a particular price, rather than moving the market in a big way at that particular price, they may break up that sizable order into numerous smaller orders and execute them at various prices near the price that was supposed to be the actual order price. Some fund managers practice this; others do not. Surprisingly, the deciding factor does not seem to be a function of the size of the fund! Some funds will throw an enormous order at a single, given price.

This same notion of staggered entries and exits is sometimes practiced in the context of multiple systems on the same market. As a simplified example, suppose I am a fund manager and I have a system that has a single parameter, and that system for today calls for me to enter a particular market at a price of 100.

Rather than operate in this fashion, it is often common to have different parameters, in effect creating different "systems," as it were, causing a staggering of entries and exits. This is a fairly easy process.

So, whereas intentional staggering of entries and exits to mitigate slippage is not universally employed, the notion of inadvertently doing this, or staggering entries and exits as a fortunate by-product of using numerous parameters on the same market system, is quite pervasive and accepted.

The concept of using an array of parameter values is also rather widely thought to help alleviate the problems of what parameter values to use in the future, based on historical testing. The thinking is that it is difficult to try to pinpoint what the very best parameter will be in the future. Since most of these systems are *robust* in terms of giving some level of positive performance against a wide spectrum of parameter values, by using an array of parameter values, fund managers hope to avoid selecting what in the future will be an outlier parameter value in terms of poor return. By using an array of parameter values, they tend to come in more toward what the typical parameter performance was in the past—which is acceptable performance.

Parameter optimization tends to be fraught with lots of questions at all levels. Though the concept of parameter optimization is, in effect, inescapable in this business, my experience as an observer here is that there is not much to it. Essentially, most people optimize over the entire data set they can get their hands on, and look at the results for parameter values over this period with an eye toward *robustness* and trying to pick that parameter value that, though not necessarily the peak one, is the one in a range of parameter values where all have performed reasonably well. Additionally, they tend to break down the long history of these runs into thirds, sometimes fourths. For example, if they have a 28-year history, they will look at things in 7-year increments—again, with the same criteria. Of note here is the question of whether they use the same parameter values from one market to another. Typically, they do not, using a different set of parameter values for different markets, though this is not universal.

Additionally, as for how frequently they optimize and reestablish parameters, this too seems to be all over the board. Some do so annually, some do it considerably more frequently than that. Ultimately, this divergence in operations also seems to have little effect on performance or correlation with one another.

There has been a trend in recent years to capture the characteristics of each individual market's prices, then use those characteristics to generate new, fictitious data for these markets based on those characteristics. This is an area that seems to hold great promise.

The notion of adding to a winning position, or *pyramiding*, is almost completely unseen among the larger fund managers. That is, there just don't seem to be any large funds out there that add to winning positions according to some schedule as a trade progresses. However, this is occluded, as many funds that employ multiple trend-following systems and/or an array of parameter values for a given market system will inadvertently add to positions. Aside from that, the concept of pyramiding is virtually unseen.

Almost as rare is the notion of taking profits. Rarely do any of these funds have a set target where they will exit a trade. Rather, there is almost always a trailing stop, whereby the position is either exited or flipped.[1]

Related to the notion of exiting a trade at a specified target is the entire concept of trying to smooth out the equity curve. These techniques have been employed with varying and, in most cases, none-too-stellar success.

Attempts to do this are often along the lines of so-called *anti-trending systems*, that is, systems that tend to profit in flat markets. Again, since these successful funds profit when there are trends, they tend to suffer in the absence of such trends. Hence the emergence of anti-trending systems along the lines of option writing (covered or uncovered, often with spreads of the butterfly type—essentially anything that takes premium at the establishment of the position), or convertible-type arbitrage, etc. (The list of anti-trending types of systems is nearly endless and unbelievably creative! There is a long list of anti-trending types of systems devised in recent years.)

---

[1]There are individual traders, however, who have had great success with taking profits on trades and are far more short-term oriented, particularly those of the bailout-type exits. The reason for this is that by being able to convert many losing trades—as well as diminishing what otherwise might be large winning trades, to effectively a scratch, the standard deviation in returns from trade to trade is tightened up. Per the Pythagorean Theorem, from previous chapters, this is effectively the same as increasing the arithmetic average trade in terms of growth on an account.

Ideally, managers would like to have trend-following systems that are uncorrelated or even perhaps negatively correlated. However, these don't seem to exist, and, quite likely if something was spotted that exhibited this characteristic, in the absence of a causal factor for the correlation, it might well be ready to turn and perform in an opposite manner.

The idea of trading anti-trending systems has been to both mitigate drawdown by attempting to smooth out the equity curve and to provide a somewhat regular return on a fund, an attempt to give a certain *buoyancy* to its performance.[2] There has been a prevailing trend in the industry that the only way you can interest the larger institutions to invest with you is if you can produce 1 to 1.5% returns per month with limited drawdowns. Fund managers have attempted to incorporate these convergent, anti-trending types of systems into the process with that goal in mind. Ultimately, however, very little (aside from the increase in automation) has changed in the way successful funds operate in terms of their market strategies. Attempts to incorporate convergent, or anti-trending, systems have shown limited success thus far.

* * *

In 1984, a group of well-trained and highly screened individuals, who were at the time nontraders, and who were subsequently dubbed the *Turtles*, laid out a lot of these basic commonalities regarding successful long-term trading when they began opening up about their training. This was the group Richard Dennis founded in a dispute with his colleague William Eckhardt over whether successful trading skills could be taught.

Since that time, supposedly, some of the original Turtles have seen great success; others, failure. The distinguishing characteristics, though speculated upon by many, aren't really known (by me anyhow). However, it should be mentioned at this point that failure, usually in a system that, in the long run, shows a profit, is solely the result of where one stops in the equity cycle. Clearly, even at a very modest level for f, the drawdowns to be expected are extreme. In a system that will, say, at the end of five years, be wildly profitable, very likely that system has had some hair-curling, greater-than-anticipated drawdowns. Quitting in such a drawdown, then, is considered failure.

---

[2]As of this writing, interest rates of any duration are and have been near their lows of four decades. In addition to protracted, multiyear drawdowns that many of these funds are experiencing as of this writing, the low interest rates seem to have created an atmosphere that may have promoted the idea of further incorporating anti-trending types of systems.

Put another way, if you own a casino, and an individual comes in, has a few plays that break in his favor, and you quit the casino business at that point, having lost money at it for the duration of being in the casino business, then, yes, you are, by definition, a failure in the casino business.

As of this writing, February 2006, the long-term trend followers are in the midst of a protracted multiyear drawdown, with many funds down well over 50%. People are saying, "Long-term trend following is dead."

As you will see in the next chapter, "The Leverage Space Portfolio Model in the Real World," this type of drawdown is absolutely expected and normal. In fact, it may well get worse before it gets any better.

* * *

Given these basic building blocks of allocation, however seemingly crude, one could (and many, in fact, have) create successful commodity funds. Merely by risking 1% of an account's capital on 20 seemingly disparate markets (or not so disparate, even, given how few markets are available for some funds because of their liquidity constraints as dictated by their size), some funds have seen wild success over the years by any measure; and, peculiarly, with no concern for correlations.[3]

One very large, very successful fund that has been around for a generation has operated that very way since inception and is known for coming through with nice returns over time, with rather small, perfectly digestible-to-most drawdowns. Another long-term, successful fund with roughly USD 1 billion currently trades only about a dozen markets with a single model and three parameter sets per market.

By contrast, one of their close competitors with a similar amount under management, and funds highly correlated to the one just mentioned, use six to eight models with dozens of parameter sets for each market and a basket of over 60 markets! As would be expected, their returns have historically been smoother, but not relatively to the extent one might expect.

With the majority of the commodity funds, however, that 20% figure could be 5% or it could be 50%, but at 20% you'll be right in the mainstream, right in or near the fattest part of the curve. As for the stop-out amount, typically this would be the lesser of 2% of an account's equity or the percentage allocated to trading (again, 20% putting you in the fat part of the curve), divided by the number of markets traded.

---

[3]This may not be such a bad approach. Given that correlations do not seem to maintain consistency with the magnitude of swings, when all cut against you, in such an allocation model, you are looking at a 20 % loss.

Does this mean that mean-variance models are not employed? Though a gross generalization, in terms of the individual funds, it most often is not. However, with the larger pools and fund of funds, it tends to be. So the general rule out there seems to be that if it is a solitary fund, a single market system, say, across markets, mean-variance is not employed most of the time, whereas, if it is a conglomeration of funds—or market systems, it tends to be used more.

This is not to say that individual funds are not looking to pair uncorrelated items together, or are not working to smooth out their equity curves via a mean-variance approach. However, and particularly more recently, the individual large funds appear to be looking more toward a *value-at-risk* means of allocation, and more toward the notion of trying to get the biggest bang they can out of their funds within the constraint of "acceptable" drawdowns.

Sometimes, they are looking at individual markets and their individual drawdowns, then the drawdowns of the portfolio as a whole. Most larger funds appear to allocate an equal amount of risk to each market and then scale the whole portfolio up to the acceptable level of risk to see what the return is.

Still others do each market individually, so in the markets that have performed better, the percentage of equity to risk on each trade in an attempt to achieve an x percent chance of a y percent drawdown is higher. Then these different percentages for each market are tested together, obtaining a single portfolio, which is then further scaled up or down to retain that x percent chance of a y percent drawdown for the entire portfolio. Note that under this method, a market that was twice as profitable will end up getting twice as much allocated to it.

The interesting aspect of this approach (versus merely allocating an equal amount of risk to each market, then scaling the portfolio up or down as a whole to achieve the desired risk level)—that is, of preprocessing by each individual market, thus, when you subsequently scale the whole portfolio, achieving different allocations to different market systems—is, in effect, you have employed mean variance indirectly. Thus, such an approach can be said to combine mean variance with value at risk.

This type of an approach is typically employed in the following manner. Let's suppose you have 25 years of historical data. For each market, then, you look through all the data, seeking to obtain that percentage of equity to risk in each market such that there are no more than $25 \times 12 = 300$ months $\times 1\% = 3$ months, where the loss was greater than 20%. This is typically regarded as the standard way to obtain value at risk from a trading study. Once you have obtained this percentage of equity to risk for this particular market system, you must decide if the return over that period justifies including it in the portfolio.

Once you have settled on the components of the portfolio and their relative percentages of risk, you then perform step 2, which is to look at the portfolio as a whole, to determine a scaling factor by which to multiply all the component risk percentages. Funds that allocate a same risk level to each market system perform only step 2 of this analysis.

In implementation, before a trade is to be initiated, the stop on the trade on a per-contract basis is determined. Now, the portfolio value (some use the current value; some, the value as of last night; others, the value at the beginning of the month) is divided by the portfolio scaling factor adjusted risk percentage for this market, and that number is then divided by the risk per contract on this trade, to determine the number of contracts to allocate.

So, if we have a $1 million account, and our stop-out on a one-contract basis on this trade is $5,000, the relative percentage of risk is 4% (the number that gave us no more than an x percent chance of a y percent drawdown over the 25 years of past data for this market), and our portfolio scaling factor is .7 (the number that gave us no more than an x percent chance of a y percent drawdown over the 25 years of past data for the portfolio as a whole). We then have:

$$1,000,000 \times .04 \times .7/5,000 = 5.6 \text{ contracts}$$

Note the .04 here, for most funds, is typically a constant from one market system to the next, but again, there are some funds that do derive this number individually for each market system.

Of note here too is the .7 portfolio scaling factor. If all markets were perfectly correlated, then this number would equal 1 divided by the number of market systems in the portfolio. Therefore, the higher you can get this number, the less correlated the constituent market systems are. If you had only two components and there was a negative correlation, your portfolio scaling factor would actually be greater than 1.

However, it may not be a bad bet to expect worst months among market systems to cluster together in the future, and therefore, may *not* be a bad bet to simply say that your portfolio scaling factor is to be 1 divided by the number of market systems in the portfolio.

* * *

These concepts aren't altogether complicated as applied in the crude ways they are being employed in the real world as outlined here. What is hard is getting software that can do this, keep track of the equity, perform the rollovers for the raw futures data rather than continuous contracts, and so on. The concepts as expressed here are actually pretty easy, but getting the tools to do it accurately is not.

Furthermore, as we are yet to see in this text, the way detailed here is far from an accurate assessment of what these fund managers are seeking to discern. The techniques shown in this chapter will give an overly optimistic assessment of the potential risk.

We've gone into greater detail here than what we really were looking for, that is the disparity between optimal f/Leverage Space Model framework and what these successful and long-standing funds do. However, we also see that they are trying to fit a mean-variance approach and a value-at-risk approach to meet the dictates placed upon managed futures of a utility preference curve that is anything but ln.

Chapter 8 explored the relationship between mean variance and optimal f. In Chapter 12, we will show how the notions of mean variance, value at risk, and the Leverage Space Model are interrelated, and how, in fact, they all work together to achieve what the fund manager seeks. It is precisely this process that is the focus of the final chapter.

The Leverage Space Portfolio Model in the Real World

n.b. The balance of this text attempts to show a viable means for applying the theories in Part 1 of the text, "The Optimal f Framework," and the resultant portfolio model of Chapter 10, "The Leverage Space Model." As such, terminologies used will reflect the new model. Rather than speak of market systems, we will refer to scenario spectrums. Rather than speak of trades or plays, or results over a certain period, we will speak of scenarios. However, the reader is alerted of the interchangeability of these terms.

In applying the concepts of the Leverage Space Model in the real world, the problems can be twofold. First, there is the computational aspect. Fortunately, this can be overcome with computer power and good software. There is no reason, from a computational standpoint, to *not* employ the Leverage Space Portfolio Model. The discernment of the scenario spectrums, their constituent scenarios, and the joint probabilities between them are no more incalculable than, say, a stock's beta or a correlation coefficient.

The second impediment to employing these concepts in the real world has been that people's utility preference curves are *not* ln. People do not act to maximize their returns. Rather, they act to maximize returns within an acceptable level of risk.

This chapter shows how to maximize returns within a given level of risk. This is a far more real-world approach than the conventionally practiced mean-variance models. Further, risk, as used in this chapter, rather than being the ersatz risk metric of "variance (or semivariance) in returns," as in classical portfolio construction, is addressed here as being risk of ruin or risk of drawdown to a certain degree. Thus, the Leverage Space Model has, as its risk metric, drawdown itself—seeking to provide the maximum gain for a given probability of a given level of drawdown.

Let us first consider the "Classical Gambler's Ruin Problem," according to Feller.[1] Assume a gambler wins or loses one unit with probability p and $(1 - p)$, respectively. His initial capital is z and he is playing against an opponent whose initial capital is $u - z$, so that the combined capital of the two is u.

The game continues until our gambler whose initial capital is z sees it grow to u, or diminish to 0, in which case we say he is *ruined*. It is the probability of this ruin that we are interested in, and this is given by Feller as follows:

$$RR = \frac{\left((1 - p)/p\right)^u - \left((1 - p)/p\right)^z}{\left((1 - p)/p\right)^u - 1} \tag{12.01}$$

This equation holds if $(1 - p) \neq p$ (which would cause a division by 0). In those cases where $(1 - p)$ and p are equal:

$$RR = 1 - \frac{z}{u} \tag{12.01a}$$

The following table provides results of this formula according to Feller, where RR is the risk of ruin. Therefore, $1 - RR$ is the probability of success.[2]

Row	p	$(1 - p)$	z	u	RR	P (Success)
1	0.5	0.5	9	10	0.1	0.9
2	0.5	0.5	90	100	0.1	0.9
3	0.5	0.5	900	1000	0.1	0.9
4	0.5	0.5	950	1000	0.05	0.95
5	0.5	0.5	8000	10000	0.2	0.8

---

[1] William Feller, "*An Introduction to Probability Theory and Its Applications*," Volume 1 (New York: John Wiley & Sons, 1950), pp. 313–314.

[2] I have altered the variable names in some of Feller's formulas here to be consistent with the variable names I shall be using throughout this chapter, for the sake of consistency.

Row	p	(1 − p)	z	u	RR	P (Success)
6	0.45	0.55	9	10	0.210	0.790
7	0.45	0.55	90	100	0.866	0.134
8	0.45	0.55	99	100	0.182	0.818
9	0.4	0.6	90	100	0.983	0.017
10	0.4	0.6	99	100	0.333	0.667
11	0.55	0.45	9	10	0.035	0.965
12	0.55	0.45	90	100	0.000	1.000
13	0.55	0.45	99	100	0.000	1.000
14	0.6	0.4	90	100	0.000	1.000
15	0.6	0.4	99	100	0.000	1.000

Note in the table above the difference between row 2, in an even money game, and the corresponding row 7, where the probabilities turn slightly against the gambler. Note how the risk of ruin, RR, shoots upward.

Likewise, consider what happens in row 6, where, compared to row 7, the probabilities p and $(1 - p)$ have not changed, but the size of the stake and the target have changed (z and u—in effect, going from row 7 to row 6 is the same as if we were betting 10 units instead of 1 unit on each play!). Note that now the risk of ruin has been cut to less than a quarter of what it was on row 7. Clearly, in a seemingly negative expectation game, one wants to trade in higher amounts and quit sooner. According to Feller,

> *In a game with constant stakes, the gambler therefore minimizes the probability of ruin by selecting the stake as large as consistent with his goal of gaining an amount fixed in advance. The empirical validity of this conclusion has been challenged, usually by people who contend that every "unfair" bet is unreasonable. If this were to be taken seriously, it would mean the end of all insurance business; for the careful driver who insures against liability obviously plays a game that is technically unfair. Actually there exists no theorem in probability to discourage such a driver from taking insurance.*[3]

For our purposes, however, we are dealing with situations considerably more complicated than the simple dual-scenario case of a gambling illustration, and as such we will begin to derive formulas for the more complicated situation. As we leave the classical ruin problem according to Feller, keep in mind that these same principles are at work in investing as well, although the formulations do get considerably more involved.

---

[3] Feller, p. 316

Let's consider now what we are confronted with, mathematically, when there are various outcomes involved, and those outcomes are a function of a stake that is multiplicative across outcomes as the sequence of outcomes is progressed through.

Consider again our two-to-one coin toss with $f = .25$:

$$+2, -1 \qquad \text{(Stream)}$$
$$1.5, .75 \qquad \text{(HPRs)}$$

There are four possible chronological permutations of these two scenarios, as follows, and the terminal wealth relatives (TWRs) that result:

$$1.5 \times 1.5 = 2.25$$
$$1.5 \times .75 = 1.125$$
$$.75 \times 1.5 = 1.125$$
$$.75 \times .75 = .5625$$

Note that the expansion of all possible scenarios into the future is like that put forth in Chapter 6.

Now let's assume we are going to consider that we are ruined if we have only 60% ($b = .6$) of our initial stake. I will attempt to present this so that you can recognize how intuitively obvious this is. Take your time here. (Originally, I had considered this chapter as the entire text—there is a lot to cover here. The concepts of *ruin* and *drawdown* will be covered in detail.) Looking at the four outcomes, only one of them ever has your TWR dip to or below the absorbing barrier of .6, that being the fourth sequence of .75 × .75. So we can state that in this instance, the risk of ruin of .6 equity left at any time is $1/4$:

$$RR(.6) = {}^1\!/_4 = .25$$

Thus, there is a 25% chance of drawing down to 60% or less on our initial equity in this simple case.

Any time the interim product $<= RR(b)$, we consider that ruin has occurred.

So in the above example:

$$RR(.8) = {}^2\!/_4 = 50\%$$

In other words, at an f value of .25 in our two-to-one coin-toss scenario spectrum, half of the possible arrangements of HPRs leave you with 80% or less on your initial stake (i.e., the last two sequences shown see 80% or less at one point or another in the sequential run of scenario outcomes).

Expressed mathematically, we can say that at any i in (12.02) if the interim value for (12.02) $<= 0$, then ruin has occurred:

$$\sum_{i=1}^{q} \left(\left(\prod_{t=0}^{i-1} HPR_t \right) * HPR_i - b \right) \tag{12.02}$$

where: $HPR_0 = 1.0$
q = The number of scenarios in multiplicative sequence (in this case 2, the same as n).[4]
b = That multiple on our stake, as a lower barrier, where we determine ruin to occur ($0 <= b <= 1$).

Again, if at any arbitrary q, we have a value $<= 0$, we can conclude that ruin has occurred.

One way of expressing this mathematically would be:

$$\text{int} \left(\frac{\sum_{i=1}^{q} \left(\left(\prod_{t=0}^{i-1} HPR_t \right) * HPR_i - b \right)}{\sum_{i=1}^{q} \left| \left(\left(\prod_{t=0}^{i-1} HPR_t \right) * HPR_i - b \right) \right|} \right) = \beta \tag{12.03}$$

where: $HPR_0 = 1.0$
q = The number of scenarios in multiplicative sequence.

$$\sum_{i=1}^{q} \left| \left(\left(\prod_{t=0}^{i-1} HPR_t \right) * HPR_i - b \right) \right| \neq 0$$

In (12.03) note that β can take only one of two values, either 1 (ruin has not occurred) or 0 (ruin has occurred).

There is the possibility that the denominator in (12.03) equals 0, in which case β should be set to 0.

We digress for purpose of clarity now. Suppose we have a stream of HPRs. Let us suppose we have the five separate HPRs of:

.9
1.05
.7
.85
1.4

---

[4]For the moment, consider q the same as n. Later in this chapter, they become two distinct variables.

Further, let us suppose we determine b, that multiple on our stake, as a lower barrier, where we determine ruin to occur, as .6. The table below then demonstrates (12.03) and we can thus see that ruin has occurred at $q = 4$. Therefore, we conclude that this stream of HPRs resulted in ruin (even though ruin did not occur at the final point, the fact that it occurs at all, at any arbitrary point, is enough to determine that the sequence ruins).

q		1	2	3	4	5
HPR		0.9	1.05	0.7	0.85	1.4
TWR	1	0.9	0.945	0.6615	0.562275	0.787185
TWR − .6		0.3	0.345	0.0615	−0.03773	0.187185
TWR − .6/[TWR − .6]		1	1	1	−1	1

Using the mathematical sleight-of-hand, taking the integer of the quantity a sum divided by the sum of its absolute values (12.03), we derive a value of $\beta = \text{int}\left(\frac{3}{5}\right) = \text{int}(.6) = 0$. If the value in column 4 in the last row is 1, then $\beta = 1$.

Note that in (12.03) the HPRs appear to be taken in order; that is, they appear in a single, ordered sequence. Yet, we have four sequences in our example, so we are calculating β for each sequence. Recall that in determining optimal f, sequence does not matter, so we can use any arbitrary sequence of HPRs.

However, in risk-of-ruin calculations, order *does* matter(!) and we must therefore consider all permutations in the sequence of HPRs. Some permutations at a given set $(b, \text{HPR}_1 \ldots \text{HPR}_n)$ will see $\beta = 0$, while others will see $\beta = 1$. Further, note that for n HPRs, that is, for $\text{HPR}_1 \ldots \text{HPR}_n$, there are n^n permutations.

Therefore, β must be calculated for all permutations of n things taken n at a time. The symbology for this is expressed as:

$$\forall nPn \tag{12.04}$$

More frequently, this is referred to as "for all permutations of n things taken q at a time," and appears as:

$$\forall nPq \tag{12.04a}$$

This is the case even though, for the moment in our discussion, $n = q$.

Notice that for n things taken q at a time, the total number of permutations is therefore n^q.

We can take the sum of these β values for all permutations (of n things taken q at a time, and again here, $n = q$ for the moment), and divide by the number of permutations to obtain a real probability of ruin, with *ruin* defined as dropping to b of our starting stake, as $RR(b)$:

$$RR\left(b, q\right) = \frac{\forall nPq \sum_{k=1}^{n^q} \beta_k}{n^q} \tag{12.05}$$

This is what we are doing in discerning the probability of ruin to a given b, $RR(b)$. If there are two HPRs. There are $2 \times 2 = 4$ permutations, from which we are going to determine a β value for each [using $RR(.6)$]. Summing these β values and dividing by the number of permutations, 4, gives us our probability of ruin.

Note the input parameters. We have a value for b in $RR(b)$—that is, the percentage of our starting stake left. Various values for b, of course, will yield various results. Additionally, we are using HPRs, implying we have an f value here. Different f values will give different HPRs will give different values for β. Thus, what we are ultimately concerned with here— and the reader is advised at this point not to lose sight of this—is that we are essentially looking to hold b constant in our analysis and are concerned with those f values that yield an acceptable $RR(b)$. That is, we want to find those f values that give us an acceptable probability for a given risk of ruin.

Again we digress now for purposes of clarifying. For the moment, let us suspend the notion of each play's being a multiple on our stake. That is, let us suspend thinking of these streams in terms of HPRs and TWRs. Rather, let us simply contemplate the case of being presented with the prospect of three consecutive coin tosses. We can therefore say that there are eight separate streams, eight permutations, that the sequence which H and T may comprise ($\forall_2 P_3$).

H H H
H H T
H T H
H T T (ruin)*
T H H
T H T
T T H (ruin)
T T T (ruin)

Now let us say that if tails occurs in two consecutive tosses, we are ruined. Thus, we are trying to determine how many of those eight streams see two consecutive tails. That number, divided by eight (the number of permutuations) is therefore our "Probability of Ruin."

The situation becomes more complex when we add in the concept of multiples now. For example, in the previous example it may be that if the first toss is heads, then two subsequent tosses of tails would not result in ruin as the first play resulted in enough gain to avert ruin in the two subsequent tosses of tails*.

We return now to assigning HPRs to our coin tosses at an optimal f value of .25 and b of .6.

Note what happens as we increase the number of plays—in this case, from two plays (i.e., $q = 2$) to three plays ($q = 3$):

$$\forall_2 P_3 =$$
$$1.5 \times 1.5 \times 1.5 = 3.375$$
$$1.5 \times 1.5 \times .75 = 1.6875$$
$$1.5 \times .75 \times 1.5 = 1.6875$$
$$1.5 \times .75 \times .75 = .84375$$
$$.75 \times 1.5 \times 1.5 = 1.6875$$
$$.75 \times 1.5 \times .75 = .84375$$
$$.75 \times .75 \times 1.5 = .84375 \qquad \text{(ruin)}$$
$$.75 \times .75 \times .75 = .421875 \qquad \text{(ruin)}$$

Only the last two sequences saw our stake drop to .6 or less at any time.

$$RR(.6) = 2/8 = .25$$

Now for four plays:

$$\forall_2 P_4 =$$
$$1.5 \times 1.5 \times 1.5 \times 1.5 = 5.0625$$
$$1.5 \times 1.5 \times 1.5 \times .75 = 2.53125$$
$$1.5 \times 1.5 \times .75 \times 1.5 = 2.53125$$
$$1.5 \times 1.5 \times .75 \times .75 = 1.265625$$
$$1.5 \times .75 \times 1.5 \times 1.5 = 2.53125$$
$$1.5 \times .75 \times 1.5 \times .75 = 2.53125$$
$$1.5 \times .75 \times .75 \times 1.5 = 1.265625$$
$$1.5 \times .75 \times .75 \times .75 = .6328125$$
$$.75 \times 1.5 \times 1.5 \times 1.5 = 2.53125$$
$$.75 \times 1.5 \times 1.5 \times .75 = 1.265625$$

$$.75 \times 1.5 \times .75 \times 1.5 = 1.265625$$
$$.75 \times 1.5 \times .75 \times .75 = .6328125$$
$$.75 \times .75 \times 1.5 \times 1.5 = 1.265625 \quad \text{(ruin)}$$
$$.75 \times .75 \times 1.5 \times .75 = .6328125 \quad \text{(ruin)}$$
$$.75 \times .75 \times .75 \times 1.5 = .6328125 \quad \text{(ruin)}$$
$$.75 \times .75 \times .75 \times .75 = .31640625 \quad \text{(ruin)}$$

Here, only the last four saw our stake drop to .6 or lower of initial equity at any time.

$$RR(.6) = 4/16 = .25$$

And now for five plays:

$$\forall_2 P_5 =$$
$$1.5 \times 1.5 \times 1.5 \times 1.5 \times 1.5 = 7.59375$$
$$1.5 \times 1.5 \times 1.5 \times 1.5 \times 0.75 = 3.796875$$
$$1.5 \times 1.5 \times 1.5 \times 0.75 \times 1.5 = 3.796875$$
$$1.5 \times 1.5 \times 1.5 \times 0.75 \times 0.75 = 1.8984375$$
$$1.5 \times 1.5 \times 0.75 \times 1.5 \times 1.5 = 3.796875$$
$$1.5 \times 1.5 \times 0.75 \times 1.5 \times 0.75 = 1.8984375$$
$$1.5 \times 1.5 \times 0.75 \times 0.75 \times 1.5 = 1.8984375$$
$$1.5 \times 1.5 \times 0.75 \times 0.75 \times 0.75 = 0.94921875$$
$$1.5 \times 0.75 \times 1.5 \times 1.5 \times 1.5 = 3.796875$$
$$1.5 \times 0.75 \times 1.5 \times 1.5 \times 0.75 = 1.8984375$$
$$1.5 \times 0.75 \times 1.5 \times 0.75 \times 1.5 = 1.8984375$$
$$1.5 \times 0.75 \times 1.5 \times 0.75 \times 0.75 = 0.94921875$$
$$1.5 \times 0.75 \times 0.75 \times 1.5 \times 1.5 = 1.8984375$$
$$1.5 \times 0.75 \times 0.75 \times 1.5 \times 0.75 = 0.94921875$$
$$1.5 \times 0.75 \times 0.75 \times 0.75 \times 1.5 = 0.94921875$$
$$1.5 \times 0.75 \times 0.75 \times 0.75 \times 0.75 = 0.474609375 \quad \text{(ruin)}$$
$$0.75 \times 1.5 \times 1.5 \times 1.5 \times 1.5 = 3.796875$$
$$0.75 \times 1.5 \times 1.5 \times 1.5 \times 0.75 = 1.8984375$$
$$0.75 \times 1.5 \times 1.5 \times 0.75 \times 1.5 = 1.8984375$$
$$0.75 \times 1.5 \times 1.5 \times 0.75 \times 0.75 = 0.94921875$$
$$0.75 \times 1.5 \times 0.75 \times 1.5 \times 1.5 = 1.8984375$$
$$0.75 \times 1.5 \times 0.75 \times 1.5 \times 0.75 = 0.94921875$$
$$0.75 \times 1.5 \times 0.75 \times 0.75 \times 1.5 = 0.94921875$$
$$0.75 \times 1.5 \times 0.75 \times 0.75 \times 0.75 = 0.474609375 \quad \text{(ruin)}$$

$$0.75 \times 0.75 \times 1.5 \times 1.5 \times 1.5 = 1.8984375 \qquad \text{(ruin)}$$
$$0.75 \times 0.75 \times 1.5 \times 1.5 \times 0.75 = 0.94921875 \qquad \text{(ruin)}$$
$$0.75 \times 0.75 \times 1.5 \times 0.75 \times 1.5 = 0.94921875 \qquad \text{(ruin)}$$
$$0.75 \times 0.75 \times 1.5 \times 0.75 \times 0.75 = 0.474609375 \qquad \text{(ruin)}$$
$$0.75 \times 0.75 \times 0.75 \times 1.5 \times 1.5 = 0.94921875 \qquad \text{(ruin)}$$
$$0.75 \times 0.75 \times 0.75 \times 1.5 \times 0.75 = 0.474609375 \qquad \text{(ruin)}$$
$$0.75 \times 0.75 \times 0.75 \times 0.75 \times 1.5 = 0.474609375 \qquad \text{(ruin)}$$
$$0.75 \times 0.75 \times 0.75 \times 0.75 \times 0.75 = 0.237304688 \qquad \text{(ruin)}$$

Now my probability of ruin has *risen* to 10/32, or .3125. This is very disconcerting in that the probability of ruin increases the longer you continue to play.

Fortunately, this probability has an asymptote. In this two-to-one coin-toss game, at the optimal f value of .25 per play, it is shown the table below:

Play #	RR(.6)
2	0.25
3	0.25
4	0.25
5	0.3125
6	0.3125
7	0.367188
8	0.367188
9	0.367188
10	0.389648
11	0.389648
12	0.413818
13	0.413818
14	0.436829
15	0.436829
16	0.436829
17	0.447441
18	0.447441
19	0.459791
20	0.459791
21	0.459791
22	0.466089
23	0.466089
24	0.47383
25	0.47383
26	0.482092

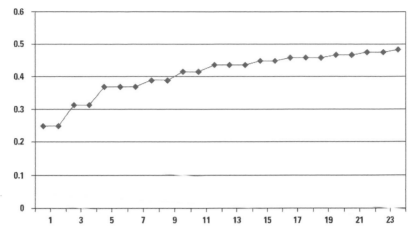

FIGURE 12.1 *RR(.6)* for two-to-one coin toss at $f = .25$

From this data, in methods to be detailed later in the text, we can determine that the asymptote, that is, the risk of ruin (defined as 60% of our initial equity left in this instance) is .48406 *in the long-run sense—that is, if we continue to play indefinitely.*

As shown in Figure 12.1, as q approaches infinity, $RR(b)$ approaches a horizontal asymptote. That is, $RR(b)$ *can* be determined in the long-run sense.

Additionally, it is perfectly acceptable to begin the analysis at $q = 1$, rather than $q = n$. Doing so aids in resolving the line and hence its asymptote.

Note that near the end of the previous chapter, a method employed in one form or another by a good deal of the larger, more successful trend-following funds, which "can be said to combine mean-variance with value at risk" was presented. Note that in the method presented—that is, in the way it is currently employed—it is akin to doing simply one run through the data, horizontally, with $n = q$ and solely for one value of k. Note that it would be if we only looked at the history of two tosses of our coin; there is no way we can approach or discern the asymptote through such a crude analysis.

Remember a very important caveat in this analysis: As demonstrated thus far it is assumed that there is no statistical dependency in the sequence of scenario outcomes across time. That is, we are looking at the stream of scenario outcomes across time in a pure sample with replacement manner; the past scenario outcome(s) do not influence the current one.

And what about more than a single scenario spectrum? This is easily handled by considering that the HPRs of the different scenario spectrums cover the same time period. That is, we may have our scenarios derived such that they are the scenarios of outcomes for the next month—or the next day, and so on.

We therefore consider each combination of scenarios for each scenario spectrum. Thus, if we are looking at two scenario spectrums of our two-to-one coin toss, we would then have the following four outcomes:

$$
\begin{array}{lcccc}
\text{Game 1} & +2 & +2 & -1 & -1 \\
\text{Game 2} & +2 & -1 & +2 & -1
\end{array}
$$

The reason we have four outcomes is that we have two scenario spectrums with two scenarios in each. Thus, n in this case will equal 4.

When you have more than one scenario spectrum:

$$ n = \prod_{i=1}^{m} \#\,scenarios_i \tag{12.06} $$

where: $m =$ The number of scenario spectrums you are including.

In other words, n is the product of all the scenario spectrums we are considering.

And in our example here, since there are two scenario spectrums ($m = 2$), each with two scenarios, we have $n = 4$.

The HPRs then for these four outcomes are the arithmetic average HPRs across scenario spectrums. The arithmetic average is used simply because an HPR represents the effect of trading one unit at a given value of f on the entire account.

So, if we assume we are going to trade at f values of .25,.25 in our example, we then have the following:

$$
\begin{array}{lcccc}
\text{Game 1} & +2 & +2 & -1 & -1 \\
\text{Game 2} & +2 & -1 & +2 & -1
\end{array}
$$

Converted to HPRs at .25,.25:

$$
\begin{array}{lcccc}
\text{Game 1} & 1.5 & 1.5 & .75 & .75 \\
\text{Game 2} & 1.5 & .75 & 1.5 & .75
\end{array}
$$

Arithmetic mean:

$$
\begin{array}{cccc}
1.5 & 1.125 & 1.125 & .75
\end{array}
$$

Thus, we have $n = 4$, and the four values are (1.5, 1.125, 1.125, .75), which we would then use in our analysis.

We digress now. To this point, we have been discussing the probability of ruin, for an aggregate of one or more market systems or scenario spectrums. Risk of ruin, $RR(b)$ represents the probability of hitting or penetrating the lower absorbing barrier of $b\times$ initial stake. Thus, this lower absorbing barrier does *not* migrate upward, as equity may increase. If an account therefore increases twofold, this barrier does not move. For example, if $b = .6$, on a \$1 million account, then the lower absorbing barrier is at \$600,000. If the account doubles now, to \$2 million, then the lower absorbing barrier is *still* at \$600,000.

This might be what many want to use in determining their relative f values across components—that is, their allocations.

However, far more frequently we want to know the probabilities of touching a lower absorbing barrier from today—actually, from our highest equity point. In other words, we are concerned with risk of drawdown, far more so in most cases than risk of ruin. If our account doubles to \$2 million now, rather than being concerned with its going back and touching or penetrating \$600,000, we are concerned with its coming down or penetrating double that, or of its coming down to \$1.2 million.

This is so much the case that in most instances, for most traders, fund managers, or anyone responsible in a field exposed to risk, it is the de-facto and organically derived[5] definition of risk itself: "The probability of drawdown," or more precisely, the probability of a $1 - b$ percentage regression from equity highs [referred to herein now as $RD(b)$].

Again, fortunately, risk of drawdown [$RD(b)$] is very closely linked to risk of ruin [$RR(b)$], so much so that we can slide the two in and out of our discussion merely by changing Equation (12.03) to reflect risk of drawdown instead of risk of ruin:

$$\text{int}\left(\frac{\sum_{i=1}^{q}\left(\min\left(1.0, \left(\prod_{t=0}^{i-1} HPR_t\right)\right) * HPR_i - b\right)}{\sum_{i=1}^{q}\left|\left(\min\left(1.0, \left(\prod_{t=0}^{i-1} HPR_t\right)\right) * HPR_i - b\right)\right|}\right) = \beta \quad (12.03a)$$

where: $HPR_0 = 1.0$.

$$\sum_{i=1}^{q}\left|\left(\min\left(1.0, \left(\prod_{t=0}^{i-1} HPR_t\right)\right) * HPR_i - b\right)\right| \neq 0$$

---

[5]All too often, the definition of risk in literature pertaining to it has *ignored* the fact that this is exactly what practitioners in the field define risk to be! Rather than the tail wagging the dog here, we opt to accept this real-world definition for risk.

Calculating in β in subsequent equations by (12.03a) will give you risk of drawdown, as opposed to mere risk of ruin.

The main difference in the mechanics of (12.03a) over (12.03) is that at any time in the running product of HPRs, if the running product is greater than 1.0, then the value 1.0 is replaced for the running product at that point.

Herein is some very bare-bones Java code for calculating equation (12.05) for one or more scenario spectrums, for determining either risk of ruin [$RR(b)$] or risk of drawdown [$RD(b)$]:

```java
import java.awt.*;
import java.io.*;
import java.util.*;


public class MaxTWR4VAR{
    String lines [];
    String msnames [];
    double f [];
    double b;
    boolean usedrawdowninsteadofruin;
    double plays[][];
    double hprs [][];
    double hpr [];//the composite (arithmetic average per
    time period) of the hprs
    int N; //the number of plays.Capital used to correspond
    to variables in the book
    long NL;// N as a long to avoid many casts
    public MaxTWR4VAR(String[] args){
        try{
            b=Double.parseDouble(args[1]);
        }catch(NumberFormatException e){
            System.out.println("Command Line format:
    MaxTWR4VAR inputfile riskofdrawdown(0.0..1.0)
    calculateRD(true/false)");
            return;
        }
        if(args.length>2){
    usedrawdowninsteadofruin=Boolean.valueOf (args[2])
        .booleanValue();
        }
        getinputdata(args[0]);
        createHPRs();
        control();
    }
```

```
public static void main(String[] args){
    MaxTWR4VAR maxTWR4VAR = new MaxTWR4VAR(args);
}

protected void getinputdata(String fileName){
    String filetext = readInputFile(fileName);
    lines = getArgs(filetext,"\r\n");
    N=lines.length-2;
    NL=(long)N;
    plays=new double[N][];
    for(int i=0;i<lines.length;i++){
        System.out.println("line "+i+" : "+lines[i]);
        if(i==0){
            msnames = getArgs(lines[i],",");
        }else if(i==1){
            f =
convertStringArrayToDouble(getArgs(lines[i],","));
        }else{
            plays[i-2]=
convertStringArrayToDouble(getArgs(lines[i],","));
        }
    }
    System.out.println("b       : "+b);
    if(usedrawdowninsteadofruin){
        System.out.println("pr of : drawdown");
    }else{
        System.out.println("pr of : ruin");
    }

}

protected void  createHPRs(){
    //first find the biggest loss
    double biggestLoss[] = new double [N];
    hprs = new double [plays[0].length][N];
    Arrays.fill(biggestLoss,Double.MAX_VALUE);
    for(int j=0;j<msnames.length;j++){
        for(int i=0;i<N;i++){
            if(plays[i][j]<biggestLoss[j]){
                biggestLoss[j]=plays[i][j];
            }
        }
    }
    //fing the hpr for each msnames for each associated
f
    for(int j=0;j<msnames.length;j++){
```

```
    for(int i=0;i<N;i++){
        hprs[j][i]= 1.0 + f[j]   (-plays[i][j] /
biggestLoss[j]);
    }
  }
  //take the arithmetic average of the hprs
  hpr = new double[N];
  for(int i=0;i<N;i++){//go through each play
      for(int j=0;j<msnames.length;j++){ //go through
each msnames
        hpr[i] += hprs[j][i];
    }
  }
  for(int i=0;i<N;i++){
      hpr[i] /= (double)msnames.length;
  }
 }

 protected String readInputFile(String fileName){
     FileInputStream fis = null;
     String str = null;
     try {
         fis = new FileInputStream(fileName);
         int size = fis.available();
         byte[] bytes = new byte [size];
         fis.read(bytes);
         str = new String(bytes);
     } catch (IOException e) {
     } finally {
         try {
             fis.close();
         } catch (IOException e2) {
         }
     }
     return str;
 }

 protected String[] getArgs(String parameter, String
delimiter){
     String args[];
     int nextItem=0;
     StringTokenizer stoke=new
StringTokenizer(parameter,delimiter);
     args=new String[stoke.countTokens()];
     while(stoke.hasMoreTokens()){
         args[nextItem]=stoke.nextToken();
```

```
            nextItem=(nextItem+1)%args.length;
        }
        return args;
}


protected double [] convertStringArrayToDouble(String
[] s){
    double [] d = new double[s.length];
    for(int i = 0; i<s.length; i++){
        try{
            d[i]=Double.parseDouble(s[i]);
        }catch(NumberFormatException e){
            d[i]=0.0;
        }
    }
    return d;
}


protected int B(double [] hprset,boolean drawdown){
    double interimHPR=1.0;
    double previnterimHPR=1.0;
    double numerator=0.0;
    double denominator=0.0;
    for(int i=0;i<hprset.length;i++){
        double useinvalue = previnterimHPR;
        if(drawdown && previnterimHPR>1.0)
            useinvalue = 1.0;

        interimHPR = useinvalue x hprset[i];
        //interimHPR = previnterimHPR x hprset[i];
        double value = interimHPR - b;
        numerator += value;
        denominator += Math.abs(value);
        previnterimHPR = interimHPR;
    }
    if(denominator==0.0){
        return 0;
    }else{
        double x = (numerator/denominator);
        if(x>=0){
            return (int)x;
        }else{
            return 0;
        }
    }
}
```

```
//n things taken q at a time where q>=n
//we really cannot use this as we get OutOfMemoryError
early on
//because we try to save the whole array. Instead, use
nPq_i()
protected double[][] nPq(int nopermutations, int q){
    double hprpermutation[][]=new
double[nopermutations][q];
    for(int column=0;column<q;column++){ // go
through column x column
        for(int pn=0;pn<nopermutations;pn++){ // go
through permutation x permutation
            if(column==0){
                hprpermutation[pn][column] = hpr[pn %
N];
            }else{
                hprpermutation[pn][column] =
hpr[(pn/(int)(Math.pow((double)N,(double)column))) % N];
            }
        }
    }
    return hprpermutation;
}


//n things taken q at a time where q>=n to return the
i'th item
protected double[] nPq_i(int q, long pn){
    double hprpermutation[]=new double[q];
    int x = 0;
    for(int column=0;column<q;column++){ // go through
column x column
        if(column==0){
            x = (int)(pn % NL);
        }else{
            x =
(int)((pn/(long)(Math.pow((double)N,(double)column))) %
NL);
        }
        hprpermutation[q-1-column] = hpr[x];
    }
    return hprpermutation;
}

protected void control(){
    int counter=1;
```

```
      while(1==1){
          long passed=0;
          long nopermutations = (long)
Math.pow((double)hpr.length,(double)counter);
          for(long pn=0;pn<nopermutations;pn++){
              double hprpermutation[]=nPq_i(counter,pn);

passed+=(long)B(hprpermutation,usedrawdowninsteadofruin);
          }
          double result=1.0-
(double)passed/(double)nopermutations;
          System.out.println(counter+" = "+result);
          counter++;
      }
  }
}
```

The code is presented "as-is," with no warranties whatsoever. Use it as
you see fit. It is merely a bare-bones implementation of Equation (12.05).
I wrote it in as generic a flavor of Java as possible, intentionally avoided
using an object-oriented approach, and intentionally kept it in the lowest-
common-denominator syntax across languages, so that you can transport
it to other languages more easily. The code can be made *far* more efficient
than what is presented here. This is presented merely to give programmers
of this concept a starting reference point.

Note that the input file format must be formatted as follows: a straight
ASCII text file, wherein the first line is the scenario spectrum name, the
second line is the f value to be used on that scenario spectrum, and all
subsequent lines are the simple stream of individual scenario outcomes.
For example:

Coin Toss 1
.25
−1
2

This shows the scenario spectrum "Coin Toss 1," at an f of .25 with two
outcomes, one of −1 and the other of +2.

For situations of multiple scenario spectrums, again the first line is
the scenario spectrum names, comma-delimited is the second line; the respective f values, comma delimited; and each line after that represents a

simultaneous outcome for both scenario spectrums, wherein each combination of scenarios from both scenario spectrums occur.

```
Coin Toss 1,Coin Toss 2
.25,.25
2,2
2,-1
-1,2
-1,-1
```

So in this file, the first outcome sees both scenario spectrums gaining two units. The next outcome sees Coin Toss 1 gaining two units, while Coin Toss 2 loses one unit (-1). Then Coin Toss 1 loses one unit (-1) and Coin Toss 2 gains two units. For the last outcome, they both lose one unit (-1). (Thus, $n = 4$ in this file. In all data files, therefore, since the first two lines are scenario spectrum name(s) and respective f value(s), n equals the number of lines in the file minus 2.)

To this point, we have not alluded to the probabilities of the scenario outcomes. Rather, as if the scenario outcomes were like a stream of trades, or a stream of coin toss results, we have quietly assumed for simplicity's sake that there has been an equal probability of occurrence with each scenario outcome. In other words, we have been inexplicitly saying to this point that for each scenario (or individual combinations of scenarios from multiple spectrums occurring simultaneously), the probability of the kth outcome among the n^q outcomes is:

$$p_k = \frac{1}{n^q} \tag{12.07}$$

Usually, however, we do not have the luxury of the convenience of all scenarios having the same probability of occurrence.[6]

To address this, we return now to Equation (12.05). We will discuss first the case of a single scenario spectrum. In this case, we not only have outcomes for each scenario [which comprise the HPRs used in Equation

---

[6]Note, however, that if we *were* talking about scenarios made up of individual coin tosses, or of results of trading a given market system over a given day, or if we did use purely empirical data in discerning our scenario spectrums and probabilities, we could use Equation (12.07) for the said probabilities. In other words, if, say, we used the last 24 trading months and examined the prices of stock ABC, we could conceivably create a scenario spectrum of 24 bins, each with an outcome of those months, each with a probability given in (12.07).

(12.03) or (12.03a) for β], but we also have a probability of its occurrence, p.

$$RX(b, q) = \frac{\forall nPq \sum\limits_{k=1}^{n^q} (\beta_k * p_k)}{\forall nPq \sum\limits_{k=1}^{n^q} p_k} \tag{12.05a}$$

where: β = The value given in (12.03) or (12.02).
           p_k = The probability of the kth occurrence.

For each k, this is the product of the probabilities for that k. That is, you can think of it as the horizontal product of the probabilities from 1 to q for that k. For each k, you calculate a β . Each β_k, as you can see in (12.03) or (12.03a), cycles through from $i = 1$ to q HPRs. Each HPR$_i$ has a probability associated with it ($Prob_{k,i}$). Multiplying these probabilities along as you cycle through from $i = 1$ to q in (12.03) or (12.03a) as you discern β_k will give you p_k in the single scenario case. For example, in a coin toss, where the probabilities are always .5 for each scenario, then however the permutation of scenarios in (12.03) or (12.03a), p_k will be .5 $\times$ .5 = .25 when $q = 2$ in discerning β_k, for each k, it will equal .25 $\times$ .25 $\times$ .25 = .015625 when $q = 3$, ad infinitum for the single scenario set case.

$$p_k = \prod_{i=1}^{q} Prob_{k,i} \tag{12.07a}$$

To help dispel confusion, let's return to our simple single coin toss and examine the nomenclature of our variables:

- There is one scenario spectrum: $m = 1$.
- This scenario spectrum has two scenarios: $n = 2$ [per (12.06)].
- We are expanding out in this example to three sequential outcomes, $q = 3$. We traverse this, "Horizontally," as $i - 1$ to q (as in [12.02])
- Therefore we have $n^q = 2^3 = 8$ sequential outcome possibilities. We traverse this, "vertically," as $k = 1$ to n^q (as in [12.04])

As we get into multiple scenarios, calculating the individual $Prob_{k,i}$'s gets a little trickier. The matter of joint probabilities pertaining to given outcomes at i, for m spectrums was covered in Chapter 9 and the reader is referred back to that discussion for discerning $Prob_{k,i}$'s when $m > 1$.

Thus, of note, there is a probability at a particular i of the manifestations of each individual scenario occurring in m spectrums together (this is a $Prob_{k,i}$). Thus, on a particular i in multiplicative run from 1 to q, in a particular horizontal run of k from 1 to n^q, we have a probability $Prob_{k,i}$. Now multiplying these $Prob_{k,i}$'s together in the horizontal run for i from 1 to q will give the p_k for this k.

$$Prob_{1,1} * Prob_{1,2} * \cdots * Prob_{1,q} = p_1$$

$$\cdots$$

$$Prob_{n^q,1} * Prob_{n^q,2} * \cdots * Prob_{n^q,q} = p_n{}^q$$

n.b. Now, when dependency is present in the stream of outcomes, the p_k values are necessarily affected.

For example, in the simplistic binomial outcome case of a coin toss, where I have two possible outcomes ($n = 2$), heads and tails, with outcomes $+2$ and -1, respectively, and I look at flipping the coin two times ($q = 2$), I have the following four (n^q) possible outcomes:

				p_k
Outcome 1	($k = 1$)	H	H	.25
Outcome 2	($k = 2$)	H	T	.25
Outcome 3	($k = 3$)	T	H	.25
Outcome 4	($k = 4$)	T	T	.25

Now let us assume there is perfect negative correlation involved—that is, winners always beget losers, and vice versa. In this idealized case, we then have the following:

				p_k
Outcome 1	($k = 1$)	H	H	0
Outcome 2	($k = 2$)	H	T	.5
Outcome 3	($k = 3$)	T	H	.5
Outcome 4	($k = 4$)	T	T	0

Unfortunately, when serial dependency seems to exist, it is never at such an idealized value as 1.0, as shown here. Fortunately, however, serial dependency rarely exists, and its appearance of existence in small amounts is usually, and typically, incidental, and can thus be worked with as being zero. However, if the p_k values are deemed to be more than merely "incidental," then they can, and in fact, must, be accounted for as they are used in the equations given in this chapter.

Additionally, the incorporation of rules to address dependency when it seems present, of the type like, "Don't trade after two consecutive losers, etc," could in this analysis be turned into the familiar tails, or T in the following stream:

HHTHTTHH

The dependency rules would transform the stream to:

$$H\,H\,T\,H\,T\,T\,H$$

Such a stream could therefore be incorporated into these equations, amended as such, with the same probabilities.

Note the nomenclature in (12.05a) $RX(b, q)$, referring to the fact that this equation can be used for either risk of ruin, $RR(b, q)$ or risk of drawdown, $RD(b, q)$.

Additionally, note that the denominator in this case is simply the sum of the probabilities. Typically, this should equal 1, excepting for floating point round-off error. However, this is often not the case when we get into some of the shortcut methods listed later, so (12.05a) will not be rewritten here with a denominator of 1.

The full equation, then, for determining risk of drawdown at a given q is then given as:

$$RD(b, q) =$$

$$\frac{\forall nPq \sum_{k=1}^{n^q} \left(\mathrm{int} \left(\frac{\sum_{i=1}^{q} \left(\min\left(1.0, \left(\prod_{t=0}^{i-1} HPR_t\right)\right) * HPR_{i-b} \right)}{\sum_{i=1}^{q} \left| \left(\min\left(1.0, \left(\prod_{t=0}^{i-1} HPR_t\right)\right) * HPR_{i-b} \right) \right|_k} \right) * \prod_{i=1}^{q} Prob_{k,i} \right)}{\forall nPq \sum_{k=1}^{n^q} \left(\prod_{i=1}^{q} Prob_{k,i} \right)}$$

$$(12.05a)$$

where: $HPR_0 = 1.0.$

$$\sum_{i=1}^{q} \left| \left(\min\left(1.0, \left(\prod_{t=0}^{i-1} HPR_t\right)\right) * HPR_i - b \right) \right| \neq 0$$

That's it. There is your equation. Solving (12.05b) will give you the probability of drawdown. Though it looks daunting, the only inputs required to calculate it are a given level of drawdown (expressed as $1 - b$; thus, if I am considering a 20% drawdown, I will use $1 - .2 = .8$ as my b value), the f values of the scenario spectrums (from which the HPRs are then derived), and the joint probabilities of the scenarios across the spectrums.

Why is (12.05b) so important? Because everything in (12.05b) you will keep constant. The only thing that will change are the f values of the components in the portfolio, the scenario spectrums from which the HPRs are derived.

Therefore, given (12.05b), one can determine the portfolio that is growth optimal within a given acceptable $RD(b)$! In other words, starting from the standpoint of "I want to have no more than an x percent probability of a drawdown greater than $1 - b$," you can discern the portfolio that is growth optimal.

Essentially then, the new model is:

Maximize TWR where $RD(b) <=$ an acceptable probability of hitting b.

$$(12.08)$$

Also expressed as:

Maximize (9.04) where (12.05b) $<=$ an acceptable probability of hitting b.

That is, whenever an allocation is measured in, say, the genetic algorithm for discerning if it is a new, optimal allocation mix, then it can be measured against (12.05b) given the f values of the candidate mix, the drawdown being considered as $1 - b$, to see whether $RD(b)$, as given by (12.05b) is acceptable (i.e., if $RD(b) <= x$).

Additionally, the equation can be looked at in terms of a fund as a scenario spectrum. We can use (12.05), (12.05a), and (12.05b) to determine an allocation to that specific fund in terms of maximum drawdown and maximum risk of ruin probabilities, rather than looking to discern the relative weightings within a portfolio. That is, in the former we are seeking an individual f value that will give us probabilities of drawdowns and ruin which are palatable to us and/or will determine the notional funding amount that accomplishes these tolerable values. In the latter, we are looking for a set of f values to allocate among m components within the portfolio to accomplish the same.

How many q is enough q? How elusive is that asymptote, that risk of drawdown?

In seeking the asymptote to (12.05), (12.05a), (12.05b) we seek that point where each increase in q is met with $RX(b)$ increasing by so slight a margin as to be of no consequence to our analysis. So it would appear that when $RX(b)$ for a given value of q, $RX(b,q)$ is less than some small amount, a, where we say we are done discerning where the asymptote lies—we can assume it lies "just above" $RX(b,q)$.

Yet, again refer to Figure 12.1. Note that the real-life gradations of $RX(b)$ are not necessarily smooth, but do go upward with spurious stairsteps, as it were. So it is not enough to simply say that the asymptote lies "just above" $RX(b,q)$ unless we have gone for a number of iterations, z, before q where $RX(b,q) - RX(b,q - 1) <= a$.

In other words, we can say we have arrived at the asymptote, and that the asymptote lies "just above" $RX(b,q)$ when, for a given a and z:

$$RX(b, q) - RX(b, q - 1) <= a, \text{ and} \ldots \text{ and } RX(b, q) - RX(b, q - z) <= a$$

$$(12.09)$$

where: $q > z$.

The problem with Equation (12.05a) or (12.05b) now [and (12.05a) or (12.05b) will give you the same answer as (12.05) when the probabilities of each kth occurrence are identical] is that it increases as q increases, increasing to an asymptote.

It is relatively easy to create a chart of the sort shown in Figure 12.1 derived from the table on page 386 to attempt to discern an asymptote when $q = 2$ as in our simple two-to-one coin-toss situation. However, when we have 26 plays—that is, when we arrive at a value of $q = 26$, then $n^q = 2^{26} = 67,108,864$ permutations. That is over 67 million β values to compute.

And that's in merely calculating the $RR(b)$ for a single coin-toss scenario spectrum! When we start getting into multiple scenario spectrums with more than two scenarios each, where n equals the results of (12.06), then clearly, computer power—speed and raw memory requirements—are vital resources in this pursuit.

Suppose I am trying to consider one scenario spectrum with, say, 10 scenarios in it. To make the pass through merely when $q = n$, I have $10 \wedge 10 = 10,000,000,000$ (ten billion) permutations! As we get into multiple scenario spectrums now, the problem explodes on us exponentially.

Most won't have access to the computing resources that this exercise requires for some time. However, we can implement two mathematical shortcuts here to arrive at very accurate conclusions in a mere fraction of the time, with a mere fraction of the computational requirements.

Now, can't I take a random sample of these 10 billion permutations and use that as a proxy for the full 10 billion? The answer is yes, and can be found by statistical measures used for sample size determination for binomially distributed outcomes (note that β is actually a binomial value for whether we have hit a lower absorbing barrier or not; it is either true or false).

To determine our sample size, then, from binomially distributed data, we will use Equation (12.10):

$$\left(\frac{s}{x}\right)^2 * p * (1 - p) \tag{12.10}$$

where: $s =$ The number of sigmas (standard deviations) confidence level for an error of x.
$x =$ The error level.
$p =$ The probability of the null hypotheses.

That last parameter, p, is circularly annoying. If I know p, the probability of the null hypotheses, then why am I sampling to discern, in essence, p?

Note, however, that in (12.10), any deviation in p away from $p = .5$ will give a smaller answer for (12.10). Thus, a smaller sample size would be required for a given s and x. Therefore, if we simply set p to .5, we are being conservative, and requiring that (12.10) err on the side of conservatism (i.e., as a larger sample size).

Simply put then, we need only answer for s and x. So if I want to find the sample size that would give me an error of .001, with a confidence to s standard deviations, solving for (12.10) yields the following:

$$2 \text{ sigma} = \left(\frac{2}{.001}\right)^2 * .5 * (1 - .5) = 1{,}000{,}000$$

$$3 \text{ sigma} = \left(\frac{3}{.001}\right)^2 * .5 * (1 - .5) = 2{,}250{,}000$$

$$5 \text{ sigma} = \left(\frac{5}{.001}\right)^2 * .5 * (1 - .5) = 6{,}250{,}000$$

Now the reader is likely to inquire, "Are these sample sizes independent of the actual population size?" The sample sizes for the given parameters to (12.10) will be the same regardless of whether we are trying to estimate a population of 1,000 or 10,000,000.

"So I need only do this once; I don't need to keep increasing q?"

Not so. Rather, you use (12.10) to discern the minimum sample size required at each q. You still need to subsequently increase q, and the answer [as provided by (12.05), (12.05a), or (12.05b)] will keep increasing to the asymptote. The reason you must keep increasing q is that at each q, the binomial distribution is different, as was demonstrated earlier in this chapter.

One of the key caveats in implementing Equation (12.10) is that it is provided for a "random" sample size. However, these minimum, random sample sizes provided for in (12.10) tend to be rather large. Thus, it's important to make sure, since we are generating random numbers by computer, that we are not cycling in our random numbers so soon that it will cause distortion in randomness, and that the random numbers generated be isotropically distributed.

I strongly suggest to the ambitious readers who attempt to program these concepts that they incorporate the most powerful random number generators they can. Over the years this has been something of a moving target, and, likely and hopefully will continue to be. Currently, I am partial to the Mersenne Twister algorithm.[7] You can use other random number generators, but your results will be accurate only to the extent of the randomness provided by them.

---

[7]Makato Matsumoto and Takuji Nishimura, "Mersenne Twister: A 623-Dimensionally Equidistributed Uniform Pseudo-Random Number Generator," *ACM Transactions on Modeling and Computer Simulation*, Vol. 8, No. 1 (January 1998), pp. 3–30.

There are additional real-world implementation issues in terms of adding floating point numbers millions of times considering the floating point roundoff errors, and so on. Ultimately, we are trying to get a "reasonable and real-world workable" resolution of the curves for RR and RD so that we can determine their asymptotes.

This particular shortcut is invoked only if the number of permutations at a given q exceeds n^q. If not, just run all the permutations. For example, where $q = 1$, where we start, there are $10^1 = 10$ permutations. Thus, we just run all 10. At $q = 2$, we have $10^2 = 100$ permutations, and again run all permutations. However, at $10^7 = 10,000,000$, which is greater than the 6,250,000 sample size required, we would begin using the sample size when $q = 7$ in this case.

Let's look at a real-world implementation of what has been discussed thus far. Consider a single scenario spectrum with the following scenarios:

Outcome	Probability
−1889	0.015625
−1430.42	0.046875
−1295	0.015625
−750	0.0625
−450	0.125
0	0.203125
390	0.078125
800	0.328125
1150	0.0625
1830	0.046875

This is a case of a single scenario spectrum of 10 scenarios. Therefore, on our $n = q$ pass through the data (i.e., $q = 10$), we are going to have $n \wedge q$, or $10 \wedge 10 = 10,000,000,000$ (ten billion) permutations, as alluded to earlier.

Now, we will attempt to calculate the risk of ruin, with ruin defined as having 60% of our initial equity left.

Running these 10 billion calculations outright gives:

$$RR(.6, 10) = .1906955154$$

at an f value of .45.

Using (12.10) with $s = 5, x = .001, p = .5$, we iterate through q obtaining quite nicely, and in a tiny fraction of the time it took to actually calculate the actual value at $RR(.6,10)$ just presented (i.e., 10 billion iterations for $q = 10$ actually versus 6,250,000! This is .000625 of the time!):

q	$RR(.6)$
1	0.015873
2	0.047367
3	0.07433
4	0.097756
5	0.118505
6	0.136475
7	0.150909
8	0.16485
9	0.178581
10	0.191146
11	0.202753
12	0.209487
13	0.21666
14	0.220812
15	0.244053
16	0.241152
17	0.257894
18	0.269569
19	0.276066
20	1

Note that at $q = 20$ we have $RR(.6) = 1$. This is merely an indication that we have overflowed the value for a long data type in Java.[8] This is still far from the asymptote.

Also note the floating point roundoff error even at $q = 1$. This value should have been 0.015625, not 0.015873.

These calculations were performed by extending the class of the previous Java program earlier in this chapter, and is included herein:

```
import java.awt.*;
import java.io.*;
import java.util.*;

public class MaxTWR4VARWithProbs extends MaxTWR4VAR{
    double probs[][];
    double probsarray[];
    double probThisB;
```

---

[8]Again, all of the code presented here can, even under present-day Java, be made far more efficient and robust than what is shown here. This is merely presented as a starting point for those wishing to pursue these concepts in code.

```
public MaxTWR4VARWithProbs(String[] args){
    super(args);
}


public static void main(String[] args){
    MaxTWR4VARWithProbs maxTWR4VARWithProbs = new
MaxTWR4VARWithProbs(args);
}


  protected void getinputdata(String fileName){
      String filetext = readInputFile(fileName);
      lines = getArgs(filetext,"\r\n");
      N=lines.length-2;
      NL=(long)N;
      plays=new double[N][];
      probs=new double[N][lines.length-2];
      for(int i=0;i<lines.length;i++){
          System.out.println("line "+i+" : "+lines[i]);
          if(i==0){
              msnames = getArgs(lines[i],",");
          }else if(i==1){
              f =
convertStringArrayToDouble(getArgs(lines[i],","));
          }else{

              plays[i-2]=
convertStringArrayToDouble(getArgs(lines[i],","),i-2);
          }
      }
      System.out.println("b : "+b);
      if(usedrawdowninsteadofruin){
          System.out.println("pr of : drawdown");
      }else{
          System.out.println("pr of : ruin");
      }


  }


  protected double [] convertStringArrayToDouble(String
[] s,int lineno){
      double [] d = new double[s.length];
      probs[lineno]= new double[s.length];
      for(int i = 0; i<s.length; i++){
          String ss[] = getArgs(s[i],";");
          try{
              d[i]=Double.parseDouble(ss[0]);
              probs[lineno][i]=Double.parseDouble(ss[1]);
```

```
            }catch(NumberFormatException e){
                d[i]=0.0;
                probs[lineno][i]=0.0;
            }
        }
        return d;
    }

    protected int B(double [] hprset,boolean drawdown){
        double interimHPR=1.0;
        double previnterimHPR=1.0;
        double numerator=0.0;
        double denominator=0.0;
        probThisB=1.0;
        for(int i=0;i<hprset.length;i++){
            double useinvalue = previnterimHPR;
            if(drawdown && previnterimHPR>1.0)
                useinvalue = 1.0;

            interimHPR = useinvalue ×  hprset[i];
            //interimHPR = previnterimHPR ×  hprset[i];
            double value = interimHPR - b;
            numerator += value;
            denominator += Math.abs(value);
            previnterimHPR = interimHPR;
            probThisB *= probsarray[i];
        }
        if(denominator==0.0){
            return 0;
        }else{
            double x = (numerator/denominator);
            if(x>=0){
                return (int)x;
            }else{
                return 0;
            }
        }
    }

    //n things taken q at a time where q>=n to return the
  i'th item
    protected double[] nPq_i(int q, long pn){
        double hprpermutation[]=new double[q];
        probsarray=new double[q];
        int x = 0;
        for(int column=0;column<q;column++){ // go through
column x column
```

```
            if(column==0){
                x = (int)(pn % NL);
            }else{
                x =
(int)((pn/(long)(Math.pow((double)N,(double)column))) %
NL);
            }
            int a = q-1-column;
            hprpermutation[a] = hpr[x];
            probsarray[a] = probs[x][0];//it's zero here
because we are only figuring one MS
        }
        return hprpermutation;
    }


    protected void control(){
        double sigmas = 5.0;
        double errorsize = .001;
        double samplesize = Math.pow(sigmas/errorsize,2.0)
x .25;
        long samplesizeL = (long)(samplesize+.5);
        int counter=1;
        RalphVince.Math.MersenneTwisterFast generator = new
RalphVince.Math.MersenneTwisterFast(System.currentTime
Millis());
        java.util.Random random = new java.util.Random();
        while(1==1){
            long permutationcount = 0L;
            double passed=0.0;
            double sumOfProbs=0.0;
            long nopermutations = (long)
Math.pow((double)hpr.length,(double)counter);
            if(nopermutations<(long)samplesize){
                for(long pn=0;pn<nopermutations;pn++){
                    double
hprpermutation[]=nPq_i(counter,pn);
                    double theB =
(double)B(hprpermutation,usedrawdowninsteadofruin);
                    if(theB>0.0){
                        theB *= probThisB;
                        passed += theB;
                    }
                    sumOfProbs += probThisB;
                    permutationcount++;
                }
            }else{
```

```
                do{
                      generator.setSeed(random.nextLong());
                      long
  pn=(long)(generator.nextDouble()*(double)nopermutations);
                      double
  hprpermutation[]=nPq_i(counter,pn);
                      double    theB =
  (double)B(hprpermutation,usedrawdowninsteadofruin);
                          if(theB>0.0){
                              theB *= probThisB;
                              passed += theB;
                          }
                          sumOfProbs += probThisB;
                          permutationcount++;

                      }while(permutationcount<samplesizeL);
                }
                double result=1.0-passed/sumOfProbs;
                System.out.println(counter+" = "+result);
                counter++;
            }
        }
    }
```

Unlike the previous code provided, this code class works only with one market system, and the format for the input file differs from the first in that in this class, each line from the third line on is a semicolon-delimited value pair of outcome;probability.

Thus, the input file in this real-world example appears as:

```
Real-world example file of a single scenario spectrum
.45
-1889;0.015625
-1430.42;0.046875
-1295;0.015625
-750;0.0625
-450;0.125
0;0.203125
390;0.078125
800;0.328125
1150;0.0625
1830;0.046875
```

The technique of using a random sample gets our first few values for the line of RX to q up and running with very good estimates in short order.

With the second technique, to be presented now, we can extrapolate out that line and hence seek its horizontal asymptote. Fortunately, lines derived from the Equations (12.05), (12.05a), and (12.05b) do possess an asymptote and are of the form:

$$RX'(b, q) = \text{asymptote-variable}A * \text{EXP}(-\text{variable}B * q) \qquad (12.11)$$

$RX'(b,q)$ will be the surrogate point, the value along the y axis for a given q along the x axis in the Cartesian plane.

We can use equation (12.11) as a surrogate for the actual calculations in (12.05), (12.05a), or (12.05b) when q gets too computationally expensive.

To do this, we need only know three values: the asymptote, variable$A$, and variable$B$.

We can find these values by any method of mathematical minimization whereby we minimize the squares of the differences between the observed values and the values given by (12.11). Those values with the minimum sum of the differences squared are those values that best fit this line, this proxy of actual $RX(b,q)$ values when q is too computationally expensive.

The process is relatively simple. We take those values we were able to calculate for $RX(b,q)$. For each of these values, we compare corresponding points derived from (12.11) and square the differences between the two. We then sum the squares.

Thus, we have a sum of the squared differences of our points to (12.11) for a given (asymptote, variable$A$, variable$B$). Proceeding with a mathematical minimization routine (Powell's, Downhill Simplex, even the genetic algorithm, though this will be far from the most efficient means— for a list and detailed explanation of these methods, see "Numerical Recipes,"[9] Press et al.) we arrive at that set of variable values that minimizes the sum of the differences squared between the observed points and their corresponding points as given by (12.11).

Returning, for example, to our two-to-one coin toss, we had calculated by equation (12.05) those $RR(.6)$ values, and these were given in Table 4.2. Here, using Microsoft Excel's Solver function, we can calculate the parameters in (12.11) that yield the best fit:

asymptote 0.48406
variableA 0.37418
variableB 0.137892

---

[9]Press, William H.; Flannery, Brian P.; Teukolsky, Saul A.; and Vetterling, William T., *Numerical Recipes: The Art of Scientific Computing*, (New York: Cambridge University Press, 1986).

These values given by (12.11) are shown in the table below.

Play#	Observed (12.05)	Calculated (12.10)
2	0.25	0.200066
3	0.25	0.236646
4	0.25	0.268515
5	0.3125	0.296278
6	0.3125	0.320466
7	0.367188	0.341538
8	0.367188	0.359896
9	0.367188	0.375889
10	0.389648	0.389822
11	0.389648	0.40196
12	0.413818	0.412535
13	0.413818	0.421748
14	0.436829	0.429774
15	0.436829	0.436767
16	0.436829	0.442858
17	0.447441	0.448165
18	0.447441	0.452789
19	0.459791	0.456817
20	0.459791	0.460326
21	0.459791	0.463383
22	0.466089	0.466046
23	0.466089	0.468367
24	0.47383	0.470388
25	0.47383	0.472149
26	0.482092	0.473683

This fitted line now, Equation (12.10), is shown superimposed as the solid line over Figure 12.1, now as Figure 12.2.

Now that we have our three parameters, I can determine for, say, a q of 300, by plugging in these values into (12.10), that my risk of ruin $[RR(.6)]$ is .484059843.

At a q of 4,000 I arrive at nearly the same number. Obviously, the horizontal asymptote is very much in this vicinity.

The asymptote of such a line is determined, as pointed out earlier by (12.09), since the line given by (12.10) is a smooth one.

Let's go back to our real-world example now, the single scenario set of 10 scenarios. Fitting to our earlier case of a single scenario set with 10 scenarios, whereby we were able to calculate the $RR(.6)$ values for $q = 1 \ldots 19$,

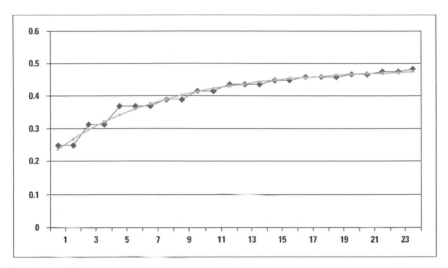

FIGURE 12.2   *RR(.6)* observed and calculated for two-to-one coin toss $f = .25$

by taking 6,250,000 samples for each q (beyond $q - 6$) , and using these 10 data points ($q = 1 \ldots 19$) as input to find those values of the parameters in (12.11) that minimize the sum of the squares of the differences between the answers given by those parameters in (12.11) and the actual values we got [by estimating the actual values using (12.10)], gives us the corresponding best fit parameters for (12.11) as follows:

$$
\begin{aligned}
\text{asymptote} &= \quad 0.397758 \\
\text{exponent} &= \quad 0.057114 \\
\text{coefficient} &= \quad 0.371217
\end{aligned}
$$

The data points and corresponding function (12.11) then appear graphically as Figure 12.3.

And, if we extend this out to see the asymptote in the function, we can compress the graphic as shown in Figure 12.4.

Using these two shortcuts allow us to accurately estimate what the function for $RX()$ is, and discern where the asymptote is, as well as how many q—which can be thought of as a surrogate for time—out it is.

Now, if you are trying to fit (12.10) to a risk of ruin, $RR(b)$, you will fit to find the three parameters that give the best line, as we have done here.

However, if you are trying to fit to risk of drawdown, $RD(b)$, you will only fit for variable A and variable B. You will *not* fit for the asymptote. Instead, you will assign a value of 1.0 to the asymptote, and fit the other two parameters from there.

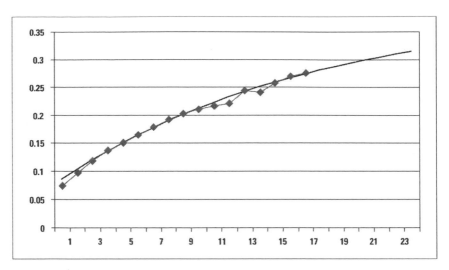

FIGURE 12.3   *RR*(.6) for real-world example at $f = .45$

To confirm the reader's burgeoning uneasiness at this point, consider the following:

In the long-run sense, the probability of hitting a drawdown (of *any* given magnitude, b) approaches 1, approaches *certainty* as you continue to trade (i.e., as q increases).

$$\lim_{q \to \infty} RD(b, q) = 1.0 \qquad (12.12)$$

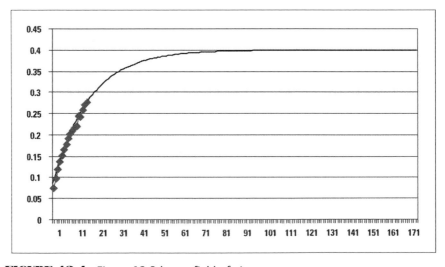

FIGURE 12.4   Figure 12.3 larger field-of-view

This is not as damning a statement as it appears on first reading. Consider the real-world example just alluded to wherein $RR(.6) = 0.397758$. Since the probability of hitting a drawdown of any given magnitude (let's say, a 99% drawdown, for argument sake) approaches 1 as q approaches infinity, yet there is only a roughly 40% chance of dropping back to roughly 60% of starting equity, we can only conclude that so many q have transpired so as to cause the account to have grown by such an amount that a 99% drawdown still leaves 60% of initial capital.

What we can know, and use, is that (12.05b) can give us a probability of drawdown for a given q. We can use it to know, for instance, what the probability of drawdown is over, say, the next quarter.

Further, since, we have a geometric mean HPR for each value of (12.05b), we can determine what T we are looking at to reach a specified growth.

$$T = \log_G \text{target} \qquad (5.07b)$$

where: target = The target TWR.
G = The geometric mean HPR corresponding to the allocation set used in (12.05b).

Thus, for example, if my target is a 50% return (i.e., target TWR = 1.5) and my geometric mean HPR from the allocation set I will use in (12.05b) is 1.1, then I will expect it to take T periods, on average, to reach my target TWR:

$$T = \log_{1.1} 1.5 = 4.254164$$

So I would want to consider the $RD(b, 4.254164)$ in this case to be below my threshold probability of such a drawdown.

Notice that we are now considering a risk of drawdown (or ruin) versus that of hitting an upper barrier [i.e., target TWR, or u from (12.01)]. Deriving T from (5.07b) to use as input to (12.05) is akin to using Feller's classical ruin given in (12.01) only for the more complex case of:

1. A lower barrier, which is not simply just zero.
2. For multiple scenarios, not just the simple binomial gambling sense (of two scenarios).
3. These multiple scenarios are from multiple scenario spectrums, with outcomes occurring simultaneously, with potentially complicated joint probabilities.
4. More importantly, we are dealing here with geometric growth, not the simple case in Feller where a gambler wins or loses a constant unit with either outcome.

Such analysis—determining T as either the horizon over the next important period (be it a quarter, a year, etc.), or backing into it as the expected number of plays to reach a given target, is how we can determine the portfolio allocation that is growth optimal while remaining within the constraints of an acceptable level of a given drawdown over such a period.

In other words, if we incorporate the concepts detailed in this chapter, we can see that the terrain in leverage space is *pock-marked*, has holes in it, where we cannot reside. These holes are determined by the utility preference pertaining to an unacceptable probability of an unacceptable drawdown.[10] We seek the highest point where the surface has not been removed under our feet via the analysis of this chapter.

The process detailed in this chapter allows you to maximize returns for a given probability of seeing a given level of drawdown over a given period—which *is* risk. This is something that has either been practiced by intuition by others, with varying degrees of success, or practiced with a different metric for risk other than drawdown or risk of ruin—often alluded to as *value at risk.*

Essentially, by seeking that highest point (altitude determined as a portfolio's geometric mean HPR or TWR) in the $n + 1$ dimensional landscape of n components, one can mark off those areas within the landscape that cannot be considered for optimal candidates as those areas where the probability of risk of ruin or drawdown to a certain point is exceeded.

---

[10]Note that as one nears the $f_1 = 0 \ldots f_n = 0$ point, as described at the end of Chapter 10, the likelihood of not being at a hole in the landscape becomes assured, without going through the analysis outlined in this chapter. However, that is a poor surrogate for *not* going through this analysis, as one would pay the consequences for deviating far left on all axes in leverage space.

Postscript

In the long run, the probability of hitting a drawdown (of any given magnitude) approaches 1, approaches certainty, as you continue to trade (i.e., as q increases).

This is not hard to see intuitively. If you can only die by being struck by lightning, eventually, you will die by being struck by lightning. Trading, like getting out of bed in the morning, flirts with disaster. I may not like this—but that's how it works, and therefore, usually, the best thing to do in life is nothing at all.

It's also the hardest thing to do.

Being aware of the risks is *enlightenment*, not a curse. Knowing that the sun will burn out someday is not a bad thing to know. Knowing that driving at 80 km/hour will cause 16 times the impact as driving one-fourth that speed is beneficial to know, in that enlightenment allows us to better assess those risks we wish to take.

RALPH VINCE

Index